Liquid Chromatography/Mass Spectrometry

Techniques and Applications

MODERN ANALYTICAL CHEMISTRY

Series Editor: David Hercules
University of Pittsburgh

ANALYTICAL ATOMIC SPECTROSCOPY
William G. Schrenk

APPLIED ATOMIC SPECTROSCOPY
Volumes 1 and 2
Edited by E. L. Grove

CHEMICAL DERIVATIZATION IN ANALYTICAL CHEMISTRY
Edited by R. W. Frei and J. F. Lawrence
Volume 1: Chromatography
Volume 2: Separation and Continuous Flow Techniques

COMPUTER-ENHANCED ANALYTICAL SPECTROSCOPY
Volume 1: Edited by Henk L. C. Meuzelaar and Thomas L. Isenhour
Volume 2: Edited by Henk L. C. Meuzelaar

ION CHROMATOGRAPHY
Hamish Small

ION-SELECTIVE ELECTRODES IN ANALYTICAL CHEMISTRY
Volumes 1 and 2
Edited by Henry Freiser

LIQUID CHROMATOGRAPHY/MASS SPECTROMETRY
Techniques and Applications
Alfred L. Yergey, Charles G. Edmonds, Ivor A. S. Lewis, and Marvin L. Vestal

MODERN FLUORESCENCE SPECTROSCOPY
Volumes 1-4
Edited by E. L. Wehry

PHOTOELECTRON AND AUGER SPECTROSCOPY
Thomas A. Carlson

PRINCIPLES OF CHEMICAL SENSORS
Jiří Janata

TRANSFORM TECHNIQUES IN CHEMISTRY
Edited by Peter R. Griffiths

A Continuation Order Plan is available for this series. A continuation order will bring delivery of each new volume immediately upon publication. Volumes are billed only upon actual shipment. For further information please contact the publisher.

Liquid Chromatography/Mass Spectrometry

Techniques and Applications

Alfred L. Yergey

National Institutes of Health
Bethesda, Maryland

Charles G. Edmonds

Battelle Pacific Northwest Laboratories
Richland, Washington

Ivor A. S. Lewis

King's College
London, United Kingdom

and

Marvin L. Vestal

Vestec Corporation
Houston, Texas

Plenum Press • New York and London

Library of Congress Cataloging in Publication Data

Liquid chromatography/mass spectrometry: techniques and applications / Alfred
L. Yergey . . . [et al.].
   p.   cm. — (Modern analytical chemistry)
   Includes bibliographical references.
   ISBN 0-306-43186-6
   1. Liquid chromatography. 2. Mass spectrometry. I. Yergey, Alfred L. II. Series.
QD79.C454L5537   1989   89-35915
543'.0894 — dc20   CIP

© 1990 Plenum Press, New York
A Division of Plenum Publishing Corporation
233 Spring Street, New York, N.Y. 10013

Printed in the United States of America

Preface

This book is intended both to be an introduction to techniques and applications of liquid chromatography/mass spectrometry and to serve as a reference for future workers. When we undertook its writing, we chose not to cover the field, particularly applications, exhaustively. Rather we wished to produce a book that would be of use to people just beginning to use the technique as well as to more advanced practitioners. In this regard, we have sought to highlight techniques and applications that are of current importance, while not neglecting descriptions of approaches that may be of significance in the future. We hope that we have succeeded in this. At the same time we hope that the bibliography, with indexes classified by author and title, will make this book of value to those who may disagree with our emphasis.

ACKNOWLEDGMENTS. One of us (C.G.E.) wishes to acknowledge the encouragment of Professor J. A. McCloskey in undertaking this project. All four of us are grateful for the continuous and expert assistance of V. A. Edmonds in the preparation of the Bibliography.

Alfred L. Yergey
Bethesda, Maryland

Charles G. Edmonds
Richland, Washington

Ivor A.S. Lewis
London, England

Marvin L. Vestal
Houston, Texas

Contents

Chapter 1

Introduction

Despite the power and usefulness of gas chromatography/mass spectrometry (GC/MS), there are many compounds that are intractable to analysis by this technique because of the difficulty or impossibility of forming satisfactory thermally stable derivatives. Certain classes of biological compounds that are precharged, phospholipids, for example, or conjugates such as glucuronides or glutathione adducts are particularly subject to this limitation. While mass spectra of such materials can be obtained using many of the newer direct sample introduction techniques such as direct chemical ionization (DCI), fast atom bombardment (FAB), and ^{252}Cf desorption, on-line chromatographic introduction of these kinds of compounds was not possible until the development of a practical liquid chromatographic/mass spectrometric interface.

The development of such a practical interface has occurred relatively recently, and in fact there are no citations in *Chemical Abstracts* under the subject heading "Liquid Chromatography *and* Mass Spectrometry" before 1971, and the number of such citations remained at a very low level for the next several years. However, it should be noted that pioneering work demonstrating the potential for such an interface was done by Tal'roze and his colleagues in 1968 [68TA28]. In the past few years, the field has grown substantially, leading to the publication of about 40 original papers a year for the last five years. To put this number in perspective, however, the recent annual publication rates for LC/MS are equal to about 25% of the papers published in GC/MS and about 4% of the total publications in mass spectrometry.

A major characteristic of any of the interfaces between a liquid chromatograph and a mass spectrometer is the elimination of the need for chemical modification of the sample by derivatization. This elimination becomes a substantial practical advantage that permits the mass spectrometric analysis of nonvolatile, thermally labile, and/or precharged molecules. On the other hand, these interfaces must be able to accommodate the greatly increased molecular flow rates associated with liquid chromatography relative to gas chromatography. Microbore liquid chromatographs operating at flow rates of about 10 μl/min

1

have molecular flows that are comparable to capillary gas chromatographs operating at flow rates of 1 atm·ml/min, as shown in Table 1.1.

Almost all such flow rates can be accommodated by single-stage mass spectrometer pumping systems, typically limited to electron ionization or other low-pressure ionization schemes, using high-vacuum pumping in the range of 300 to 400 liter/s to achieve analyzer pressures of about 3×10^{-6} torr. Certainly, the entire molecular flow equivalent to 10 µl/min can be accommodated by a CI source; in fact, a "makeup gas" is generally required for good CI spectra. On the other hand, Table 1.1 shows that the molecular flow rate developed by a packed column LC operating in the conventional flow range, e.g., 1 ml/min, is about 1000 times greater than that for a capillary GC. As might be expected, these molecular flow rates are about 100 times greater than the pumping capacity of a well-designed, differentially pumped chemical ionization mass spectrometer. In fact, such conventional LC flow rates are even about 10 times greater than the typical 6000 liter/s pumping speed of a cryopumped system. These pumping considerations and the past, present, and possibly future dominance of packed column liquid chromatography make it clear that any practical design of an LC/MS interface must take into account these great differences in molecular flow rates.

In addition to these physical considerations, one might wish to describe other ideal characteristics of a potential LC/MS system. Snyder and Kirkland indirectly suggested many of these characteristics in their 1974 monograph *Introduction to Modern Liquid Chromatography,* by describing the "ideal detector for liquid chromatography," which is summarized in Table 1.2. Synder and Kirkland felt that it would be impossible to attain all nine of these desired features in any one detector, but we suggest that, aside from the destruction of the solute that typically occurs in mass spectrometric analysis, all of these characteristics are well satisfied by modern LC/MS systems.

A similar set of characteristics could be imagined for the "ideal inlet to a mass spectrometer," which might include sequential delivery of samples requiring minimum preparation, and could strongly suggest liquid chromatography as this inlet. Clearly, then this combination of ideal components would require linkage. That is the purpose of this monograph.

TABLE 1.1. Molecular Flow Rates for Various Chromatographies

Input	Molecular flow rate (molecules/s)
Capillary GC, 1 atm · ml/min	4.5×10^{17}
Microbore LC, 10 µl/min	5.6×10^{18}
Packed column LC, 1 ml/min	5.6×10^{20}

TABLE 1.2. Characteristics of the "Ideal Detector for Liquid Chromatography"[a]

The ideal detector should . . .
• Have high sensitivity and predictable response.
• Respond to all solutes, or else have a predictable specificity.
• Be unaffected by changes in temperature and carrier flow.
• Respond independently of the mobile phase.
• Not contribute to extracolumn broadening.
• Be reliable and convenient to use.
• Have a response which increases linearly with the amount of solute.
• Be nondestructive of the solute.
• Provide qualitative information on the detected peak.

---

[a]L. R. Snyder and J. J. Kirkland, *Introduction to Modern Liquid Chromatography*, John Wiley and Sons, New York (1974), p. 136.

There have been a number of successful approaches to the solution of this problem. These range from mechanical transport of solute to the mass spectrometer after solvent removal external to the vacuum system (belt and wire systems) to bulk solution introduction with or without splitting [direct liquid introduction (DLI) or nebulizing] to direct ionization methods involving the entire solvent stream (thermospray, electrospray). These methods are discussed in detail in the following chapters. Applications of LC/MS methods to specific compound types are presented in subsequent sections. The final portion of this volume is a classified bibliography of the LC/MS literature through May 1988. It should be noted that all references to the LC/MS literature that are cited in the individual chapters are given in the format of the codens used in the bibliography and are set off in square brackets ([ ]).

Chapter 2

Direct Liquid Introduction Interfaces

2.1. INTRODUCTION

The most obvious method for connecting a liquid chromatograph and mass spectrometer is to couple them directly, i.e., direct liquid introduction (DLI). Tal'roze *et al.* [68TA28] were the first to publish a method for accomplishing this using nanoliter/minute flow rates from a glass capillary tube. Baldwin and McLafferty [73BA71] showed that a capillary tube containing a solute in a small volume of solvent (~10 µl) could be placed directly in a mass spectrometer source and used to generate chemical ionization spectra of the solute. Neither of these papers resulted in a practical LC/MS interface, but both demonstrated the attractiveness of the principle of direct coupling, giving rise to a literature rich in variations on this apparently simple and direct idea.

The factor common to all the direct introduction approaches is that the molecular flow rate into the ion source is limited to what can be accommodated by a typical pumping capacity as shown in Table 1.1, i.e., 10–20 µl/min. Although this limit can be extended to about 100 µl/min by the addition of a cryopump operating in the 6000-liter/s range, it is clear that all DLI interfaces with conventional LC flow rates may admit only a small fraction of the total flow into the mass spectrometer. On the other hand, the entire effluent of microbore columns operating up to 20 µl/min could be accommodated by typical differentially pumped instruments. In their early stages of development, microbore systems did not provide satisfactory separations in reasonable times. However, the simplicity of a direct interface coupled with the rapid development of microbore and open tubular column liquid chromatography technologies has led to a recent growth in this approach to interfacing [83BR11, 85NI37].

Table 2.1 lists the various types of physical configurations that have been used to effect DLI interfaces, together with selected references that describe important aspects of each approach. Aside from the use of atmospheric pressure ionization (API) with a nebulizer and cryopumping or the more recent use of continuous flow fast atom bombardment (FAB), the mass spectrometric instrumentation is typically a combination of a chemical ionization source pumped

TABLE 2.1. Characteristics of DLI Interfaces

Interface	Liquid to MS (μl/min)	Split ratio	Pumping configuration	Orifice radius (μm)	Source temperature (°C)	Selected references
Conventional LC[a]						
Capillary and	2–20	100:1	DP	30–40	280–320	78HE07
splitter	10	100:1	DP	5	250	81EV86
Diaphragm and	25	20:1	Cryopump	5	300	81DI5′
splitter	10	100:1	DP	5	210	82VO43
	100	10:1	Cryopump	5	800	82HE41
Nebulizer	2–20	100:1	DP	30–45	280–320	81EV86
	~1000	API	Cryopump (& Nozzle/ skimmer)			83TH2′
Microbore LC[c]						
Direct connec-	<20	N/A[d]	DP	100	200–250	81SC65
tion	40–70	N/A	DP	25	270	85HI39
Nebulizer	2–16 (+ 50 (ml/min gas))	N/A	DP	150	120–200	79TS16
Continuous	2[e]	N/A	DP	300	200–250	85IT61
flow FAB						86CA89
SFC	20–40[f]	N/A	DP	100	200–250	82SM43

[a]Liquid flow rates: 0.5–2 ml/min.
[b]DP: diffusion pumping, with differential barrier.
[c]Liquid flow rates: 3–60 μl/min.
[d]N/A: not applicable.
[e]Glycerol + water.
[f]Supercritical fluid.

with either diffusion or turbomolecular pumps and a quadrupole mass filter. The definition of split ratio, the third column of Table 2.1, is the inverse of the ratio of the flow into the mass spectrometer to the total flow, both in ml/min.

The major differences among the several types of DLI interfaces are due principally to orifice size as summarized in the fifth column of Table 2.1. The design of these orifices varies from using a section of narrow-bore capillary, with or without a splitter, to using a diaphragm with a small orifice to using nebulizers. These different approaches represent attempts to improve operating stability and reduce plugging. All of these approaches are somewhat subject to plugging as attested by the provision for rapidly changing orifice or capillary in all of them. Nebulizing interfaces tend to reduce the tendency to plugging some-what, but they are more complex to fabricate than the capillary/orifice interfaces.

The source operating temperature given in the sixth column of Table 2.1, and the interface operating temperature are among the most important considera-tions of DLI interfaces. In general, the portion of these outside of the ion source proper must be cooled, or at least not heated. If the capillaries become too hot, they dry out and plug. In the case of diaphragm interfaces, cavitation associated with boiling interrupts steady flow, and the high temperatures lead to plugging. Although higher liquid flow rates would probably reduce capillary drying at a particular temperature, limitations of gas flow into the mass spectrometer prohibit this. Source temperatures must therefore be set to a level that will allow solvent to be completely removed from the droplets formed at the tubing orifice without drying the direct input capillary or causing boiling in the diaphragm interfaces.

Ionization in DLI interfaces is generally accomplished by using the vaporized solvent as a chemical ionization (CI) reagent gas. Both normal and reversed phase solvent systems have been used with DLI, but solvent systems are typically chosen for their separation power, not for their reagent gas proton affinity. Almost all of the organic solvents that are appropriate for performing a desired separation will be acceptable as a CI reagent gas. Reversed phase solvents have included binary mixtures such as methanol/water or acetonitrile/water mixtures up to a limit of about 60% water as well as quaternary mixtures of hexane/methanol/cyclohexanol/water [85TS70]. In general, it is not possible to include any buffering salts in DLI solvent systems due to the tendency of the capillaries to plug when heated, although Covey and Henion [83CO55] have reported the use of 0.05 M ammonium acetate buffers in methanol/water systems. The interface in the latter case is, however, a combination of DLI and thermospray devices and, as such, may be more immune to plugging than normal for DLI systems. Normal phase solvent systems are chosen for separation power with the expectation of good performance as a CI reagent gas mixture; a typical example of such a system is 50:50 acetonitrile/tetrahydrofuran [85HI39].

A general observation about any of the DLI configurations is that they seem to yield a high degree of chromatographic fidelity. That is, for the cases in the literature which present mass chromatograms in conjunction with UV or refractive index chromatograms, the two sets of curves are indistinguishable. The exception to this at the present time is in the use of the continuous flow FAB interfaces. These devices appear to provide degraded chromatographic peaks from the mass spectrometers, most likely due to tailing effects of the glycerol matrix at the tip of the interface.

2.2. OPERATING PRINCIPLES

The physical basis for operation of the DLI interfaces is a combination of thermal energy input and liquid flow rate. Liquids enter the interface at rates limited only by the pumping capacity of the mass spectrometer. Sufficient thermal energy must be supplied to vaporize them in a manner that yields good vaporization without drying out the interface.

In the larger-flow-rate DLI systems, liquid leaving the orifice of either a diaphragm or capillary inlet breaks into droplets with a size distribution determined by capillary diameter, solvent flow rate, and temperature. At a given temperature, such a system of droplets surrounded by a gas is unstable with respect to the distribution of droplet size, as shown in Kelvin's equation describing the effect of curvature on the equilibrium state[1]:

$$\ln \frac{p'}{p^0} = \frac{v^0}{RT} \frac{2\sigma}{r} \left(1 - \frac{\kappa\sigma}{r} \right) \tag{1}$$

where r is droplet radius; p^0 and p' are the vapor pressure of the liquid and the droplet, respectively, at temperature T; σ, κ, and V^0 are the surface tension, compressibility, and molar volume of the droplet; and R is the gas constant. That is, in a system at equilibrium, as the droplets begin to evaporate and become smaller, the rate of evaporation of small droplets increases, and the vapor accumulates onto the bigger droplets. Clearly, the DLI desolvation process is not an equilibrium one, and the growth of larger droplets is not expected. On the other hand, one can expect accelerating solvent losses from small droplets (<1 μm) which will lead to their evaporative cooling. A suggestion made by Bruins and Drenth [83BR11] to explain their observation of droplet streams changing direction under vacuum is consistent with equation (1).

Conductive heat transfer from the ion source must be sufficient to reduce the droplets to vapor by the time they reach the region of the source in which ions are formed. If the droplets were allowed to evaporate completely in the absence of further thermal input, the desolvation could be considered a completely adiabatic process. A purely adiabatic system would involve long times or distances from the formation of droplets to the production of vapor for ionization, i.e., very large desolvation chambers. In all of the sources, many of which use desolvation chambers, high temperatures in the source and desolvation chamber are used to speed droplet evaporation. Systems such as these, however, cannot be considered completely adiabatic although adiabatic droplet cooling probably exists in the region immediately around the final inlet orifice. Of course, in the absence of additional thermal energy input to the system, freezing of the droplet stream might be expected and, in fact, has been observed by both Arpino and Beaugrand [85AR45] and Bruins and Drenth [83BR11].

The situation in open tubular column systems is different, and there are two current points of view on how best to operate these interfaces. Bruins and Drenth [83BR11] advocate balancing thermal energy input with solvent flow rate so that all evaporation occurs just at the tip of the capillary. Too little heat into the interface leads to liquid flowing into the mass spectrometer. Too much heat leads to drying of the capillary. These points are shown in Figure 2.1. In their use of 25–50-μm capillaries, these workers have found that the balanced heat/solvent flow approach leads to very stable source pressures and hence to very stable mass spectra. On the other hand, Niessen and Poppe [85NI37] generate liquid jets in

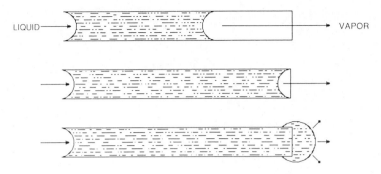

LIQUID →   → VAPOR

FIG. 2.1. Three possible situations during evaporation of a liquid from a capillary into a vacuum. (Reproduced from 83BR11 with permission.)

their open tubular column system. The jet was formed by spraying effluent through a small orifice. They found that at typical flow rates of 0.006–0.6 μl/min through a 5-μm i.d. column, orifice sizes (0.04–0.8 μm i.d.) which gave liquid jets were too subject to plugging. They generated liquid jets by adding solvent to the column effluent to give a total flow of about 20 μl/min. Under these conditions, 4-μm i.d. diaphragms were used to give stable flow. An advantage to this approach is that the makeup liquid can be chosen for its chemical ionization reagent gas properties since it is not involved in the separation.

Nebulizing interfaces use a stream of gas at the end of the interface capillary to assist in breaking the liquid stream into droplets for subsequent desolvation. These interfaces fall into two classes, vacuum and atmospheric nebulizers. Vacuum nebulizers deliver their droplets into a subatmospheric pressure region, often to the ion source itself, and were first reported by Tsuge, Hirata, and co-workers, who used a modified GC jet separator as the basis of their interface [79TS16, 79HI46]. The atmospheric nebulizers deliver their droplets at atmospheric pressure, and molecules or ions of interest must be then coupled into the mass spectrometer. Atmospheric pressure ionization sources are typically used for this. The atmospheric nebulizer applications reported to date use nebulizer and pressure coupling technology developed by Sciex Corporation.

Nebulizing gas is introduced through a coaxial tube surrounding the capillary containing LC effluent. The gas tube is terminated in a manner to give the best droplet formation. The molecular ratio of liquid to nebulizing gas is about 1 : 1000 at typical flow rates (10–40 μl/min liquid, 10–50 atm·ml/min gas) for a vacuum nebulizing system. It is this high gas/liquid flow ratio that breaks the liquid stream into droplets than can then be desolvated. Vacuum nebulizer systems are cooled in the region where the gas and liquid streams are in separate tubes, but heated after nebulization in order to augment desolvation. It is clear that, while the molecular flow into the low-pressure region of vacuum nebulizing systems is substantially greater than that encountered with a simple DLI microbore system, the gas load is within the range of the pumping capacity of a well-designed CI instrument, or a CI system equipped with a jet separator. In the case of the atmospheric nebulizer system, the use of cryogenic pumped nozzle/skimmer systems has been the only way of maintaining satisfactory mass spectrometer pressures.

2.3. SPECIFIC DLI INTERFACES

2.3.1. Capillary Inlets

Direct coupling of LC effluent into a mass spectrometer source was done first by the simple expedient of using a capillary tube between the two devices. This approach is shown schematically in Figure 2.2. The capillary can be a commercial heavy-walled glass capillary of about 6.25 mm o.d. which is ground down so that the tip is about 5 mm o.d. with an orifice estimated to be about 5 μm [78HE07] as shown in Figure 2.3. The heavy-walled tube is used as a modified solids inlet probe. An alternative approach is to draw down an 8-mm o.d. × 1-mm i.d. capillary so that the final tube is a 0.25–0.3-mm o.d. × 30–40-μm i.d. orifice [81EV86]. In the latter case, the capillary is mounted inside a modified solids inlet probe. The value shown in Figure 2.2 is required only when working with conventional flow rates in order to adjust the split ratio of the system.

When microbore systems are used, the lower gas loads are compatible with typical CI pumping capacities, and thus such systems can couple the LC effluent directly into the source by using a straight connection from the microbore col-

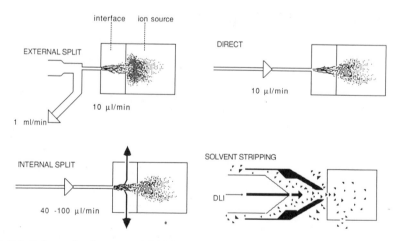

FIG. 2.2. Schematic diagram of DLI in split and direct modes. (Reproduced from 83SU47 with permission.)

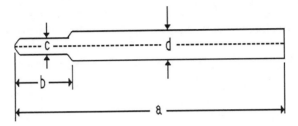

FIG. 2.3. Schematic of an all-glass interface probe for direct LC/MS coupling. Dimensions are: a, 25 cm; b, 2.5 cm; c, 5 mm; d, 6.25 mm. The capillary i.d. is 0.075 mm and the orifice at the probe tip is approximately 5 μm in diameter. (Reproduced from 78HE07 with permission.)

umn. A number of different strategies have been used to effect the coupling of the microbore or open tubular columns to the mass spectrometer. In general, they all use a capillary that is smaller than the column i.d. but is coaxial with the last segment of the column. Schaeffer and Levsen [81SC65], as shown in Figure 2.4, used a 0.1-mm i.d. stainless steel capillary (the smallest available at the time) nested inside a larger capillary to make the connection, while Bruins and Drenth [83BR63] used a larger, 0.25-mm i.d. capillary. Eckers *et al.* [82EC82] used a 35-cm long × 10-μm i.d. glass capillary. The steel capillary has the advantage of ruggedness but requires coupling the column to the interface with a fitting of PTFE polymer or similar material. The small-bore glass tubing interface is less rugged but is much simpler to use. More recently, a number of authors [85HI39, 85TS70], have emphasized the importance of 25-μm i.d. fused silica capillary. Hirter *et al.* [85HI39] coupled their 1-mm i.d. microbore column to the 25-μm fused silica through a standard zero-dead-volume fitting butted against a column end frit. The final assembly, including tip heaters, fits insider their standard 1/2-inch o.d solids probe. A potentially significant aspect of the paper by Tsuda *et al.* [85TS70] is that, by adjusting split ratios to 1 : 1000–1 : 1500, they obtained EI

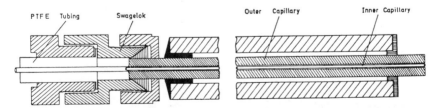

FIG. 2.4. Microbore LC/MS interface. Stainless steel inner capillary: 0.1 mm i.d., 0.2 mm o.d., 30 cm long. Stainless steel outer capillary: 0.25 mm i.d. Outer tube: 5 mm o.d. Connectors shown. (Reproduced from 81SC65 with permission.)

spectra at the 10-ng level of compounds injected on column. Niessen and Poppe have demonstrated an interface using 5-μm i.d. columns at flows of 0.006–0.6 μl/min accompanied by makeup solvent flows for a total flow of about 20 μl/min into the souce [85NI37], but as noted above, their system is really a diaphragm interface.

2.3.2. Diaphragm Interfaces

Diaphragm interfaces employ a small orifice (ca. 5 μm) at the end of a large-bore capillary (ca. 100 μm i.d.) to introduce liquids into the mass spectrometer ion source. This type of inlet, shown in Figure 2.5, was first announced by Serum and Melera [78SE5'] with important contributions within a short time by Arpino *et al.* [79AR59, 79AR6'] and Henion [80HE815]. An important consequence from the developments of these groups is that they led to commercial products, thereby increasing the rate at which LC/MS techniques spread.

Figure 2.5 shows the salient features of diaphragm interfaces. Particularly important are the need for cooling the capillary prior to the diaphragm to prevent solvent boiling before ejection, the capability for removing plugged diaphragms, and the splitter valve required when working at conventional flow rates. Cooling is typically done with water flowing within an enclosed jacket, but radiative cooling was used in one of the later diaphragm sources [82DE42]. The ranges of flow rate and split ratio that have been used in diaphragm interfaces are summarized in Table 2.1.

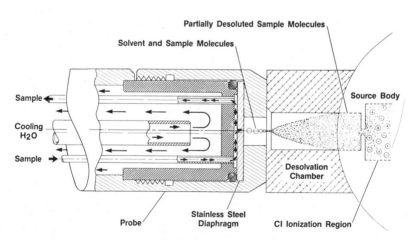

FIG. 2.5. Schematic diagram of diaphragm LC/MS probe inlet showing desolvation chamber. Diaphragm orifice: 5 μm. Capillary i.d.: 100 μm. (Reproduced from 81DI5' with permission.)

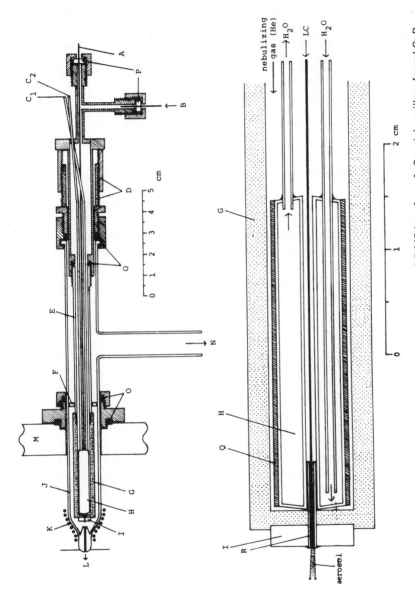

FIG. 2.6. Schematic of improved vacuum nebulizing microbore LC/MS interface. A, Coaxial capillary from LC; B, nebulizing gas entry; C, cooling water in and out; D, probe position adjuster; E–H, mounting tube and cooling waterjacket (G is of Macor); I, Cu disk orifice; J, Pyrex liner; K, heater. (Reproduced from 82YO14 with permission.)

2.3.3. Nebulizing Interfaces

Tsuge, Hirata, and co-workers' first vacuum nebulizing interface [79TS16, 79HI46] was an extensively modified GC/MS jet separator. A thin-walled stainless steel capillary, 150 μm i.d., with a 130-μm central wire was placed on the central axis of a Pyrex nebulizing tip drawn down to 400 μm i.d. at the end. This assembly was placed on the central axis of the high-pressure side of the separator. Gas at 50 atm·ml/min was passed through the nebulizing tip while LC effluent, flowing at 2–16 μl/min, was passing through the central capillary. The normal action of the jet separator removed the bulk of the nebulizing He, and the remaining material entered the heated, low-pressure side of the jet separator. Later versions of this interface [82YO14], shown in Figure 2.6, used a less elaborate nebulizing tip that permitted water cooling the high-pressure side of the system.

Somewhat simpler nebulizing interfaces have been described by Evans and Williamson [81EV86] and Apffel and co-workers [83AP50]. Both systems work in the flow ranges of the device described above, although Evans and Williamson used a 100 : 1 split ratio to reduce liquid flows from conventional to microbore levels. The simplifying aspect of these two interfaces is that the interface is essentially a modified solids probe that avoids the need for the more complex Pyrex pumping apparatus described by Tsuge, Hirata, and co-workers. A capillary of about 100 μm i.d. and 35 cm long is placed inside a larger tube through which nebulizing gas passes. Evans and Williamson pumped the tip of the device in addition to inserting it directly into the ion source. Apffel *et al.* dispensed with this additional pumping.

2.4. INTERFACES USING UNUSUAL FLUIDS

Both supercritical fluids and glycerol can be introduced continuously into mass spectrometer sources. Supercritical fluids can be used for chromatography (SFC) but are liquids only because of the constraints of high temperatures and pressures. Glycerol is too viscous to be used as the sole mobile phase for liquid chromatograpy but, when diluted, can be used for this purpose. Continuously flowing dilute glycerol may then serve as the matrix for fast atom bombardment. While the physical configurations for both of these interfaces are similar to typical microbore capillary interfaces, their increasing importance requires that they be discussed separately.

2.4.1. Continuous Flow FAB Interfaces

The ability to obtain mass spectra of high-molecular-weight compounds is one of the great advantages of FAB mass spectrometry. A device that would couple this ability with direct liquid chromatographic sampling would be of considerable importance. Ito *et al.* [85IT61, 86IT81, 86MI67'] were the first to describe such a continuous flow FAB interface. Their device used a syringe pump-based microbore LC system operating at 2 μl/min. Their liquid phase was 10% glycerol in a 50:50 mixture of acetonitrile and water. The entire effluent was passed through a glass frit at the source end of the capillary. The frit served as the FAB target and could be positioned in three dimensions within the source. These authors showed very reproducible spectra that showed virtually no evidence of compound memory.

Caprioli and co-workers [86CA1', 86CA89] are the other principal exponents of this technique. This group does not employ the frit at the source end of the capillary. Their interface consists simply of a 75-μm i.d. fused silica capillary passed through a hollow probe and terminated about 0.2 mm beyond a 0.3-mm hole drilled in the copper FAB probe tip placed at its normal position in the source. This interface is shown in Figure 2.7. The interface typically operates at 2–10 μl/min of a 2:8 glycerol/water mixture containing 0.2% FTA. The effluent flow rate must be adjusted so that there is a constant ion signal. This is achieved by balancing the flow rate of material into the source with the system pumping speed. Pumping speed is best controlled by controlling evaporation rate at the probe tip by varying tip temperature and liquid volatility. Such a balanced input leads to no excess material at the tip which would lead to contamination in the spectra of subsequent compounds.

In a report [86CA1'] of the direct injection spectrum of substance P, a peptide with a molecular weight of 1347 daltons, Caprioli and Fan noted that the background of glycerol matrix peaks was reduced substantially from the typical FAB observation. This higher than normal signal to noise was also seen in the spectra of oxidized bovine insulin (MW 3493) and the B chain of intact bovine insulin (MW 5730) [86CA89]. Published results from the use of this technique to achieve real separations before obtaining mass spectra, as opposed to the demonstration of spectra from the flow injection of single compounds, have only recently appeared [87HU24].

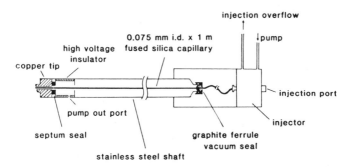

FIG. 2.7. Schematic of continuous flow FAB probe. Orifice in Cu tip is 0.3 mm i.d. (Reproduced from 86CA89 with permission.)

2.4.2. SFC Interfaces

Supercritical fluids are materials that are constrained by pressure and temperature to exist in a single, fluid state. As chromatographic solvents, they offer the advantage of controllable solvating power by means of pressure variation. An additional advantage is the potential of greater separatory power than possible with liquids as a consequence of greater solute diffusivity in the supercritical fluid. Supercritical fluids are solvents for high-molecular-weight and heterofunctional compounds and have been used extensively for separations of hydrocarbons, substituted aromatics, and polymers. While chromatography with supercritical fluids is carried out at temperatures considerably below those used for GC, the temperatures are elevated appreciably over the ambient temperatures typically used for LC. A capillary column SFC interfaced to a mass spectrometer offers the potential of a simple DLI interface that operates at low mobile phase flow rates using high-volatility solvents.

Randall and Wahrhaftig [78RA03, 81RA23] were the first to report an SFC/MS interface. In that work, they used packed column SFC and a supersonic nozzle/skimmer arrangement for pressure coupling the effluent into the mass spectrometer. The lack of chromatographic efficiency of the packed column coupled with the complexity of the interface delayed further work on SFC/MS until capillary column techniques were developed.

Smith and co-workers have published extensively on the development and applications of capillary column SFC/MS since announcing their first straightforward, and essentially unchanged to the present, interface [82SM43, 83SM56]. The interface, illustrated in Figure 2.8, is the diameter probe that has interface heaters, and in later versions cooling capability as well, running axially. The interface is maintained at the same temperature as the column. Typically flow rates of 20–40 µl/min of supercritical fluids at 20–40 atm (equivalent to 10–15 µl/min of liquid) are used. Column temperatures vary with mobile phase critical temperature, and of course must be at least equal to that temperature. This interface does not require a desolvation chamber since cluster sizes are

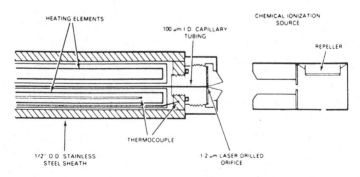

FIG. 2.8. Schematic of SFC interface. (Reproduced from 83SM56 with permission.)

10^6–10^9 smaller than obtained with liquid interfaces having similar dimensions, as calculated using expressions developed by Randall and Wahrhaftig [81RA23].

ADDITIONAL REFERENCES

1. W. Thompson, *Phil. Mag. 42*, 448 (1871).

Chapter 3

Mechanical Transport Devices

3.1. INTRODUCTION

Mechanical transport devices have in common the feature that the reduction in solvent volume and concurrent concentration of solute is performed in a region physically removed from the ion source of the mass spectrometer. Solutes are subsequently sequentially transported by some physical means into the ion volume for ionization and analysis. Devices of this type may be conveniently divided into two distinct categories: on the one hand, those which purport to offer a continuous representation of solute composition at the exit of the chromatograph, termed continuous monitors, and, on the other hand, those which, by design, segment eluent into fractions which, following concentration, are presented to the spectrometric ion source. The latter, termed storage devices, may broadly be classified as mechanized fraction collection systems and are therefore not, in the strictest sense, direct liquid chromatographic–mass spectrometric unions. In either case, within tolerable limits, systems have been demonstrated to exercise the requisite pressure reduction between chromatograph and spectrometer. The requirements for pressure reduction are variable, being a function of the rate of liquid flow, the speed of transportation, the nature of the enrichment device used, and the pumping efficiency of individual systems. In any event, both of these approaches offer the significant theoretical advantage that the ionization of solutes can be made independent of any influence of the mobile phase. This virtue should not be underestimated. By way of emphasis, it is worth making the point that transport devices offer the possibility of measurement of true electron impact, chemical ionization (with free selection of reagent gas), fast atom bombardment, and other spectra. That is, the spectra are not modulated by the mobile phase, the advantage being that the measured spectra are comparable with those acquired by direct means. These devices therefore offer the analyst spectra with which he or she is likely to be familiar and, in particular, which are comparable with library spectra, where they exist.

Of the two approaches, that which yields a continuous report on the elution of material from the column is idealistically the more attractive. Nevertheless, both warrant discussion.

3.2. CONTINUOUS TRANSPORT SYSTEMS

3.2.1. Moving-Wire Interfaces

Scott and his colleagues [74SC95] modified a moving-wire gas chromatography detector, as manufactured at the time by Philips Chromatography, such that it entered the source region of their mass spectrometer. In doing so, they engineered a system which satisfied the minimal requirements of a combined liquid chromatography/mass spectrometry system. That is to say, the device permitted the introduction of sample from solution at atmospheric pressure into a mass spectrometric ion source at high vacuum. They successfully demonstrated the principle, by the enrichment, transportation, and ionization of several materials including diazepam, phenobarbital, and vitamin A. For a time, these researchers continued to develop the system and report on its potential [77SC13, 79SC95]. They were, however, certainly aware of its limitations in two major respects: firstly, its lack of sensitivity, and, secondly, its limited ability to successfully vaporize sample, albeit relatively efficiently introduced into the ion source. In suggesting the use of heater coils, infrared radiation, and lasers as possible solutions to the latter problem, and consequently possibly the former, they managed to presage much of the subsequent development of continuous transport systems.

3.2.2. Moving-Belt Interfaces

3.2.2.1. Introduction

Current manifestations of this type of interface owe their existence to Scott's pioneering work, in the midseventies, on the moving-wire interface. In a remarkably short period, through the work of McFadden and co-workers, the Finnigan Corporation, which had supported Scott and his collaborators, developed [76MC29, 77DA54], patented [77MC87'], promoted [77MC7', 77MC56], and commercialized a transport interface. The interface, schematically represented in Figure 3.1, introduced, in place of the moving wire, a moving belt in the form of a band, which was made to continuously cycle within an enclosed housing. Chromatographic eluent is deposited on the belt, which is supported on a pulley driven by means of a variable-speed motor. The band passes below an infrared heater, whose output is adjusted to evaporate most of the mobile phase before the belt reaches the entrance slit to the first vacuum lock. The pressure drop between atmosphere and high ion source vacuum is maintained through stepwise reduction by means of two differentially pumped vacuum locks. The

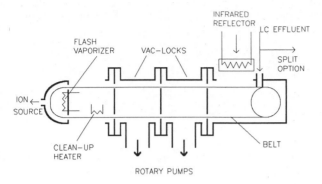

FIG. 3.1. The original Finnigan Corporation moving-belt interface, presented in schematic form, by kind permission of the Finnigan Corporation.

belt in turn passes into the ion volume through the locks, over a heater which vaporizes sample into the ion source; then over a further heater which cleans the belt by evaporation of any residue, and then out through the locks to recycle. The introduction of a belt instead of a wire presented the user with a flat, relatively broad surface area for sample deposition. Such a surface naturally enough is capable of higher, and more efficient, sample loading than the equivalent moving wire. The efficiency of sample delivery, measured as the percentage parent ion current afforded after transportation with respect to that obtained from the solids probe, was shown to be 30–40% [77MC7'], an efficiency, albeit compound dependent, comparable to that of jet-separator GC/MS interfaces. In part thereby, the inventors overcame one of the problems associated with the moving wire, i.e., its limited sensitivity. The basic design of this form of interface, irregardless of source of origin, remains essentially unchanged to date. In 1979, shortly following the introduction of the original Finnigan system, VG Micromass introduced a comparable device [79BA9', 79YO4'] as did, a little later, in 1981, Finnigan MAT [81BR0']. No other manufacturer has since entered the field; however, a number of research groups [80BE59, 81HA32, 81SM33, 85ST73] have since constructed similar devices in satisfaction of particular requirements.

To date, over 200 papers and reports have been presented that either cite examples of the use of the interface or report its continuing development. Two research groups, in particular, have maintained a constant output. They are those of Games at University College Cardiff, beginning in 1977 [77GA5'], and of Karger of Northeastern University in Boston, which became involved in 1979 [79KA6', 79KA14]. Until recently, manufacturers have, in general, albeit perhaps a little reluctantly, consistently tried to improve their systems in the light of both academically inspired advances and developments in other areas of mass spectrometry. Between them, they have produced and sold well over 100 interfaces. Superficially, these facts may lead the reader to believe that this type of interface has achieved universal acceptance and that it is widely used. Closer examination, however, reveals this not to be the case.

3.2.2.2. Early Applications: Volatiles by EI and CI

The number of groups reporting the use of the belt has from the time of its introduction remained small. A large majority of the reports emanate either from the two groups mentioned above or from representatives of one or another of the manufacturers. The generally expressed explanations for the apparent lack of use of the belt are that the system is relatively difficult to set up and that it can be unreliable. Therefore, it appears, the moving-belt interface is rarely, if ever, fitted, particularly in laboratories required to carry out a wide range of analyses of various other types. Nevertheless, the moving belt has been successfully applied to a wide range of problems. It has been refined over the years, being made simpler to use. Its potential has also been considerably extended. Despite the fact that it remains in use only in the hands of a relatively small number of researchers, it continues to occupy an important position as an LC/MS interface.

At the time of the introduction of the first moving-belt interface in 1976, few institutes were equipped with field desorption (FD),* the only viable desorption ionization technique then commercially available. Consequently, virtually all mass spectrometry laboratories were limited to the analysis of relatively volatile, thermally stable compounds using electron impact (EI) or chemical ionization (CI), regardless of the means of sample introduction. Understandably, the moving belt was also, therefore, restricted to EI and CI, and consequently to compounds amenable to those ionization techniques. The principle of a belt interface in combination with FD ionization was patented [78BR7'] but never pursued. In principle, therefore, the system was superficially restricted to compounds of chemical classes and in the range that might equally as well have been analyzed by combined gas chromatography/mass spectrometry (GC/MS). At first sight, given the fact that by then GC/MS was fully established as a reliable and sensitive combined chromatographic spectrometric technique, the moving belt might easily have been considered a technique without a clearly defined, if any, role. On closer examination, however, the early decision to develop and make available the moving-belt system, was laudable. The logic behind its introduction was quite straightforward and remains perfectly valid. The fact that a mixture may in theory be analyzed by GC/MS does not preclude the desire or indeed the requirement for analysis by LC/MS. Liquid chromatography is more common than gas chromatograpy. Chromatographers most often require direct one-to-one correspondence between spectra and chromatographic components which have been separated by established liquid methods. They are often justifiably unwilling to develop a corresponding GC/MS method, which invariably

*H. D. Beckey, *Principles of Field Ionisation and Field Desorption Mass Spectrometry*, Pergamon, London (1977).

leads to scrambling of the order of compound elution, with respect to that obtained with liquid methods. Other significant factors that might incline an analyst to favor LC/MS over GC/MS are the length of time spent in workup, the relative speed of analysis, the greater separative versatility of LC/MS, and an individual analyte's on-column behavior, be it derivatized or not. Of course, these arguments are applicable in favor of any LC/MS system. However, as stated above, one significant and particular advantage offered by the moving belt that does set it apart from other available LC/MS systems is its ability to consistently yield true EI data. One other system, based on the generation of a monodisperse aerosol [84AB6', 87WI72], i.e., MAGIC, claims this advantage. Recent results and the growing commercial importance of these interfaces suggest this is true. Further discussion of these interfaces can be found in Chapter 5.

The moving belt, in its original form, however, rapidly established a track record in the analyses, by electron impact or chemical ionization, of a wide range of materials. In a short interval following its introduction, it was demonstrated practicable in the study of alkaloids [80EC86, 82EC92, 82EC93], antibiotics [80GA17, 80RO5', 80EC86, 81KE6', 81TW8', 82AB47], pesticides [76MC29, 78WR7' 79KA14, 80GA6', 80GA17, 80EC86,], polynuclear aromatics [77DA54, 78DA69, 79HI5', 79DE8', 80ST87], chlorinated aromatics [78DY9', 79DY95, 79WR2', 81WR85], herbicides [80SK88], petroporphyrins [78MC2', 79MC78, 80EG23], pharmaceuticals [80GA17, 77MC56, 80KI95, 80RO5', 78GA5RE', 80GA83, 81MA11, 82MA15], nucleosides [80GA17, 81GA31, 80GA73, 80QU2'], peptides [80GA17, 81GA31, 81YU89], coumarins [80GA17, 80GA72, 80EC86, 78GA5RE', 78GA51], steroids [80GA17, 80MI89, 78GA5RE'], sugars [81GA84, 81GA31, 80GA73], lipids [77MC56, 77ER27, 78PR11], aromatic acids [79KA14, 80KA81], aflatoxin [79MC72, 77MC56], bile acids [80GA17, 80QU2'], phenolics [80TH4', 81TH87], liquid crystal mixtures [80MA95, 82MA34, 81MA33], rotenoids [81GA31, 81WE43], and waxes [79MC72, 79MC78].

As a consequence of the lessons learned during these early studies, and in the period immediately following them, the belt was gradually improved. It was made more reliable and more versatile. Incidental advantages, which followed from the system's design, were quickly appreciated and used to advantage. For example, it was observed [78GA51] that involatile or labile compounds which failed to yield abundant parent ion currents by conventional direct insertion methods would do so when measured from the belt. The rapid volatilization of sample in transit across the tip heater was offered as an explanation for this phenomenon. A steady progress in the system's ability to yield spectra from relatively labile and involatile species followed and was demonstrated [84GA17] consequent to the introduction of the belt's sample evaporator into the ion volume [80MI89], and then into the electron beam [81BR0']. These improvements expanded the range of materials amenable to analysis, permitting the analysis, for example, of trisaccharides [84GA17], improving in many cases on direct probe sample volatilization. Nevertheless, the gain was a limited one. Difficulties encountered in sample evaporation undoubtedly contributed to the system's reputation of being unreliable. Given the absence of any viable alternative when confronted with involatile materials, it was the analyst's natural tendency to drive the sample heater over and above the limit of the belt's tolerance.

3.2.2.3. Development

The original means of solvent removal also contributed to the system's fragility and inefficiency. As stated above, all commercial examples of this type of interface were equipped with differentially pumped regions, between the point of solvent application and the spectrometric ion source, which easily yielded the requisite pressure reduction between atmospheric pressure and the high-vacuum source. All early models, however, required the direct application of solvent to the belt and solute enrichment by indirect irradiation of the belt, typically using an infrared source. In operation, they therefore required careful balance between the rate of solvent deposition and speed of transportation in order to maintain that pressure differential and, thereby, successful solute ionization. Interfaces of this type, quite tolerant of relatively high organic solvent flows, were barely able to effectively accomplish desolvation of substantially aqueous phases, even at reduced flow through conventional columns. The positioning of additional heaters in the vacuum locks was proposed [80DY4'] as an aid to more effective desolvation. Nevertheless, in the main the analyst had little choice other than to run the lamp at high power, thus risking damage to the Kapton polyimide belt, which is fitted to all conventional systems. In certain cases, the use of segmental flow extractors, placed between the chromatograph and the interface [79KA6', 79KA14, 80KI4', 80KA81, 80KI95, 81KI39, 82KI25, 82VO6', 82VO15, 85VO7', 85VO9'], which extract solute or ion pairs thereof into an organic phase, was shown to effectively bypass the problem, in that the interface was thereby required to contend with the removal of the organic phase, rather than the much more difficult aqueous phase. The simple expedient of reduction in bulk liquid flow to the interface was employed, either by splitting the volume of eluent or by the use of microbore columns, [81GA6', 82AL99, 82GA02, 82TA0', 82WE91, 82GA95, 83LA69, 83AL83, 84KA2', 85HA53', 85HA53]. The latter of these solutions confers its own particular advantages, i.e., narrowed chromatographic peaks, a subject reviewed by Bruins [85BR39]. The use of mobile phases constituted of high proportions of water was restricted in those early interfaces in a further respect, which is that the belt material does not lend itself to wetting and, therefore, to even distribution of solute. It was shown that the addition of water-miscible solvents, e.g., 2-propanol [82AL15], or detergents [81KE6'] improved the wetting characteristics of the belt, resulting in smoother deposition from aqueous phases. All of the above present themselves as partial solutions to the problems of effective desolvation and even deposition. One approach, however, would appear to address both problems directly, namely, the use of spray deposition devices. They range from gas-assisted nebulization devices, introduced [81SM39] by Smith and Johnson, to the use of thermospray [80HA6', 81HA32, 82HA0', 83HA5', 87KR49] and electrospray [83HA5'], the latter applied to a metal-ribbon system. The devices vary in their ability to handle bulk flows. Thermospray and nebulization devices are capable of handling relatively high volume flows, the former (*vide infra*) of conventional column eluents and the latter reportedly [81SM39] of up to 0.7 ml/min of aqueous phase. Electrospray remains a low-flow device ($\sim$10 μl/min). All these devices vary in their efficiencies of deposition, a dependency not only related to the nature of the mobile phase, but also very much related to the nature of the sample. Clearly, volatile species may be lost along with the mobile phase, and labile materials may suffer decomposition. Gas-assisted nebulizers are capable of high deposition efficiencies. They result in smooth sample deposition and have been demonstrated to be significantly preferable to direct deposition in that they contribute very little to band broadening [83HA55, 83HA1']. A number of devices modeled on the Smith and Johnson approach have been incorporated into commercially available systems [85KI7']. They have effectively replaced the indirect heating method of solvent removal. They significantly simplify the use of the belt; the analyst is no longer required to balance solvent flow rate and belt speed with the heat output of the solvent evaporator. They contribute greatly to the device's reliability, in that they allow the removal of the infrared lamp, a source of heat damage to the belt.

A considerable number of LC/MS analyses using moving-belt systems, incorporating one or another of the above-mentioned improvements, have been reported. It is clear that any material that can be analyzed by EI or CI via direct probe methods, or indeed by GC/MS, will consistently yield interpretable spectra from the moving-belt system, with little loss in sensitivity. The loss in absolute response is more than compensated for, in the analysis of complex mixtures, by the gain in specificity, provided adequate, but not necessarily complete, chromatographic separation is possible. In most cases, indeed, the systematic loss of sensitivity can be compensated for by the modification of chromatographic conditions so as to minimize the bandwidth of analytes. The reliability of modern moving-belt interfaces provides for the straightforward, systematic, and routine analysis of such materials. Aromatic hydrocarbons and their derivatives clearly fall into this classification, and there are plentiful examples of their analysis [82GA95, 82TA0', 82WE91, 83FO08, 83SM87, 85KR73, 85KR72, 86MO31, 86QU5', 86HE11]. A considerable number of analyses of lipid materials have been reported, from early studies using in-line catalytic reactors [77ER27, 78PR11] to analyses of derivatized sphingoid bases [84JU59], derivatized fatty acids [85BA25], intact free glycosphingolipids [83EV3', 87EV49] and their derivatives [83MC1'], intact phospholipids [84JU58, 83EV0'], triglycerides [81PR26, 85HO2'], and acetylated monoglycerides [87FI8']. Analyses of pesticides of the carbamate and urea class [83CA04, 82WR01] and of the relatively high molecular weight perchloro cage class [82CA43] by chemical ionization have been reported. The latter report, demonstrating residual detection limits at the 1-ppm level, emphasizes the advantage of the freedom to select reagent gases that this system provides. It compares the effects of the use of ammonia and methane as reagent gases, reporting the loss of molecular weight information with the former. The detection of aldicarb at the sub-ppb level by a method requiring no derivatization and minimal cleanup [82WR80] and a direct comparison favoring LC/MS over GC/MS for organophospate insecticides [83WH68] confirm the potential sensitivity and utility of the belt system. Cautionary evidence as to the method's applicability to quantitative studies is offered [83CA03] by the confirmation of a concentration-dependent competition between thermal degradation and sample vaporization for thermally labile materials. Nevertheless, the use of the system in quantitative studies, using labeled analogues, has been demonstrated. Examples are in the identification and quantitation of diketopiperazines [87VA47] and of a number of drugs and their metabolites [77GA5', 80MI89, 82MA15], including steroids [85VA37]. The use of labeled compounds in the detection and identification of metabolites is illustrated in drug studies as exemplified in the following. Several amphetamines have been analyzed [87HA89] following their separation on chiral phases. Comparison of detection limits by low- and high-resolution selected ion monitoring (SIM) for a number of drugs, including bromazepam, favored the high-resolution method, which was also shown to permit the determination on-line of the mass of the drug, at residue levels, to within 5 ppm [86TA06]. Few other examples of accurate mass determination have been reported since the earliest on record [82BA91], although the precision and accuracy of the system is clearly demonstrated in a study by Missler *et al.* [86MI6'].

This survey by no means represents a comprehensive listing of the use to which the belt has been put but is rather intended to illustrate a number of the belt's particular characteristics. It does, however, give some indication of the belt's utility and serves to illustrate that a considerable body of analyses may be performed by means of the belt interface in its electron impact/chemical ionization form. Analyses of the more important chemical classes are discussed elsewhere, and in those cases comparisons between individual LC/MS techniques are made.

3.2.2.4. The Analysis of Involatiles: Alternative Ionization

All the examples presented above involve the investigation of relatively volatile species. As stated above, the analysis of involatile materials using the belt system, in its earliest form, is severely restricted. In particular, the analysis of biopolymers, i.e., glycosides, glucuronides, saccharides, peptides, and nucleotides, is limited to the smallest in each class. A number of investigations have established the optimization, defined spectral characteristics, and outlined the analytical benefits of the device in each case. For example, a study of the fragmentation behavior of a number of relatively small, underivatized glucuronides [82CA46], using NH_3 and ND_3 as reagent gases, led to the determination of the composition of ions common to their spectra. These ions may therefore be taken as characteristic of structure. Conditions under which a number of glycosides may yield intense negative ion spectra by chloride attachment have been described [81GA31, 84LE18]. Early studies of amino acids and peptides, cited above, have in time been supplemented by a number of others in which the device's capabilities have been assessed [81GA84, 83YU03, 83RO87, 84YU15]. Most of the latter are referred to in an investigation [84GA4'] reporting the analysis of phenylthiohydantoin (PTH) and dansyl amino acids, as well as BOC- and N-acetyl-N, O-permethylated peptides up to and including examples of tetrapeptides, which concludes that the interface is not competitive with established peptide sequencing methods. The latter argument, incidentally, is only applicable when the peptide in question is composed of the more common amino acids. As might be expected, nucleoside [81GA84] and nucleotide [83GA93] analyses using the belt interface in its electron impact or chemical ionization form fare little better in the investigation of higher-order oligomers.

It will be as clear to the reader as it was to a number of research groups that the dependence of any system on the thermal evaporation of samples would preclude its use in the analysis of plainly involatile materials. Lines of investigation of means whereby involatile compounds might be analyzed using this type of interface were researched as soon as any viable possibility arose. They include the use of in-line chemical reactors [77ER27, 78PR11, 81PR26, 85VO6', 85VO9', 86TS82], which suffer from the disadvantage that they are not generally applicable; the application of particle beams, i.e., either conventional fast atom bombardment (FAB)* [82KE6', 83DO65] or secondary ion (SIMS) systems [79BE6', 80BE59]; and the use of lasers, i.e., laser desorption (LD).

The application of laser desorption, patented [80NO9', 80RE2', 80WE5'] and concurrently applied [80HA6', 80HU0', 80SM0'], has in essence been pursued by one group, that of Vestal at the University of Houston. This group built a system housing a stainless steel band from which spectra could be measured following its exposure to pulses from a Nd:YAG laser [81HA9', 81HA32]. In their studies [82HA0', 83HA5'], which were subsequently more fully reported [84FA60], the analysis of amino acids, nucleosides, nucleotides, and small peptides with detection limits of approximately 1 ng/cm² of belt cover was demonstrated. Spectra were given for a number of relatively involatile materials, for example, cytidine, erythromycin, and gramicidin S, for which a spectrum obtained from one microgram was presented. They reported interesting matrix

---

*In this context, the term *FAB* is used to describe the technique developed by Barber *et al.* [M. Barber, R. S. Bordoli, R. D. Sedgwick, and A. N. Tyler, *J. Chem. Soc. Chem. Commun.* **1981**, 325 (1981)], which involves the use of energetic atoms as primary impact particles. Distinction between it and SIMS, liquid or otherwise, is purely on this basis. Discussion as to whether the process might be more properly named [A. Benninghoven, F. G. Rudenauer, and H. W. Werner, *Secondary Ion Mass Spectrometry*, John Wiley and Sons. New York (1987)] is intentionally avoided.

effects on the overall ion yield of these compounds and directly compared LD to SIMS [84FA0', 84FA60], acknowledging that, in principle, the continuous nature of SIMS ionization favors it over the pulsed laser system. They did, however, detail a number of instances in which the absolute parent ion current yield is greater by LD than by SIMS. The inevitable conclusion, however, is that the LD method, effective for the vaporization of involatile species, but not necessarily for their ionization, provides no startlingly obvious advantages over particle beam methods.

The principle of the use of a primary ion beam to yield secondary ions from chromatographic analytes, using a moving-belt system, was patented [79BE6'] and discussed [80BE59] by Benninghoven, but apart from the work of the group mentioned above, its application has been almost exclusively pursued by Smith and co-workers [80SM0', 80SM97, 81SM2', 81SM5', 81SM07], who reported [81SM30, 81SM33] the design and performance of an elegant instrument capable of such analyses. Unfortunately, we are left with little evidence as to the capabilities of such SIMS devices in direct LC/MS studies. The prospect of high secondary-ion beams from involatile materials by direct SIMS are nowhere supported in the current literature for combined moving-belt SIMS analyses.

An alternative particle bombardment technique, fast atom bombardment, has also been applied to moving-belt analysis, primarily in the expectation of achieving the analysis of involatile materials. In the first instance, Lewis and co-workers [82KE6'] and Dobberstein et al. [83DO65] simply adapted existing moving-belt interfaces to instruments with FAB capabilities, by assuring passage of the belt into the fast atom beam. In both cases, preliminary results were encouraging in that peptides with molecular weights in the range of 700–1200 daltons were shown to yield intense molecular ions. The incorporation of a spray deposition device as well as a solvent wash bath as aids to desolvation and even deposition, as well as to more effective belt cleanup, followed [83LE0', 83LE20']. A preliminary report [84ST97] on the use of one such interface discusses the application of LC/FAB/MS to the analysis of the products of peptide hydrolysis and reports the on-line measurement of molecular ions from peptides with molecular weights of the order of 2000. A fuller report [85ST75] details the detection and structural determination of a number of products of the hydrolysis of the hexadecapeptide antiamoebin I, the detection of previously unknown antiamoebins, the measurement on-line of the emerimicins IIA and IIB, and the spectrum of the cardiac glycoside digitonin. A direct deposition device, suitable for microbore (1 mm i.d.) columns, is proposed [85ST3'] as an alternative to the spray and its efficacy in LC/FAB/MS studies demonstrated. Its use [85ST4', 86ST88] in the investigation of the leucinostatins A and B revealed four previously unknown analogues. It has been applied [86BO4'] as an aid in the intended structural determination of an oligopeptide (MW > 20.000), following enzymatic hydrolysis. It is reported effective in the study of complex carbohydrate mixtures [86HE34'], yielding molecular information to 4000 daltons. Its promising early performance is remarkable considering that no investigation, or modification, of the belt itself has been undertaken. The majority of the data cited above were obtained with systems in which the size of the belt, its angle of entry, and its position in the device are those dictated by mechanical design considerations determined 10 years ago, indeed before the introduction of FAB ionization.

3.2.2.5. Future Prospects

The moving-belt FAB interface is the only one of the prospective solutions to the analyses of involatile materials discussed above that has thus far become commercially available. This approach, as well as all the others, warrants further development, not only mechanically but methodologically. The emergence of the continuous FAB probe, and the steady improvement of thermospray, may put that development in jeopardy. The belt has, however, as discussed, a clear and established role in analysis by electron impact and chemical ionization. As well as performing in the FAB mode, it has a promising future in the interfacing of supercritical fluid chromatography with mass spectrometry. An early study performed on certain ergot alkaloids using conventional liquid chromatography [80EC86, 82EC92, 82EC93] has provided a vehicle for comparison of the advantages of supercritical fluid chromatography/mass spectrometry via the belt interface [86BE1', 86BE31, 86BE37, 87BE0']. In all these respects, the belt is truly a versatile and valuable interface.

3.2.3. The Heated-Wire Concentrator

The heated-wire concentrator is a transport device, but it is not, strictly speaking, a mechanical interface, in that the transport medium, the heated wire, is static. Instead, transportation of solute to the mass spectrometric ion source is performed by the mobile phase itself. The effluent from column flows down the wire, which is electrically heated so as to concentrate solute. It is in this regard, i.e., that solvent is removed in a region separate from the ion source, that the heated-wire concentrator may be classified as a transport system. At the base of the wire, the concentrated phase is admitted to the ion source, though a needle valve, which can be used to control the rate of flow. It may, therefore, also be considered a direct liquid introduction (DLI) device. Its conception, design, construction, and application are the preserve of one group, that of Christensen, of the National Institute of Standards and Technology [formerly the National Bureau of Standards], Washington. They demonstrated its practicability in a number of early reports [79CH0', 80CH8RE', 80CH1RE', 80HE08], patented it [81WH6'], and have clearly described the device and a number of applications in sevral more recent studies [81CH31, 81CH7KI', 83CH7', 83CH11, 84CH8', 85CH336].

3.3. MECHANIZED FRACTION COLLECTORS

3.3.1. Discussion

Mechanized fraction collectors may be divided into two distinct classes: those in which collection and analysis are connected as part of the same physical system, requiring analysis during the course of chromatographic separation, and those in which the two functions are physically separate so that analysis may be performed after separation. An example of the former is the system of Lovins *et al.* [73LO53, 74LO67], which is in essence an automated probe. A valve system is inserted in the liquid line permitting fractions of interest to be selected, which fractions are forced down the center of a hollow probe for concentration and collection at the tip. The probe is then driven into the ion source, after conventional rough pumping and valve opening to high vacuum, whereupon the sample is analyzed. The collection volume and cycle time of each sampling limit the number of analyses that may be performed during the course of each chromatographic separation. In the system under discussion, the probe cycle time is quoted as between 3 and 5 min and clearly limits the number of spectra that can be recorded during any analysis to multiples of that interval. Fraction collection systems of the latter type, in which fractions are collected on discrete regions of some form of storage device which is then transferred to the analyzer, are also limited to a restricted number of samples, dependent on the physical form of the collector. However, analysis is temporally independent of the collection process. Examples of this type are the systems of Huber *et al.* [83HU17], Beavis *et al.* [85BE2', 86BE99], and Jungclas *et al.* [82JU76, 82JU79], which have been coupled with laser desorption, secondary ion, and ^{252}Cf ionization spectrometers, respectively. Jungclas *et al.* quote a limit of 12 samples per analysis [83JU67] with their system. The system described by Beavis *et al.* consists of an aluminum foil mounted on a drum which is rotated by means of a stepping motor in front of a focused electrospray of the column eluate. Mass spectral analysis is performed by driving the strip though a beam of primary ions. The spot size of the spray is quoted as 1 mm^2, as is the area of the ionizing primary beam. The high definition of both permits the loading and analysis of a large number of fractions onto a foil of manageable proportions, and in the extreme this type of fraction collector could permit a number of samplings from each eluent in fairly complex separations. The performance of the device has been demonstrated in the separation and analysis of peptides; as an example, the spectrum of leucine enkephalin at 40 picomol injected on column was given.

3.3.2. Future Prospects

In our opinion, the practical difficulties and the time involved in the use of these interfaces will, in the future, restrict them to analyses and instruments for which there are no viable alternatives, an increasingly reducing option.

Chapter 4

Thermospray

4.1. INTRODUCTION

The technique known as "thermospray" has emerged from efforts to develop a practical liquid chromatography–mass spectrometry (LC/MS) technique applicable to nonvolatile samples in aqueous effluents at conventional LC flow rates on the order of 1 ml/min [84VE65]. It is now widely recognized as a practical and efficient method for interfacing these two powerful techniques and is applicable to a wide range of samples including volatile and nonvolatile, polar and nonpolar, labile and stable, under a variety of chromatographic conditions [85GA34]. Early in this work, it was discovered that ions were produced even though the filament normally used was unheated [80BL21]. Initial measurements of mass spectra produced from nonvolatile compounds such as peptides, nucleosides, and nucleotides showed that the spectra were quite different from those obtained by chemical ionization and were, in fact, most similar to those obtained by field desorption. Subsequent work has focused on elucidating the mechanism of this "soft" ionization.

The word *thermospray* was derived by combining thermo with spray to provide a descriptive term for "production of a jet of fine liquid or solid particles by heating." It is not true that the term comes from early efforts which involved equal applications of heat and prayer that were occasionally successful. Briefly, thermospray is a method for directly coupling liquid chromatography (LC) with mass spectrometry (MS) which involves controlled, partial vaporization of the LC effluent before it enters the ion source of the MS. This is accomplished by heating the capillary tube connecting the two devices, and it is vital that the heat input be properly controlled so that premature complete vaporization does not occur inside the capillary. As a result of heating, the liquid is nebulized and partially vaporized, and any unvaporized solvent and sample are carried into the ion source as micro droplets or particles in a supersonic jet of vapor. Nonvolatile samples may be ionized by direct ion evaporation from highly charged droplets or particles, and more volatile samples may be ionized directly or by ion–molecule reactions in the gas phase.

The original work which led to the development of thermospray used laser heating to rapidly vaporize both the sample and the solvent and employed molecular beam techniques to transport and ionize the sample while minimizing contact of the sample molecules with solid surfaces [78BL81]. This "crossed-beam" LC/MS used a large and expensive vacuum system and a 50-W CO_2 laser. Later, a simplified version was developed which used oxyhydrogen flames to vaporize the LC effluent and a substantially less elaborate vacuum system [80BL86]. These early systems were designed to efficiently transfer the sample to either an electron impact (EI) or chemical ionization (CI) source while vaporizing and removing most of the solvent.

These systems gave satisfactory performance for a number of relatively nonvolatile samples and were used successfully with reversed phase separations employing aqueous buffers at flow rates as high as 1 ml/min. The major problem was that the vaporizer was difficult to control properly, which often caused uncontrolled fluctuations in performance and sometimes frustrated attempts at application to real analytical problems. This problem was subsequently solved by development of an electrically heated vaporizer which allowed more precise control of the solvent vaporization [83BL50]. This inexpensive vaporizer consisted of a few centimeters of stainless steel capillary tubing brazed into a copper block heated by commercial cartridge heaters. This electrically heated vaporizer greatly improved the stability and reproducibility of the vaporization of solvents and nonvolatile samples and made it possible to predict with some confidence the conditions appropriate for a particular solvent and flow rate. Also, the vacuum system was greatly simplified with only a single-stage mechanical vacuum pump being required in addition to the standard vacuum system on commercial mass spectrometers equipped for chemical ionization.

All of this earlier work was based on the premise that very rapid heating over a short length of the capillary was required to vaporize the sample without pyrolyzing it, but more recent work has shown that this premise was false. The history of work on capillary vaporizers leading to the development of a practical thermospray system is summarized in Table 4.1. At each step in the development, the heated length has increased and the surface temperature has decreased.

TABLE 4.1. Summary of the Characteristics of Different Thermospray Vaporizers

Method	Surface temperature (°C)	Heated length (cm)	Energy flux (W/cm²)	Total power (W)
CO_2 laser	>1000	0.03	30,000	25
Oxyhydrogen flame	1000	0.3	5,000	50
Indirect electrical	250	3	700	100
Direct electrical	200	30	70	150

With the longer-heated lengths, it is most convenient to heat the capillary by passing an electrical current through the capillary itself [85VE73]. Essentially all of the thermospray systems presently in use now employ direct electrical heating of the capillary.

4.2. APPARATUS FOR THERMOSPRAY

A schematic diagram of a recent version of the thermospray LC/MS interface is shown in Figure 4.1. This device employs direct electrical heating of the capillary and electrical heating of the ion source to convert a liquid stream into gas phase ions for introduction into a conventional mass analyzer. It appears that any LC/MS interface must accomplish nebulization and vaporization of the liquid, ionization of the sample, removal of the excess solvent vapor, and extraction of the ions into the mass analyzer, although different techniques may involve variations in the methods by which these steps are accomplished.

In the thermospray system depicted in Figure 4.1, the LC effluent is partially vaporized and nebulized in the directly heated vaporizer probe to produce a supersonic jet of vapor containing a mist of fine droplets or particles. As the droplets travel at high velocity through the heated ion source, they continue to vaporize due to rapid heat input from the surrounding hot vapor. A portion of the vapor and ions produced in the ion source escapes into the vacuum system through the sampling cone, and the remainder of the excess vapor is pumped away by a mechanical vacuum pump. A description of the major components of the thermospray LC/MS system and their primary functions are given below.

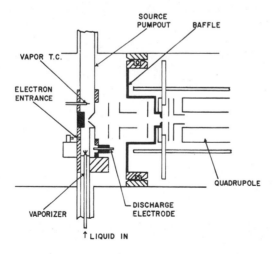

FIG. 4.1. Schematic diagram of thermospray ion source.

4.2.1. The Vaporizer Probe

The vaporizer is the heart of the thermospray system. In most presently used systems, it consists of a stainless steel tube 0.1 to 0.15 mm i.d. by 0.8 to 1.6 mm o.d. The heated length does not appear to be a very critical parameter and is usually chosen for mechanical convenience or to match the characteristics of the heater power supply. Heated lengths of between 15 and 50 cm are common. The vaporizer is heated by passing either an AC or DC current through the heated length, and the power input is controlled by a thermocouple attached near the input end. A second temperature sensor is usually attached near the exit from the vaporizer. The latter is very useful for sensing the state of the fluid exiting the vaporizer and determining that complete vaporization or "takeoff" has not occurred, but this temperature depends on several parameters and is generally not useful for direct feedback control. The vaporizer may be mounted in a probe for insertion through a vacuum lock as indicated in Figure 4.1 or it may be coiled up inside the vacuum system.

4.2.2. The Thermospray Ion Source

The vapor produced by vaporizing one ml/min of solvent is about one hundred times the amount that can be accommodated by a typical mass spectrometer equipped for chemical ionization. The thermospray ion source allows high mass flows to be vaporized by using a very tight ion chamber similar to that used in CI but with a mechanical vacuum pump (ca. 300 liters/min) attached directly to the source chamber through a port opposite the vaporizer. Orifices for ion exit and electron entrance are typically about 0.5 mm in diameter, and only about 1% of the total vapor exits through these to the source housing vacuum manifold. Ions may be produced by the direct ion evaporation from charged droplets or by CI initiated either by an electron beam or a low current discharge. The source block is strongly heated by high-capacity electrical cartridge heaters (ca. 50–100 W), and a thermocouple is normally positioned to monitor the vapor temperature just downstream of the ion sampling orifice in addition to a temperature sensor to monitor the source block. In the most recent version of the thermospray ion source shown in Figure 4.1, a separately heated block surrounds the vaporizer tip. This allows the region cooled by the rapid adiabatic expansion to be properly heated without overheating the portions of the ion source further downstream. Recent versions of thermospray ion sources are often equipped with repeller electrodes which, under some circumstances, can be used either to enhance high mass sensitivity or to produce increased fragmentation for qualitative identification [86RO0', 86ZA3'].

4.2.3. Electron Gun

The electron gun serves the same function as in a conventional CI source, but some minor changes are generally required. Thoriated iridium filaments such as those used in nonburnout ionization gauges are used in place of tungsten or rhenium filaments to achieve satisfactory operating life in the presence of high pressures of water and methanol vapors. Also, the pressures inside the ion source tend to be somewhat higher than in conventional CI (ca. 10 torr at 1 ml/min); thus, electron energies of 1 keV or more are required for efficient penetration into the ion source. The high velocity of the vapor through the ion source requires that the electron beam entrance be displaced upstream from the ion sampling cone; otherwise, ions are swept downstream and few diffuse to the ion sampling cone and into the mass analyzer.

4.2.4. Discharge Electrode

Even with the thoriated iridium filaments, it is difficult to maintain satisfactory emission at high flows of water; the discharge electrode overcomes this limitation. The lifetime of the discharge electrode is determined by the time required for it to be fouled by conductive deposits to the point that electrical leakage paths prevent the discharge from striking [85VE1']. With water, fouling is not a problem, and the lifetime is virtually infinite. Running the discharge in the presence of organic solvents produces carbonaceous deposits which will eventually cause fouling. Running for a few hours in water vapor can sometimes clean a partially fouled discharge electrode. The electron beam and the discharge generally give very similar results in thermospray.

4.2.5. Vacuum System

A typical vacuum system for a thermospray LC/MS is shown schematically in Figure 4.2. Except for the added mechanical pump connected directly to the

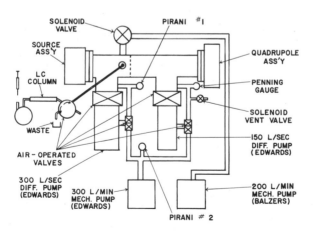

FIG. 4.2. Schematic diagram of typical vacuum system required for thermospray on a quadrupole mass spectrometer.

ion source, this system is similar to those normally provided with commercial mass spectrometers equipped for CI. The pumping speeds indicated in the figure are sufficient for LC flows of at least 1.5 ml/min. By increasing the capacity of the diffusion pump on the source housing and its associated mechanical pump, higher flow rates can be accommodated. Since commercial MS systems rarely are designed to operate at maximum throughput, the mechanical pumps backing the diffusion pumps are often the limiting factor, and replacing the mechanical backing pump with one of larger capacity is sometimes required. Earlier versions of thermospray systems used a refrigerated cold trap between the ion source and the source pump. To be effective, this trap must be cooled to liquid nitrogen temperature since otherwise the organic solvents are not efficiently trapped and the lifetime of conventional mechanical vacuum pumps may be seriously shortened. The new "hot" pumps which operate with oil reservoirs at ca. 80°C, provided they are always used with full gas ballast, appear to perform satisfactorily without use of a cold trap.

4.2.6. Thermospray Control System

Precise temperature control, particularly of the vaporizer, is required for the best performance. The control system should normally be capable of maintaining the control temperature to within one degree. The source block and tip heater must also be controlled for reproducible results, but the control requirements are not as stringent. In order to reproduce conditions, it is necessary to monitor both the vaporizer exit temperature and the vapor temperature, since these temperatures which are not directly controlled are the best indicators of system performance.

4.2.7. Ion Optics and Mass Analyzer

Thermospray LC/MS can, in principle, use any type of mass analyzer. Most of the work to date has employed quadrupoles, but thermospray is now commercially available on magnetic deflection instruments as well. The high voltage employed on the ion source of magnetic instruments requires that the thermospray ion source and its associated control electronics also be at high potential. The LC can be electronically isolated from the vaporizer probe by using one or two meters of fused silica in the flow path [84VE60], and several satisfactory solutions to other problems associated with operating thermospray at high voltage have been developed. In both magnetic and quadrupole instruments, it is desirable to employ ion lenses between the source and analyzer which are as open as possible to allow efficient pumping of this region.

4.3. MECHANISMS INVOLVED IN THERMOSPRAY VAPORIZATION AND IONIZATION

In this section an attempt is made to summarize the present understanding of the mechanisms involved in thermospray with a view toward optimizing performance, particularly for samples containing large, nonvolatile, thermally labile, or otherwise difficult molecules.

4.3.1. Vaporization and Nebulization in the Thermospray Capillary

When liquid is forced at high velocity through an unheated capillary tube, a solid jet issues from the tube and breaks up into regular droplets according to Rayleigh's well-known theory of liquid jet stability [45RA0']. According to this theory, the length of the liquid column which separates to form a liquid droplet is 4.508 times the diameter of the solid jet. This breakup leads to droplets with uniform diameters approximately 1.9 times the diameter of the nozzle. A train of droplets of uniform size and velocity is produced as shown in Figure 4.3a. When the tube is heated gently, the properties of the jet are modified slightly by the drop in surface tension accompanying the increase in temperature, but very little change is observed visually. When enough heat is applied to produce significant vaporization inside the capillary, the appearance of the jet changes drastically as shown in Figure 4.3b. Further increase in the applied heat reduces the visibility of the jet as the number and size of the droplets decrease until eventually the jet is invisible, as shown in Figure 4.3c, where the only evidence of the jet is downstream condensation which occurs due to cooling in the atmosphere.

Some information on the processes occurring when a flowing liquid is vaporized as it is forced through a heated capillary tube is provided by measurement of the temperature profiles produced by direct electrical heating of the capillary. This method provides uniform heating along the length of the capillary, and by using a relatively long heated zone (ca. 30 cm), the temperature drop between the outside of the capillary and the fluid inside is at most a few degrees. Examples of such temperature profiles are given in Figure 4.4 [85VE73]. In this experiment, 53 W were dissipated in the capillary with water flowing through at rates in the range from 0.7 to 4.0 ml/min as indicated on the figure. The experimental result at 0.7 ml/min is compared with a calculated profile in Figure 4.5, where the processes assumed to occur in different regions of the heated capillary are represented schematically. At the inlet end, the flowing water is heated until vaporization begins. Downstream from this point, the temperature remains nearly constant until vaporization is complete since the heat flux is used to provide the latent heat of vaporization. The slight decrease in temperature along the vaporization region results from the fact that the pressure is decreasing toward the exit end. At the point corresponding to complete vaporization, the

FIG. 4.3. Flash photographs of jets issuing from a thermospray vaporizer. (Top) Thermospray capillary heated to below the threshold for formation of the thermospray jet; (middle) thermospray capillary heated to just above the onset of thermospray; (bottom) fully developed thermospray jet corresponding to ca. 98% vaporization. (Photographs by G. J. Fergusson.)

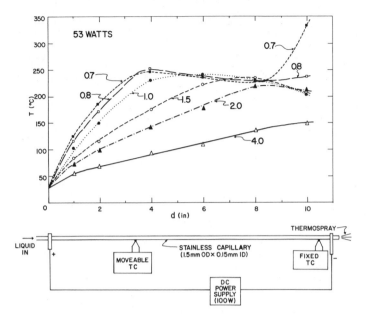

FIG. 4.4. Temperature measured along the capillary tube in a directly heated thermospray vaporizer. The parameter labeling the curves is the water flow rate in ml/min.

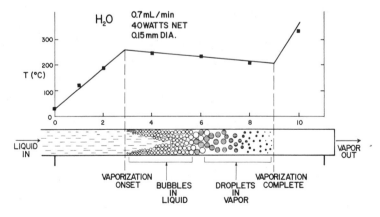

FIG. 4.5. Comparison between calculated (solid line) and measured (points) temperature profile at 0.7 ml/min together with a schematic representation of the model used in the calculation.

temperature again rises rapidly since only the heat capacity of the vapor is available to absorb the input heat.

4.3.1.1. Control of the Heat Input to the Thermospray Vaporizer

In these experimental studies, a manually controlled DC power supply was employed. More recently, a triac-controlled AC supply has been used for supplying the power to the directly heated thermospray vaporizer. A block diagram of this controller is shown in Figure 4.6. The power output to the capillary is controlled by feedback so as to maintain the temperature, T_1, constant as indicated by a thermocouple attached to the capillary near the inlet end, where no vaporization occurs.

The total power which must be coupled into the flowing liquid to vaporize a fraction, f, of a given mass flow F (g/s) is

$$W = fF\Delta H_v + F(1-f)C_L(T_2-T_0) \tag{1}$$

where ΔH_v is the total specific enthalpy (J/g) to convert liquid at the entrance temperature, T_0, to vapor at the exit temperature, T_2, and C_L is the specific heat capacity of the liquid. The total power coupled into the liquid in the region between the entrance to the vaporizer and the location of the thermocouple monitoring T_1 is given by

$$W_1 = C_L(T_1-T_0) \tag{2}$$

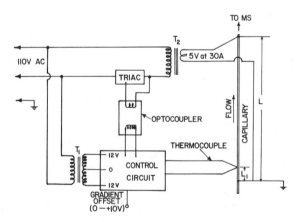

FIG. 4.6. Block diagram of the triac-controlled AC supply used for direct heating of capillary vaporizers.

where T_1 is the temperature at the control point. If the tube is mounted and insulated so that essentially all of the power dissipated in the tube is coupled into the flowing fluid, then uniform heating of the tube implies that

$$W_1/W = L_1/L \tag{3}$$

where L is the total heated length and L_1 is the length of the heated portion up to the point monitored by T_1. Combining equations (1) through (3) and solving for T_1 gives

$$
\begin{aligned}
T_1(0) &= T_0 + (L_1/L)(T_2 - T_0) \\
T_1(f) &= T_0 + (L_1/L)\langle f\,H_v/C_L + (1 - f)(T_2 - T_0)\rangle \\
-T_1(1) &= T_0 + (L_1/L)\langle (C_v/C_L)(T_2 - T_0)\rangle
\end{aligned}
\tag{4}
$$

where C_v is the specific heat of the vapor, and the last term in $T_1(1)$ represents the heat input required to heat the dry vapor.

Since the flow rate, F, does not appear explicitly in equation (4), this result implies that the indicated temperature, T_1, is linearly related to the fraction vaporized, f, and independent of flow rate. Thus, by controlling the power input so as to maintain a constant T_1, the fraction vaporized may be maintained at a constant selected value even though the flow rate may change. Furthermore, this equation predicts how the set point required to maintain a certain fractional vaporization depends on the composition of the fluid. As the fluid composition changes, for example, in gradient elution, the ratio $\Delta H_v/C_L$ changes somewhat. If the set point T_1 is changed accordingly, then the fraction vaporized may be maintained constant even though the solvent composition may vary from pure water, for example, to pure methanol. As discussed below, this compensation can be accomplished automatically by sensing the change in heat capacity which accompanies a change in composition at constant flow rate. To a good approximation, T_1, given by equation (4), is independent of solvent flow rate but significantly dependent on solvent composition. A hidden dependence on flow results from the fact that ΔH_v depends on the temperature of the vapor at the exit, which in turn depends on the flow velocity.

The rate of vaporization of a liquid at temperature T is given by

$$Z = \frac{P_v(T) - P_a}{(2\pi m k T)^{1/2}} \tag{5}$$

where $P_v(T)$ is the equilibrium vapor pressure at temperature T, P_a is the ambient pressure of the vapor, m is the molecular mass, and k is the Boltzmann constant. This expression gives the net flux (no. of molecules/cm$^2 \cdot$ s) evaporating. It can be transformed into an effective vaporization velocity by multiplying by the molecular mass and dividing by the density of the liquid to give

$$v_v = \frac{P_v(T) - P_a}{\rho_L{}^0} \left(\frac{m}{2\pi kT} \right)^{1/2} \tag{6}$$

If the liquid is completely vaporized, then the vaporization velocity must be at least equal to the liquid flow velocity, which is given by

$$v_L{}^0 = F/\rho_L{}^0 A \tag{7}$$

where F is the mass flow (g/s) and A is the cross-sectional area of the flow channel in the capillary tube. For complete vaporization, conservation of mass requires that

$$\rho_L{}^0 v_L{}^0 = \rho_e v_e \tag{8}$$

where ρ_e and v_e are, respectively, the density and the velocity of the vapor at the exit from the capillary, and $\rho_L{}^0$ is the density and $v_L{}^0$ the velocity of the liquid at the entrance to the vaporizer. The maximum velocity with which the vapor can exit the tube is that of the local speed of sound in the vapor, given by

$$v_s = (\gamma kT/m)^{1/2} \tag{9}$$

where $\gamma = C_p/C_v$ is the specific heat ratio for the vapor. Combining equations (8) and (9) and approximating the behavior of the vapor by the ideal gas law, the pressure of the vapor at the exit from the capillary is given by

$$P_e = v_L{}^0 \rho_L{}^0 v_s/\gamma \tag{10}$$

Substituting the exit pressure for P_a in equation (6) and solving for the liquid velocity gives

$$v_L{}^0 = \frac{P_v(T)}{\rho_L{}^0 v_S(T)} \left(\frac{\gamma}{(2\pi\gamma)^{1/2} + 1} \right) \qquad \text{at } f = 1 \tag{11}$$

The temperature at which this equation is satisfied correponds to the minimum temperature of the fluid at which complete vaporization occurs. This equation can be inverted (at least numerically) to give the minimum temperature for complete vaporization as a function of liquid flow. Results for several common solvents are summarized in Figure 4.7. If the heat supplied to the liquid is more than that required to reach this temperature (for a given flow rate), vaporization will occur prematurely and superheated, dry vapor will emerge from the capillary. If the heat supplied is slightly less than the critical value, then a portion of

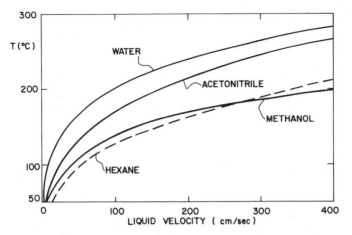

FIG. 4.7. Maximum exit temperature (at which complete vaporization occurs) as a function of liquid velocity for some common LC solvents. For a typical capillary (ca. 0.15 cm i.d.), a flow rate of 1 ml/min corresponds to a velocity of about 100 cm/s.

the liquid is not vaporized and will emerge along with the vapor jet as small entrained droplets. For most applications of thermospray, it appears that the best operating point corresponds to a fluid temperature at which partial but nearly complete vaporization occurs. In this range, the residual droplets tend to be relatively small and are accelerated to high velocities by the expanding vapor. Since the vapor pressure is a very steep function of the temperature, it is essential to have very precise control of the temperature if a stable fraction vaporized is to be maintained at nearly complete vaporization.

The equations derived above can be used to calculate, for a given position of the control thermocouple (L_1/L), both the temperature at the control point, T_1, and the temperature at the exit, T_v. Results for water and methanol flowing through a 0.15-mm diameter capillary with $L_1/L = 6$ are given in Figure 4.8. As can be seen from the figure, the dependence of T_1 on flow rate is rather small, being on the order of 1°C for a 1-ml/min change in flow rate of water, but the dependence on composition is fairly substantial, corresponding to about 15–20°C difference between water and methanol at the same flow rate. On the other hand, the exit temperature, T_v, is a fairly strong function of both flow rate and composition. Most other common solvents (for example, acetonitrile and hexane) fall in the range between water and methanol. For many purposes, the flow rate dependence of T_1 may be unimportant; however, if 99% or more vaporization is required in a given application, then the set point must be adjusted whenever large changes in flow rate are made. It is clear, however, that this method of control provides a basis for correcting for the small flow fluctuations introduced by typical LC pumps, provided the overall response of the control system and vaporizer is sufficiently fast.

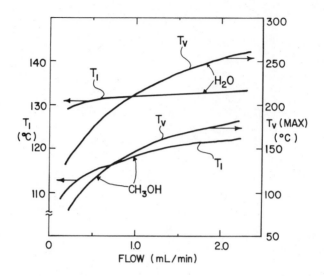

FIG. 4.8. Control point temperature, T_1, and vapor exit temperature corresponding to complete vaporization at the exit, calculated for a capillary inside diameter of 0.15 mm and a control ratio of $L_1/L = 6$.

Two examples of results on the variation of vaporizer exit temperature with control point temperature, T_1, are shown in Figure 4.9 for water at 1 ml/min being vaporized at atmospheric pressure. These results were obtained with capillary tubing whose internal diameter, according to the manufacturer's specifications, was between 0.10 and 0.15 mm (Handy and Harmon, Morristown, Pennsylvania). In Figure 4.9a, the sharp upward break in exit temperature at $T_1 = 150°C$ corresponds to complete vaporization ($f = 1$) at the exit. The point at which $T_v = 100°C$ indicates the onset of vaporization ($f = 0$). With this particular vaporizer probe, the fraction vaporized is given by $f = T_1 - 50$ for temperatures between 50 and 150°C. After correction for convective losses to the atmosphere, these results are generally in good agreement with theoretical expectations, except that the exit temperature (170°C) at which $f = 1$ corresponds to a liquid velocity of only 50 cm/s. This is about half the velocity expected for 1 ml/min flowing through a 0.15-mm i.d. capillary. Subsequent measurements of the capillary showed that the internal diameter was, in fact, about 0.2 mm as indicated by the above results.

Results for a vaporizer with an authentic i.d. of 0.15 mm are shown in Figure 4.9b. In this case, the exit vapor temperature at $f = 1$ is about 200°C, in agreement with theoretical expectations based on equation (11). It should be noted that the critical values of the control temperature, T_1, are different for these two vaporizers. This occurs primarily as the result of different locations for the thermocouple with L_1/L being smaller in the case of Figure 4.9b than in that of Figure 4.9a. Performance obtained with an oversize capillary such as depicted in Figure 4.9a is generally inferior to that obtained with those with smaller i.d.

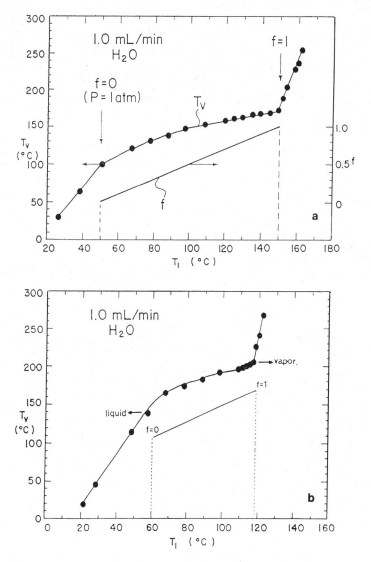

FIG. 4.9. Exit temperature and fraction vaporized as a function of control temperature for an oversized capillary (a) and for a 0.15-mm i.d. capillary (b).

Reasons for this dependence on capillary diameter are discussed in a following section.

4.3.1.2. Extension of Thermospray to Lower Flow Rates

Recent work [87OS4′] has focused on establishing a practical approach to thermospray in the lower flow rate ranges appropriate for use with smaller-bore columns. The theoretical treatment given above suggests that capillary diameters in the range of 25–50 μm should give satisfactory results. The initial approach to thermospray at lower flow rates employed 30-gauge stainless steel needle tubing which is available with nominal internal diameters of 50, 100, and 150 μm. The vaporizer was heated by passing a current through the tube and controlling the power with a thermocouple attached a short distance from the entrance end of the vaporizer. A second thermocouple was attached within a few millimeters of the exit to give an approximate indication of the vapor temperature at the exit. Plots of exit temperature as functions of control temperature or input power were prepared for flows in the range from 0.05 to 2 ml/min using these capillaries. Most of these studies employed deionized water, but a more limited range of conditions was also investigated using methanol and acetonitrile. These results were generally in reasonable agreement with the theoretical model for the larger-diameter capillaries, but significant disagreement was observed with the smaller diameters. In particular, the exit temperature at the point of complete vaporization was substantially lower in the experimental results than predicted.

Some additional experiments were conducted which have significantly clarified the effect of capillary diameter on the properties of the thermospray. In this work, a series of thermocouples were attached at intervals along the capillary as shown in Figure 4.10, and these temperatures were monitored as functions of the control temperature for several different flow rates. As has been discussed previously [85VE73], the thermospray controller used in this work provides a constant power per unit flow at a given value of control temperature, T_1. Results obtained for several different flows of water through a 50-μm i.d. stainless steel capillary are summarized in Figure 4.11. From these data, approximate profiles of the tube temperatures along the heated portion can be constructed as shown in Figure 4.12. The calculated exit temperature for each flow is also indicated on the figure. From these results, it appears that the calculated values are more

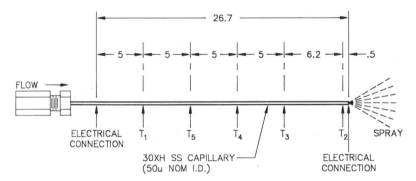

FIG. 4.10. Schematic diagram of experiment for determining temperature along vaporizer capillary. The arrows indicate the locations of thermocouples and the electrical connections to the vaporizer power supply.

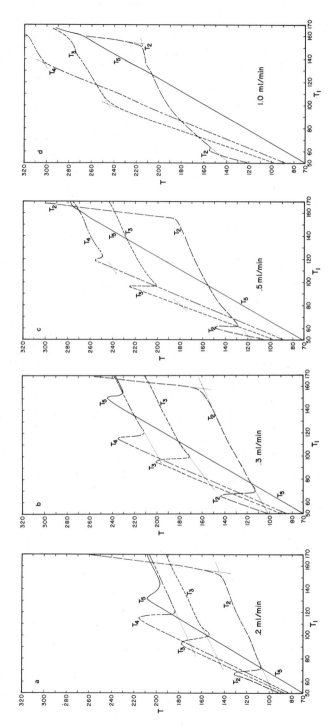

FIG. 4.11. Measurement of temperatures at selected points (as indicated in Figure 4.10) along the thermospray capillary for water through a 50-μm i.d. capillary at various flow rates. The initial rise corresponds to heating liquid, the flatter portion to vaporization, and the sharp upward break in T_2 to complete vaporization.

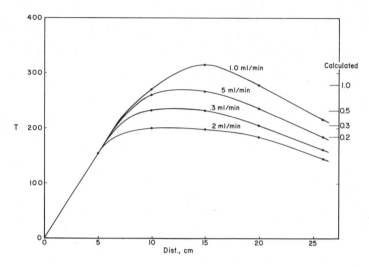

FIG. 4.12. Temperature versus distance along the capillary for complete vaporization ($f = 1$) at the exit, taken from the data in Figure 4.11.

closely related to the average temperature along the capillary in the region where vaporization is occurring than to the temperature at the exit.

After considering this discrepancy, it can be concluded that the model used in calculating exit temperatures at complete vaporization as a function of liquid velocity contains an unstated approximation which becomes poorer at smaller capillary diameters. In the earlier treatment [85VE73], the exit temperature at complete vaporization was calculated by setting the liquid velocity equal to the vaporization velocity at the exit to derive equation (11). By numerically inverting this equation, the exit temperature can be calculated as a function of liquid velocity. After considering the recent results, it is clear that this equation should be written as

$$v_L{}^0 = v_v = \frac{1}{\rho_L{}^0} \left(\frac{m}{2\pi k} \right)^{1/2} \int_0^L \frac{P_v[T(x)] - P(x)}{[T(x)]^{1/2}} \, dx \qquad (12)$$

where the integral is taken over the length of the heated capillary in which vaporization is occurring. If the pressure of the vapor inside the capillary is constant, then for uniform heating as employed here, the vaporization occurs at constant pressure and temperature and equation (12) reduces to equation (11). However, for higher flows through smaller-bore tubes, this condition is increasingly violated.

Both the exit pressure and the pressure drop in the tube are proportional to the mass flow of vapor, but the exit pressure is proportional to the square of the tube diameter while the pressure drop goes as the fourth power of the diameter.

Thus, equation (11) is a very poor approximation for flows of even 0.2 ml/min through a 50-μm tube, but it is reasonably satisfactory for a flow of 1 ml/min through a 150-μm tube.

Another problem which becomes increasingly severe at smaller capillary diameters is that of heat transfer to the flowing liquid. For tubes of similar characteristics (e.g., surface roughness), the heat transfer efficiency is proportional to the surface area in contact multiplied by the time of contact. This product is proportional to the length of the heated portion and to the third power of the diameter. With 150-μm capillaries, it is found that heated lengths of only about 3 cm can provide sufficient heat transfer without large temperature differentials. If the capillary diameter is reduced by a factor of 3, then the heated length must be increased by a factor of 27 to maintain the same heat transfer efficiency.

Another problem with smaller-diameter capillaries is caused by the pressure drop itself. For example, the total pressure required to drive the 50-μm capillary at a flow of 1 ml/min near complete vaporization is about 200 bar. Thus, a substantial part of the LC pump capacity is required to drive the vaporizer, and the range of pressure drop available for the column is proportionally less. Also, this high head pressure makes the use of in-line auxiliary detectors rather questionable.

From these studies it has become increasingly obvious that merely decreasing the capillary tube diameter is not a practical way to increase thermospray performance at lower flow rates. It is really the exit diameter that is important in determining the vaporizer exit conditions. If the exit diameter is significantly smaller than the tube diameter in the heated region, then equation (11) is a good approximation. The temperature inside the vaporizer is then never significantly higher than that at the exit and most of the pressure drop occurs at the exit, causing the total pressure for a given flow and exit diameter to be very much lower than in the case of the straight tube. The size of the capillary in the heated region must be small enough to avoid band broadening but is otherwise not critical.

Several approaches to forming a reduced-diameter "nozzle" at the end of the capillary are presently being studied. The ultimate goal is an exit nozzle diameter optimized for a particular flow rate regime, solvent composition, and type of sample. This appears to be the most important technical problem remaining to be solved for thermospray to become more generally accepted.

4.3.1.3. Incomplete Vaporization

Equations (5) through (12) allow the state of the exiting vapor to be calculated for the critical case corresponding to just achieving complete vaporization ($f = 1$) at the exit from the vaporizer. Calculation of exit state for the more interesting cases involving partial, but incomplete, vaporization ($0 < f < 1$) is complicated by the fact that the velocity of the liquid droplets exiting the vaporizer is generally unknown. This velocity depends upon the acceleration of liquid droplets by the rapidly expanding sonic vapor, and the result clearly depends on the geometry of the vaporizer.

For incomplete vaporization, conservation of mass [see equation (8)] requires that

$$\rho_L{}^0 v_L{}^0 = f\rho_e v_e + (1-f)\,\rho_L v_L \tag{13}$$

where ρ_L is the density of the liquid droplets. The exit velocity of the vapor, v_e, can be calculated as a function of the exit temperature, but the liquid droplet velocity, v_L, depends on the dynamics of the two-phase flow through the vaporizer. While the final velocity of the liquid droplets can, in principle, be calculated from well-known aerodynamic equations, these calculations require a more detailed knowledge of the two-phase flow parameters than is presently available. For the present, it is assumed that v_L depends linearly on f according to an equation of the form

$$v_L = v_L{}^0(1 + \alpha f) \tag{14}$$

where α is a positive number which depends only on the geometry of the vaporizer. Using this approximation in equation (12) together with the velocity of sound [equation (9)] and the ideal gas law, and neglecting the weak temperature dependence of γ, yields

$$P_e = \frac{v_S \rho_L{}^0 v_L{}^0}{\gamma}\, [\alpha f - (\alpha - 1)] \tag{15}$$

Substituting this for the ambient pressure, P_a, in equation (6) gives

$$v_v = f v_L{}^0 = \frac{P_v(T)}{\rho_L{}^0 v_s(T)} \left[\frac{\gamma}{(2\pi\gamma)^{1/2} + \alpha - (\alpha - 1)/f} \right] \tag{16}$$

For the special case $\alpha = 1$, equations (15) and (16) reduce to particularly simple forms, namely,

$$P_e(f) = f\,P_e(1) \tag{17}$$

and

$$v_v = f v_L{}^0 = \frac{P_v(T)}{\rho_L{}^0 v_s(T)} \left[\frac{\gamma}{(2\pi\gamma)^{1/2} + 1} \right] \tag{18}$$

While it is difficult to determine the liquid droplet velocity and α directly, it is possible to determine them approximately by comparing calculations of exit temperature as a function of fraction vaporized with experimental results. In the preceding derivation, the final state of the fluid has been determined by considering only the exit conditions even though vaporization is occurring over a significant length of the capillary tube. This is permissible at low liquid velocities because the uniform heating employed assures that the vaporization rate along the tube is nearly uniform even though the pressure and temperature vary slightly along the tube.

4.3.1.4. Automatic Compensation for Changes in Solvent Composition

As discussed above, the control temperature must be programmed with changes in solvent composition such as occur in gradient elution if proper thermospray operation is to be maintained. This can be accomplished automatically by a fairly simple modification to the basic thermospray controller [85VE7']. This modification is illustrated in the block diagram given in Figure 4.13. The additional circuitry over that shown in Figure 4.6 allows the specific heat of the solvent actually present in the thermospray vaporizer to be determined by measuring the power input required to maintain the preset control temperature. As there is a direct relationship between the specific heat of the solvent mixture and the optimum control temperature, a signal to adjust the control temperature set point can be derived from the specific heat measurement and fed back to the temperature control circuit.

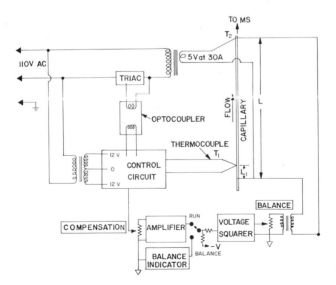

FIG. 4.13. Schematic diagram of thermospray vaporizer controller with added feedback system for automatic gradient compensation.

   The power input to the capillary is determined by using a four-quadrant multiplier as a voltage squarer, to produce an output proportional to the square of the instantaneous voltage across the capillary. The power input to the capillary could also be determined from the square of the current through the capillary. The most accurate measurement of power would result from the product of voltage and current as this is independent of the resistance of the capillary. However, as the change in resistance of the capillary over the temperature range involved with different solvents is very small, adequate accuracy results from voltage squaring, and this method allows the simplest interconnecting wiring. Heater power for the vaporizer may be either alternating or direct current.

   To set up the gradient controller, the set point temperatures for the desired degree of vaporization of two solvent compositions at the given flow rate must first be determined. These usually may be taken as 100% solvent A (for example, water) and 100% solvent B (for example, methanol or acetonitrile). The proper set point temperatures can be determined experimentally by measuring the vaporizer exit temperature, T_v, as a function of control temperature, T_1, to determine, as illustrated in Figure 4.14, the point at which complete vaporization occurs ($f = 1$). As can be seen from the figure, the dependence on composition is fairly substantial, corresponding to about a 20°C difference in control temperature, T_1, and about a 70°C difference in exit temperature between water and methanol at a 1.0-ml/min flow rate.

   The desired operating temperature (f slightly less than 1) can be determined

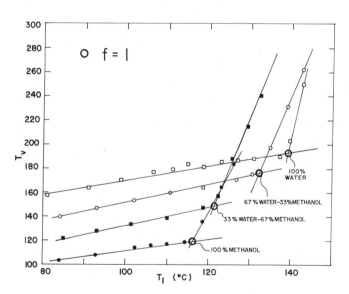

FIG. 4.14. Plots of exit temperature as a function of control temperature for water/ methanol mixtures. The large circles indicate the experimentally determined "takeoff" points at which complete vaporization occurs just at the vaporizer exit.

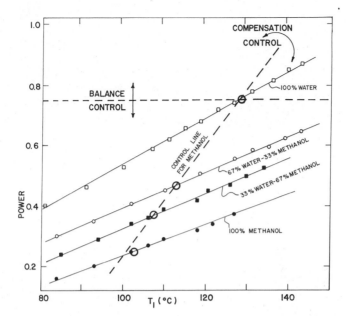

FIG. 4.15. Measured vaporizer power (arbitrary units) as a function of control temperature for water/methanol mixtures. The inclined dashed line corresponds to the control line established by automatic gradient compensation; the large circles represent the experimentally determined operating control temperature for each composition.

directly from Figure 4.14. Generally, operating at a set point, T_1, about 5% below that corresponding to $f = 1$ gives satisfactory results. With solvent A flowing through the vaporizer and the controller set at the desired operating point for solvent A, the balance control is adjusted until the balance meter is zeroed. At balance, the reference level is set so that the input to the compensation amplifier is zero. The liquid flow is then switched to solvent B (e.g., methanol) and the temperature, T_1, for the desired f is determined as described above. The controller is then switched to the gradient mode and the gain of the compensation is adjusted until the correct T_1 for solvent B is obtained. The system may now be operated with any mixture of A and B with the vaporizer controller automatically tracking to maintain a constant fraction vaporized.

The operation of the controller is illustrated in Figure 4.15, where the measured vaporizer power (arbitrary units) is plotted as a function of control temperature, T_1, for water, methanol, and two intermediate compositions. Because the specific heat of methanol is so low compared to that of water, it is somewhat better to adjust the output of the compensation amplifier using a mixture of something like 33% water and 67% methanol, rather than pure methanol. The desired operating points for 100% water and the mixture were determined as described above to be 129°C and 108°C, respectively. Setting the balance and compensation as described causes the control system to operate along the broken line shown in Figure 4.15. In this way, vaporizer power is programmed continuously so that the fraction vaporized is maintained constant to within about 1% of the desired value as the composition varies anywhere from pure water to pure methanol. In the absence of gradient compensation, the power would vary along the vertical corresponding to the particular value of T_1 selected. In going from water to methanol, the power input would be decreased by about a factor of two (corresponding to the heat capacity ratio of the two liquids), but the final power for methanol would still be about 70% too high. This would cause premature vaporization and drastic overheating of the exit portion of the vaporizer, accompanied by sample decomposition and plugging of the capillary tube.

4.3.1.5. Thermospray Nebulization

The thermospray vaporizer shares many of the properties of the concentric pneumatic nebulizers used in atomic spectroscopy in that a high-velocity gas is used to shatter a liquid stream into a fine jet of droplets swept along in the gaseous stream. A unique feature of thermospray is that the nebulizing gas is generated *in situ* by partial vaporization of the liquid. Attempts have been made to directly measure the droplet size distribution produced by thermospray [86KA5', 86KO88], but these efforts have met with limited success, primarily because a very large number of small, high-velocity droplets are produced. In the absence of meaningful experimental results, the droplet diameter can be estimated using the equation of Nukiyama and Tanasawa. According to this equation, the Sauter mean diameter (i.e., the diameter of the drop whose volume-to-surface area ratio is the mean of the distribution) produced by pneumatic nebulizers is given by

$$d_s z \frac{585}{V} \left(\frac{\gamma}{\rho} \right)^{0.5} + 597 \left(\frac{\eta}{(\gamma\rho)^{0.5}} \right)^{0.45} \left(\frac{10^3 Q_L}{Q_g} \right)^{1.5} \tag{19}$$

where d_s (μm) is the Sauter mean diameter, V is the velocity difference between gas and liquid flows (m/s), γ is the surface tension of the liquid (dyn/cm), η is the liquid viscosity (poise), and Q_L and Q_g are the volume flow rates of liquid and gas, respectively. While some doubts have been expressed about the ability of this equation to accurately describe aerosols generated by all pneumatic nebulizers,[1,2] it has been shown to have considerable value, at least for predicting trends for aerosol generation in atomic spectroscopy.

Mean diameters calculated for thermospray vaporization and nebulization using equation (19) are summarized in Figure 4.16, where the mean droplet diameter is given as a function of fraction vaporized and flow rate for water in a typical 0.15-mm capillary. In these calculations, the velocity of the exiting vapor

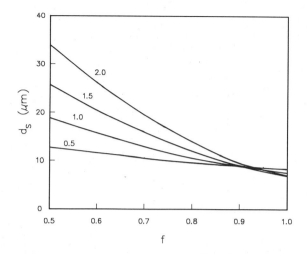

FIG. 4.16. Sauter mean droplet diameters produced by thermospray vaporization and nebulization calculated [see equation (19)] for water at the indicated flow rates (ml/min) through a 0.15-mm capillary vaporizer as a function of fraction vaporized.

has been taken as sonic and the state of the exiting vapor determined using equations (16) and (17). With these approximations, the volume flow rate of the vapor is independent of the fraction vaporized, and the volume flow rate of the liquid corresponds to the unvaporized fraction.

4.3.1.6. Droplet Charging

When a liquid surface is disrupted by spraying or bubbling, the droplets produced are often electrically charged [58LO00]. Various "static electrification" phenomena may be involved in this charging, but for liquids containing substantial concentrations of dissolved ions and in the absence of an applied electrical field, the statistical mechanism described by Dodd [53DO43] appears to be dominant. According to this model, the net charge on the droplets is due to the statistical fluctuations accompanying the rapid separation of a droplet from the bulk liquid. If the separation occurs sufficiently rapidly that conduction in the liquid can be neglected, then the net charge on the droplet is determined from simple statistics. For a large population of such droplets, the resulting charge distribution is Gaussian with zero mean and standard deviation equal to the square root of the total number of ions (of either sign) within the volume of the droplet. This charge distribution is given by

$$f(q) = \frac{1}{(2\pi\sigma)^{1/2}} \, e^{-q^2/2\sigma^2} \tag{20}$$

with

$$\sigma = (2VN)^{1/2}$$

where V is the volume of the droplet and N is the average number of ions of each sign per unit volume. The average charge of either sign is obtained by averaging over the appropriate half of the distribution:

$$\langle q^+ \rangle = \langle q^- \rangle = \langle |q| \rangle = \int_0^\gamma qf(q)dq = \left(\frac{2}{\pi} \right)^{1/2} \gamma = \left(\frac{4vN}{\pi} \right)^{1/2} \quad (21)$$

If the force exerted on the surface by the coulomb repulsion produced by an excess of charge of one sign exceeds that supplied by surface tension, then the droplet is unstable with respect to subdivision into smaller droplets. This limit on the maximum charge that a droplet can sustain was originally derived by Lord Rayleigh [45RA0'] and is given by

$$q_R = 8\pi(\gamma\epsilon_0)^{1/2}r^{3/2}\rho \quad (22)$$

with r the radius of the droplet, γ_0 the surface tension, and ϵ the permittivity of free space.

Another limit on the droplet charge is imposed by the field-assisted ion evaporation mechanism involved in production of ions in thermospray. This "field desorption" limit can be expressed as

$$q_E = 4\pi\epsilon_0 E r^2/e \quad (23)$$

where E is the minimum electric field required for ion emission. The value of E depends, in general, on the properties of the ions desorbed and on the time scale of the observation.

The skeleton of a mechanism for production of molecular ions in thermospray can be constructed from the concepts described above. Charged droplets are produced by the combination of thermospray vaporization and nebulization. The size and excess charge on the droplets are given by equations (19) and (20), respectively. If the droplets are completely vaporized and no ions are lost, for example by recombination, then all of the excess charge will be used to produce molecular ions. The charge on the droplets is determined during the nebulization from the bulk liquid; subsequent vaporization or Rayleigh subdivision should have little effect on the total ion current since the available charge will be conserved in these processes.

This model can be used to calculate the total ion current produced by thermospray as functions of liquid flow rate, fraction vaporized, and electrolyte concentration. For a given fraction vaporized, the nominal diameter of the particles is given by equation (19). The total number of droplets produced per second, n_d, is given by the volume of unvaporized liquid divided by the average volume of the droplets at the instant of nebulization from the bulk liquid. This may be expressed as

$$n_d = \left(\frac{10^{11}}{\pi} \right) \frac{E(1 - f)}{d^3} \tag{24}$$

where F is the liquid flow rate (ml/min), f is the fraction vaporized, and d is the mean droplet diameter (μm). The standard deviation of the droplet charge distribution is

$$\sigma = (2\pi)^{1/2} \times 10^4 d^{3/2} \left(\frac{M}{1 - f} \right)^{1/2} \tag{25}$$

where M (mol/liter) is the concentration of electrolyte in the liquid. The maximum total ion current of either sign is then given by

$$I_{max} = \left(\frac{2}{\pi} \right)^{1/2} n_d \sigma = \left(\frac{2}{\pi} \right)^{1/2} \times 10^{15} F(1 - f)^{1/2} M^{1/2}/d^{1/2} \tag{26}$$

Examples of the variation in droplet diameter, charge, and number of droplets as functions of fraction vaporized are given in Figure 4.17. The maximum ion current calculated as a function of fraction vaporized and flow rate through a 0.15-mm capillary (water containing 1 mol/liter of electrolyte) is shown in Figure 4.18. These results are qualitatively in good agreement with experiment, as illustrated by one such comparison in Figure 4.19.

This simple model appears to account satisfactorily for most of the experimental observations relevant to production of stable, relatively volatile ions from aqueous solutions of volatile salts. A favorite example is ammonium acetate. On the other hand, any attempt to understand the wide variations in ion production efficiencies which are sometimes observed requires probing more deeply into the kinetics of the processes involved in molecular ion production from charged liquid droplets.

4.3.1.7. Vaporization of Liquid Droplets

The rate of vaporization of a spherical liquid droplet can be expressed by

$$\frac{dr}{dt} = -v_v \tag{27}$$

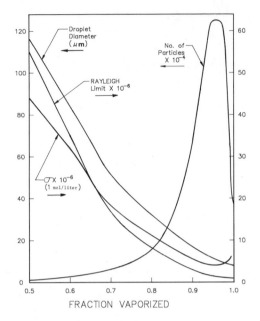

FIG. 4.17. Calculated droplet diameter, charge, Rayleigh limit, and number of droplets produced per second as functions of fraction vaporized for 1 ml/min of 1-mol/liter aqueous electrolyte through a 0.15-mm thermospray capillary.

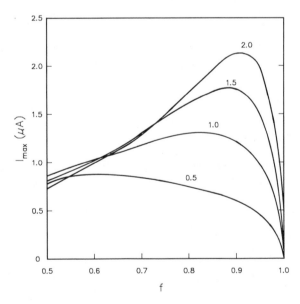

FIG. 4.18. Calculated maximum thermospray ion current produced as a function of fraction vaporized and flow rate through a 0.15-mm capillary (water containing 1 mol/liter of volatile electrolyte).

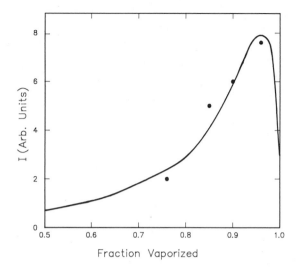

FIG. 4.19. Comparison of calculated thermospray ion current as a function of fraction vaporized (solid line) for 1 ml/min of 0.1 M ammonium acetate through a 0.15-mm capillary compared with experimental results (points) under the same conditions.

where r is the radius of the droplet and v_v is the net vaporization velocity given by equation (6). Since this rate is independent of r, the isothermal lifetime, τ, of a droplet can be expressed as

$$\tau = \frac{r_0}{v_v} \tag{28}$$

For water at 200°C in the presence of water vapor at 1 atm, the net velocity of vaporization is about 100 cm/s. Thus, the isothermal lifetime of water droplets under these conditions is about 1 μs per micron radius. At 100°C, water and its vapor at 1 atm are in equilibrium and the net rate of vaporization goes to zero. The very strong dependence of vaporization rates on temperature implies that the thermal environment of the droplets and particles must be properly controlled for efficient vaporization and ion production.

It should be noted that this treatment contains several unstated approximations and assumptions which are probably not strictly valid. For example, it is assumed throughout that a single temperature characterizes the walls of the vaporizer, the vapor, and the liquid at any point along the vaporizer. Since the heat is conducted from the walls of the vaporizer through the vapor to the liquid droplets, it is obvious that significant temperature differences must be involved.

4.3.2. Free Jet Expansion

Since the thermospray ion source is normally operated at reduced pressure (ca. 1–10 torr), a free jet adiabatic expansion occurs at the exit from the thermospray vaporizer. According to the work of Ashkenas and Sherman,[3] the Mach number in a free jet downstream from the nozzle is given by

$$M = A \left(\frac{x - x_0}{d} \right)^{\gamma - 1} - \frac{1}{2A} \left(\frac{\gamma + 1}{\gamma - 1} \right) \left(\frac{d}{x - x_0} \right)^{\gamma - 1} \tag{29}$$

where $A = 3.82$, $x_0 = 0.6d$, and d is the nozzle diameter, with x being the distance downstream from the nozzle in units of nozzle diameters. The Mach number is defined as the ratio of the axial velocity u_g to the local speed of sound, given by equation (9). For an adiabatic expansion, this gives

$$C_p T_0 = \tfrac{1}{2} m u_g^2 + C_p T \tag{30}$$

which can be inverted to give the temperature downstream as

$$T = T_0 \left(1 + \frac{\gamma - 1}{2\gamma} M^2 \right) \tag{31}$$

The pressure of the adiabatically expanding vapor is given by

$$P = P_0 \left(\frac{T_0}{T} \right)^{\gamma/\gamma - 1} \tag{32}$$

where T_0 and P_0 are the temperature and pressure, respectively, of the vapor at the nozzle, which in this case is the exit from the thermospray vaporizer.

The adiabatic expansion continues until the pressure in the jet is comparable to the pressure of the background gas. This is the location of the so-called Mach disk relative to the nozzle. Beyond this point, heat transfer between the jet and the background gas becomes important, and the jet undergoes a normal shock with the velocity distribution rapidly relaxed back to the local thermal distribution. Between the nozzle and the Mach disk, the supersonic vapor stream controls the environment and any evaporation of the remaining solvent. Downstream from the Mach disk, the surrounding atmosphere may be controlled and additional heat may be coupled into the droplets to assist in further vaporization.

During the early stages of the free jet expansion, the entrained liquid droplets are highly superheated (relative to the surrounding vapor) and undergo adiabatic vaporization in which the heat required is supplied by the internal enthalpy of the droplets. For this adiabatic process,

$$dH = (-\Delta H v \rho)\, dV + C_L \rho V\, dT = 0 \tag{33}$$

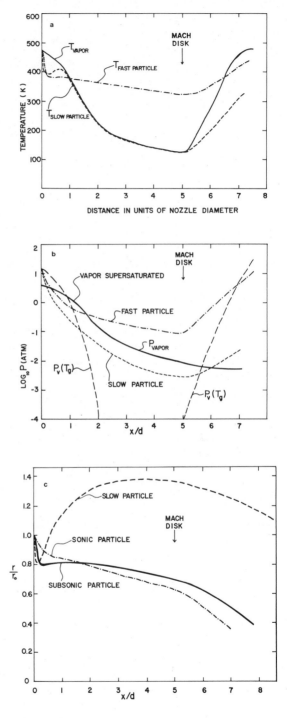

FIG. 4.20. Calculated properties of an aqueous thermospray jet downstream from the vaporizer: (a) temperature of vapor and particles along the axis of the jet; (b) pressure of the vapor and vapor pressure of the droplets or particles; and (c) radius of the particles relative to their initial radius at the exit of the vaporizer.

which can be integrated to give

$$T_0 - T = 3 \left(\frac{\Delta H_v}{C_L} \right) \ln \left(\frac{v_0}{v} \right) \qquad (34)$$

Equations (31) and (32), together with conservation of mass in the jet, allow the pressure and temperature of the adiabatically expanding vapor to be calculated as functions of the distance downstream from the nozzle, and equation (34) gives the temperature of the droplet as a function of its radius under adiabatic conditions. If the velocity of the droplet is known (or can be estimated), then these values can be used in the expression for vaporization velocity, equation (6), to determine the droplet size as a function of distance traveled from the nozzle.

Rapid vaporization of the superheated droplets exiting the vaporizer capillary causes the temperature of the droplets to initially fall more rapidly than the temperature of the vapor in the jet. Within a short distance downstream (about one nozzle diameter or less), conditions are such that recondensation on the droplets may occur. However, at low temperatures both vaporization and condensation rates are low, and during the short time the particles spend in this region, the amount of recondensation is expected to be rather small under normal operating conditions. Examples of calculated temperature and pressure profiles for a typical free jet expansion in a thermospray source are shown in Figure 4.20, together with some examples of calculated particle radii within the free jet expansion. Downstream from the Mach disk, the temperature of the vapor and of the droplets or particles is determined by heat transfer from the wall of the ion source.

4.4. IONIZATION MECHANISMS

In the preceding section, the skeleton of a mechanism for production of ions in thermospray was described. Briefly, charged droplets are produced as the result of rapid nebulization and vaporization in the thermospray vaporizer. As these droplets fly through the ion source, they continue to vaporize and produce molecular ions as a consequence of the high local electric fields generated on the surface of a small, highly charged droplet. This model accounts satisfactorily for the production of ions from solutions of volatile electrolytes in water, but to understand the wide variations in ion production rates which are often observed with less volatile samples and buffers, it is necessary to probe more deeply into the dynamics of the processes involved.

4.4.1. Production of Molecular Ions

The dominant mechanism for production of molecular ions from charged liquid droplets appears to involve field-assisted ion evaporation. This mechanism, originally described by Iribarne and Thomson [76IR47], is very similar, if not identical, to that involved in the early stages of conventional field desorption (FD), which has been extensively studied by Roellgen and co-workers.[4]

The rate of vaporization of molecules from a surface can be represented approximately by an Arrhenius equation of the form

$$Z = vC_s \, e^{-Q/kT} \tag{35}$$

where Q is the activation energy for vaporization, C_s is the surface concentration of the molecular species of interest, v is a frequency factor, T is the absolute temperature, and k is the Boltzmann constant. For applications to neutral molecules, Q can be taken as the enthalpy of vaporization and the product vC_s can be estimated from kinetic theory and thermodynamics. For example, neglecting the temperature dependence of the enthalpy of vaporization, ΔH_v, the rate of evaporation of a pure liquid is given by

$$Z = \frac{P_v(T)}{(2\pi mkT)^{1/2}} = \frac{P_v(T_0)}{(2\pi mkT)^{1/2}} \, e^{\Delta H_v/kT_0} \cdot e^{-H_v/kT} \tag{36}$$

Thus,

$$vC_s = \frac{P_v(T_0)}{(2\pi mkT)^{1/2}} \, e^{\Delta H_v/kT_0} \tag{37}$$

Here, $P_v(T)$ is the vapor pressure of the liquid at temperature T, the molecular mass is m, and T_0 is a convenient reference temperature, such as the boiling point. By independently estimating C_s from the molecular diameter, the frequency factor can be calculated and is generally found to correspond to a typical vibrational frequency of ca. 10^{14} s^{-1}.

Equation (35) appears to provide a satisfactory approximation to the rate of vaporization of any substance provided the necessary parameter values are available. The preexponential factors can be estimated satisfactorily from the molecular diameter and mass, and errors introduced by uncertainties in these values will generally be negligible compared with the overriding effect of the exponential term. The activation energy for evaporation of a positive ion from the surface of a pure metal is given by

$$Q = e(V_I + W) + \Delta H_v \tag{38}$$

where V_I is the ionization potential, ΔH_v is the enthalpy of vaporization, and W is the work function of the metal. For metal surfaces, the major short-range force between an ion and the surface can be represented by the Coulomb attraction between the ion and its image on the surface. In the simplest case, in which only the image force is present, the effect of applying an electric field to accelerate ions away from the surface is the well-known Schottky lowering of the potential barrier by an amount proportional to the square root of the field. This gives for the activation energy

$$Q = e(V_I - W) + \Delta H_V - \left(\frac{eE}{4\pi\epsilon_0} \right)^{1/2} = \Delta H_I - \left(\frac{eE}{4\pi\epsilon_0} \right)^{1/2} \quad (39)$$

where ϵ_0 is the permittivity of free space, and ΔH_I is the enthalpy of ionization of the ion. Thus, the activation energy for evaporation of metal ions from a metal surface can be obtained from well-known properties of the metal. Unfortunately, for evaporation of ions from liquid solution, the parameter values are not known with accuracy. Furthermore, serious questions may be raised concerning the applicability of this model to the physical situation involved.

Iribarne and Thomson [76IR47] have discussed the energetics of ion evaporation from aqueous solution. Enthalpies and free energies of vaporization can be calculated for ions using thermochemical cycles from tabulated thermochemical data, provided the absolute values for solvation of the proton are known. In addition, since the ions leaving the liquid surface are almost certainly partially solvated, it is necessary to know the enthalpy of vaporization of the hydrated ions. Recently, Meot-Ner[5] has published an extensive compilation of heats of hydration of organic ions. His results include enthalpies of vaporization of a number of different types of ions clustered with up to four water molecules. These studies encompass a variety of organic ions including protonated alcohols, ethers, aldehydes, ketones, esters, and amines as well as the alkali ions. Interestingly, the values for vaporization of the quadruply hydrated ions all fall in the range of 2.7 to 3.2 eV, with the small alkali ions being at the top end of this range. For most of these ions, clustering by four water molecules brings the energies for further clustering down to a value similar to the enthalpy of condensation of neutral water. Therefore, these values should be reasonable estimates of the activation energies for vaporizing ions from liquid droplets.

4.4.2. Ion Evaporation from Charged Liquid Droplets

The rate equation for ion evaporation can be rewritten in the form

$$\frac{dn_I}{dt} = C_s A v \, e^{-(\Delta H_I - 1.2E^{1/2})/kT} \quad (40)$$

where ΔH_I is the activation energy for ion evaporation (eV) and E is the electric field strength (in units of 10^9 V/m) at the surface of the droplet. As E increases, the exponent approaches zero and the rate of ion evaporation becomes large at all temperatures.

The field produced by the excess charge on a spherical droplet is given by

$$E = \frac{eq}{4\pi\epsilon_0 r^2} \tag{41}$$

Thus, for any amount of excess charge, q, as the particle vaporizes, E increases until eventually the rate of ion evaporation becomes large enough that the excess charge is dissipated by release of molecular ions into the gas phase. So long as the particle continues to vaporize, the process will continue as r decreases, until essentially all of the excess charge has been released as molecular ions into the gas phase.

This process will terminate, at least from the point of view of sampling the ions into a mass spectrometer, when the droplet or particle passes the MS sampling orifice or when it reaches a condition in which vaporization effectively ceases. The latter can occur if the temperature of the droplet becomes too low or if the droplet has ejected essentially all of the solvent molecules and has become a dry particle of nonvolatile sample or buffer.

If it is assumed that molecular ion production ceases when the droplet charge reaches some minimum value, then the total ion current (of either sign) is given by

$$I = n_P \int_{q_0}^{\infty} q\, e^{-q^2/2\sigma^2}\, dq = \left(\frac{2}{\pi}\right)^{1/2} \sigma\, e^{-q_0^2/2\sigma^2} \tag{42}$$

where n_p is the number of particles or droplets. Equation (42) can be used to calculate the ion intensity as a function of electrolyte concentration with various assumptions about the minimum charge required. One interesting case, which should correspond to the use of volatile buffers, is that in which a definite minimum droplet size is reached by the time that the droplets fly past the sampling orifice. In this case,

$$q_\infty = \frac{4\pi\epsilon_0 E_c r^2}{e} \tag{43}$$

where E_c is the critical field strength required for producing ions at an observable rate. Some sample calculations for $E_c = 1 \times 10^9$ are shown in Figure 4.21. If the particle is completely vaporized, then the ion current is expected to vary as the square root of the electrolyte concentration. As shown in the figure, the deviation expected from this square root law is negligible for final droplet sizes of 0.1 μm or less. On the other hand, final droplet sizes on the order of 1 μm give a behavior qualitatively similar to that observed in earlier experiments [83BL50].

When the solvent contains significant concentrations of nonvolatile sample or buffer, the vaporization may be slowed or halted when all of the solvent has

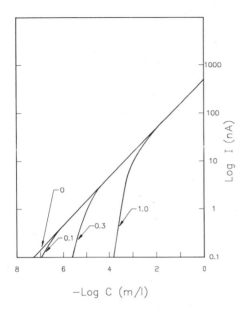

FIG. 4.21. Ion current calculated as a function of volatile electrolyte concentration assuming that the final droplet size is dependent on heat transfer and independent of electrolyte concentration. The parameter labeling the curves corresponds to the final droplet diameter in microns.

evaporated and an essentially dry particle of nonvolatile material is formed. The volume of the dry particle produced by thermospray nebulization and vaporization is given by

$$V_d = \frac{V_0}{1 - f}\left(\frac{\rho_0}{\rho_d}\right) C_0 \qquad (44)$$

where V_0 and ρ_0 are the initial volume and density of the droplet, respectively, ρ_d is the density of the dry particle, c_0 is the concentration of dissolved solid (g/g) in the solution, and f is the fraction of the liquid vaporized when the droplet is initially formed. If ρ_0 is approximately equal to ρ_d, and the concentration is expressed in moles/liter, the diameter of the dry particle is given by

$$d_p = d_0 \left(\frac{c_0}{1 - f}\right)^{1/3} \qquad (45)$$

For cases involving nonvolatile buffers, or nonvolatile samples in the absence of buffer, it is expected that ion production will stop, at least by the ion evaporation mechanism, when the particle reaches its "dry particle" diameter. The residual charge remaining on the particle can be calculated from equation (43) for any particular value of the critical electric field strength, E_c. The ion current as a function of nonvolatile electrolyte concentration and E_c can then be calculated using equation (42). The results of one such calculation are shown in Figure 4.22. In this case, it was assumed that the liquid was 99% vaporized at the point

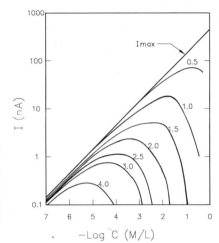

FIG. 4.22. Ion current calculated as a function of nonvolatile electrolyte concentration assuming that the final particle size is determined by concentration of the electrolyte in the solution. The parameter labeling the curves is the assumed critical electric field strength for ion evaporation in units of 10^9 V/m.

of nebulization and that 8-μm diameter droplets were initially produced. These results are qualitatively in accord with several experiments involving nonvolatile samples or buffers.

Roellgen and co-workers [86SC02] have suggested that formation of solid particles has a significant influence on thermospray mass spectra. Their results on the $(M + Na)^+$ ion intensity produced from glucose using a sodium acetate buffer as a function of buffer concentration are represented by the points in Figure 4.23. The solid curve is the calculated intensity taken from Figure 4.22 for a critical field strength of 1.5×10^9 V/m. At buffer concentration above $0.005\ M$, the calculated intensities are in good agreement with the experimental

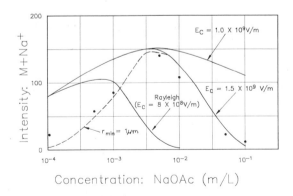

Concentration: NaOAc (m/L)

FIG. 4.23. Comparison of calculations of ion intensity as a function of NaOAc concentration (curves) with the experimental results (points) of Roellgen and co-workers [86SC02]. The best fit to the experimental results is obtained using a critical field strength of 1.5×10^9 V/m and a minimum final particle radius of 1 μm.

points, but at lower concentrations this calculation significantly overestimates the intensity. However, if the minimum droplet size attained is limited to 1 μm in diameter due to incomplete vaporization, then the dashed curve is obtained, and this provides a reasonable fit to the experimental data. From these results, it is clear that the deviation from the square root dependence at high buffer concentrations is due to the formation of solid particles which vaporize slowly, if at all, while the deviation at lower concentrations is due to incomplete vaporization of the liquid droplets.

Several studies have shown that addition of volatile buffers enhances the

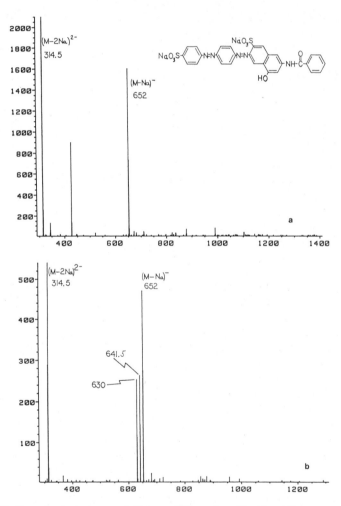

FIG. 4.24. Negative ion spectra of direct red 81 produced by direct thermospray ionization in (a) water and (b) 0.001 M ammonium acetate. Averages of (a) 0.435 to 2.170 min and (b) 2.089 to 3.775 min. Data from [87MC31].

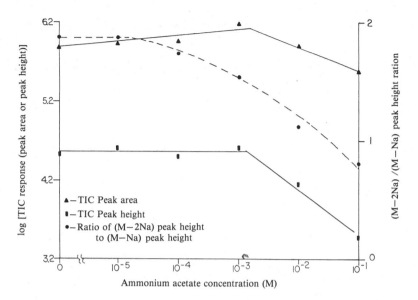

FIG. 4.25. Intensities of the major negative ions from direct red 81 as a function of ammonium acetate concentration.

ionization of many volatile and nonvolatile samples [85VE73], but this is not true in every case. One particularly striking example is that of the polysulfonated azo dyes. These exist as multiply charged anions in solution, but initial efforts to detect these compounds in 0.1 M aqueous ammonium acetate were unsuccessful. Recently, these compounds were detected in the negative ion mode by injecting them in water without a buffer [87MC31]. The spectrum of one of these (direct red 81) is shown in Figure 4.24. The dianion is the base peak, and the only other significant intensities arise from attachment of Na^+ to the dianion and from a triply charged ion corresponding to the Na^+-bound dimer of two dianions.

The effect of adding ammonium acetate buffer is summarized in Figure 4.25. At low concentrations there is essentially no effect, but at concentrations above ca. 0.001 M the sample ion intensity falls inversely with increasing buffer concentration. Also, the mass spectrum changes qualitatively as the negative charge is increasingly reduced by protonation of the anions. The mass spectrum at an intermediate buffer concentration is shown in Figure 4.24. The triply charged dimer has been converted to a protonated doubly charged dimer at m/z 641.5, and some of the dianion has been converted to the protonated anion at m/z 630. In this case, the addition of buffer does not enhance ionization but rather appears to neutralize the negative ions present in solution. It is possible that some of this neutralization is the result of gas phase recombination, but it appears that it is more likely due to the shift in ionic equilibria which occurs in the vaporizing droplet. This point will be discussed in more detail in a later section.

By carefully matching the conditions assumed for the calculations with actual experimental conditions, it may be possible to obtain useful values for the critical field strengths required for evaporation of various ions. Some recent work [87KA86] has focused on this problem using a drift tube apparatus. In these experiments, the mobility of the charged droplets or particles was determined after they had traveled several centimeters through a mixture of solvent vapor and nitrogen at pressures of a few torr and temperatures on the order of 100°C. If the particle diameter is small compared to the mean free path for the bath gas, the mobility is proportional to the charge on the particle divided by the cross section for collision with neutral molecules. Since the particles in question are large compared to molecular dimensions, the collision cross section can be approximated by the physical diameter of the particle. Thus, the mobility is proportional to the electric field strength at the surface of the particle. The experiments generally gave rather narrow single peaks in the mobility spectra for several volatile and nonvolatile electrolytes thermosprayed in water. Reduced mobilities (at STP) fell in the range 0.44 to 0.53 $cm^2/V \cdot s$, corresponding to surface field strengths of 7.6×10^9 to 9.0×10^9 V/m. Since these peaks are quite sharp and remarkably independent of operating conditions, these values appear to represent the first direct measurement of the field required to produce ions from liquid droplets. The precision of these determinations is estimated at about 5%, and the only systematic error which is difficult to estimate is in the approximation of the collision cross section by the physical size of the particle. For particles in the 0.01-μm range and higher, which is the range of main interest here, this does not appear to be too serious, since molecular scale effects such as the ion-induced dipole attractive force should not greatly affect the collision cross section.

4.4.3. Ion Production from Charged Liquid Droplets—A Summary

In a broad sense, the above treatment accounts satisfactorily for many of the qualitative features of the direct ionization process occurring in thermospray, but some of the more detailed observations do not yet fit smoothly together. For example, the data of Roellgen and co-workers [86SC02] on glucose in sodium acetate buffer, given in Figure 4.23, fit with the simple model, assuming a critical field strength of 1.5×10^9 V/m and a final droplet size of about 1 μm. However, the mobility measurements of Katta [87KA86, 87KA6'] indicate a critical field strength about one-half as large. Furthermore, the Rayleigh theory predicts that the larger droplets (greater than ca. 0.1 μm in diameter) should subdivide into smaller droplets well before the critical field strength is reached, unless it is about an order of magnitude smaller than that inferred from these experiments.

The situation is summarized in Figure 4.26, where droplet charge is given as a function of droplet radius. The lower line represents the approximate mean charge on the droplets as they are initially produced by statistical charging for a 0.1 M solution. The parallel band indicates the Rayleigh limit for water droplets

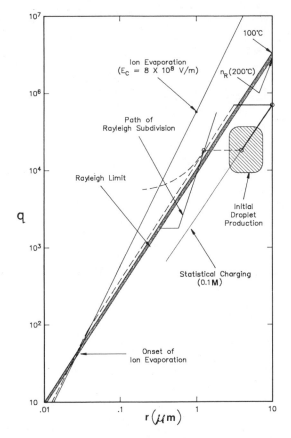

FIG. 4.26. Plot of droplet charge versus radius, summarizing possible mechanisms for producing ions by thermospray.

at temperatures between 100 and 200°C. The steeper line (slope of 2 on this log–log plot) represents the critical field for ion evaporation of about 8×10^8 V/m as determined by Katta [87KA86]. While detailed experimental measurements of the distributions of charge and size of the droplets produced by thermospray are not yet available, all of the present evidence suggests that the droplets are initially formed with radii of a few microns and charges given, at least approximately, by the statistical charging model. As indicated on the figure, these droplets are stable with respect to both limits. As these droplets vaporize, their charge should remain constant until the first instability is encountered, which, under the conditions indicated, corresponds to the Rayleigh limit. At this point, subdivision into smaller charged droplets can occur spontaneously. By successive processes of subdivision and vaporization, the droplets will increase in number and reduce in size along a path close to the Rayleigh limit, and no molecular ion production is expected until the Rayleigh limit intersects critical field line.

These processes of subdivision followed by ion production should continue until a solid particle is formed, the droplet stops evaporating or passes the sampling point, or the droplet is completely vaporized. The problem is that if the values indicated in Figure 4.26 are used, droplets containing high concentrations of nonvolatile buffers should produce fewer ions than are observed experimentally. On the other hand, if the Rayleigh limit is ignored, then a critical field of ca. 1×10^9 V/m implies that ion production should be much more efficient at high concentrations of nonvolatile electrolytes than is observed. Results of some model calculations are compared with the experimental results of Roellgen and co-workers [86SC02] in Figure 4.23.

It appears likely that the observed discrepancy may be due to the simple treatment of the droplets as having the properties of bulk water. The properties of the droplets must change as the concentration of less volatile solute increases during vaporization. Before reaching the final state of a dry, solid particle, the droplet may pass through a transition state in which it is still a liquid but with much higher surface tension and viscosity than bulk water. As a result, Rayleigh subdivision will become more difficult and the onset of ion evaporation will occur at larger droplet sizes than predicted by Figure 4.26. These considerations provide a plausible explanation for the observed discrepancies, but more definitive work on the properties of microscopic, charged droplets and particles will be required for a more quantitative treatment.

4.4.4. Ion Production from Charged Solid Particles

While ion evaporation from charged liquid droplets appears to be the major mechanism by which primary ions are produced directly by thermospray, it appears that, in some cases, ions may be produced from the residual dry particles. When essentially all of the solvent molecules have evaporated from the particle, further reduction in particle diameter will become quite slow even at elevated temperatures. As a result, continued production of ions by the field-assisted evaporation mechanism cannot be very important. However, at least two other mechanisms can be conceived which could contribute significantly. These are collision of high-velocity charged particles with surfaces and collisions between the massive charged particles and gas phase ions with the opposite charge.

4.4.4.1. Collision of Massive Charged Particles with Surfaces

Depending on the initial concentration of nonvolatile material in the solution, the final diameter of the residual particles may typically be in the range between 0.1 and 1 μm, with a charge determined by the limiting field strength of about 1×10^9 V/m. These particles are accelerated by the rapidly expanding vapor and may achieve velocities on the order of 10^4 cm/s. A 1-μm particle moving with this velocity has a kinetic energy on the order of 30 MeV. Krueger and co-workers [83KR83] have shown that dust particles accelerated to these energies can efficiently produce secondary ions upon collisions with surfaces. Moreover, the ions produced show the characteristics of "soft ionization" seen in other techniques involving collisions of high-energy particles with surfaces.

4.4.4.2. Collision of Massive Charged Particles with Oppositely Charged Gas Phase Ions

Both positive and negative gas phase ions are produced in the thermospray ion source. Ion densities are typically on the order of 10^8 ions/cm^3. Ions with the opposite charge from that of the particle will be accelerated strongly toward the particle when they are within a few mean free paths away. To a first approximation, the energy of the resulting collision will be given approximately by the electric field strength at the surface of the particle multiplied by the mean free path for collision of the gas phase ion with the molecules in the ion source. At a typical source pressure of 10 torr, the mean free path is approximately microns. Thus, the ion impacts the surface of the particle with an energy on the order of 10,000 eV, which is sufficient to efficiently produce secondary ions. As the particle traverses the ion source, approximately 1000 such collisions may occur. If the efficiency of secondary ion production is on the order of 0.1 or more, a significant number of secondary ions are produced. It appears that under the conditions normally existing in the thermospray ion source, this may be an important mechanism for producing molecular ions, as well as fragment ions from the solid particles.

4.4.5. Ionic Equilibria in a Vaporizing Droplet

It is frequently found that the ions observed in the thermospray spectrum do not correspond to those which are the dominant equilibrium species in solution. For example, in recent work on peptides it has been advantageous to work in the negative ion mode even though many of the peptides analyzed were basic species which are clearly positively charged in solution [88ST58]. Bursey and co-workers [86PA814] have presented a detailed study of the pH dependence of thermospray mass spectra for several molecules and concluded that the ions were produced by gas phase processes rather than by direct ion evaporation. It is quite clear that gas phase ion–molecule reactions play an important role in determining the final identity of the ions observed in the mass spectrum, but for truly non-volatile samples, such as large peptides, which, for all practical purposes, do not exist in the gas phase as intact neutral molecules, gas phase processes cannot be involved in producing protonated or deprotonated molecular ions. Rather, it is necessary to consider the shifts in equilibria which may occur in an isolated vaporizing droplet.

Consider a zwitterionic molecule which may exist in solution as a positive ion, a neutral, or a negative ion depending on the pH of the solution. If the solution is sufficiently acidic, the molecule should be positively charged in solution, and positively charged, protonated molecular ions should be produced as the result of ion evaporation from positively charged droplets. However, thermospray produces essentially equal intensities of positive and negatively charged droplets; thus, the sample molecule or ion is just as likely to be found in a negatively charged droplet. In this case, it cannot be vaporized as a positively charged gas phase ion, but may, in some cases, be converted to a negatively charged ion and ejected into the gas phase.

On a negatively charged droplet containing ammonium acetate buffer, the important equilibria are

$$Ac^- + MH^+ = M + HAc$$

$$Ac^- + M = (M - H^+)^- + HAc$$

(46)

These reactions are driven strongly to the right by the axcess Ac^- that necessarily exists on a negatively charged droplet and by the depletion of the relatively volatile acetic acid as the droplet vaporizes. Thus, if the sample M has a proton which is more acidic than that of acetic acid, it can be converted to a negative ion in a negatively charged droplet even though it may exist as a positive ion or neutral in the neutral bulk solution. Conversely, on positively charged droplets, the pertinent equilibria are

$$NH_4^+ + (M - H^+)^- = M + NH_3$$

$$NH_4^+ + M = MH^+ + NH_3$$

(47)

and compounds with a site which is more basic than ammonia can be converted to positive ions even though they may be negative or neutral in solution.

These shifts in equilibria accompanying vaporization of the droplets appear to adequately account for most of the differences between observed spectra for peptides and the ionic state in solution. For the polysulfonated azo dyes, the protonation reactions may be occurring in the bulk solution since the addition of an excess of ammonium acetate would be expected to at least partially neutralize the multiply charged anions. In addition, the loss of ammonia during vaporization of the droplets will tend to cause further neutralization during the course of vaporization even on the negatively charged droplets.

Polar neutral molecules which are less basic than ammonia often may be detected as adducts with the ammonium ion. Many polar molecules displace water from the inner solvation shell of ions. For example, in water/methanol or water/acetonitrile mixtures, with ammonium acetate present, the clusters of the organic species with ammonium ion such as $NH_4^+(CH_3OH)$, $m/z = 50$, or $NH_4^+(CH_3OH)_2$, $m/z = 82$, are always much more intense in the thermospray mass spectrum than are the analogous hydrated species. Similarly, nonvolatile molecules, such as sugars, which are sufficiently polar to be strongly associated with ions in solution may be detected as adducts such as MNH_4^+ in the mass spectrum. On the other hand, less polar samples which also are neither more basic than ammonia nor more acidic than acetic acid will not be ionized by thermospray in ammonium acetate solution.

4.4.6. Gas Phase Ion–Molecule Reactions

In many respects, the thermospray ion source is quite similar to a conventional "chemical ionization" (CI) source (see Munson, Ref. 6). The major differences are that the pressure in the thermospray source may be somewhat higher (typically 2–20 torr) compared to typical CI (ca. 0.2–2 torr) and the gas throughput is very much higher. In CI, the source is made as tight as possible consistent with injecting an electron beam to ionize the gas and with extracting the ions. Typical orifices for electron beam entrance and ion exit are on the order of 0.5 mm. In thermospray, a similar tight source is used, but a vacuum pump is connected opposite the vaporizer to exhaust the copious vapor that is produced by vaporizing LC effluent at liquid flows of 0.5–2 ml/min. In conventional CI, the gas flows involved are typically in the range of 1–20 ml/min (STP), while in thermospray, the vapor flow is in the range of 0.5–2 liter/min (STP). The gas flow through the thermospray source at operating pressure may be on the order of 1–2 liter/s and local gas velocities may approach sonic velocity at restricted cross sections.

In a conventional CI source, ionization is initiated by high-energy electrons either from an electron beam or a Townsend discharge [85VE1']. If water vapor is used as the CI reagent gas, the initial process is

$$H_2O + e^- = H_2O^+ + 2e^- \tag{48}$$

The water molecular ions react on virtually every collision to produce the hydronium ion by the proton transfer reaction

$$H_2O^+ + H_2O = H_3O^+ + OH \tag{49}$$

The hydronium ion does not react further with water except to form clusters by the reaction

$$H_3O^+ + 2H_2O = H_3O^+(H_2O) + H_2O \tag{50}$$

where the second H_2O molecule acts as a third body to carry away the excess energy and stabilize the complex. Clustering may continue by a similar set of reactions

$$H_3O^+(H_2O)_n + 2H_2O = H_3O^+(H_2O)_{n+1} + H_2O \qquad n = 0,1,2,\ldots \tag{51}$$

The distribution of hydrated hydronium ions observed in the CI mass spectrum is determined by the equilibrium constants for reactions (51) and the temperature and pressure in the ion source. These ions do not react further with water vapor, but if samples with higher proton affinity than water are present, they will react by proton transfer reactions such as

$$H_3O^+ + M = MH^+ + H_2O \tag{52}$$

Polar samples of lower proton affinity may react by displacing a water molecule from a cluster, for example,

$$H_3O^+(H_2O) + M = H_3O^+(M) + H_2O \tag{53}$$

Samples which are both less basic and less polar than water will not react and will not be detected in the positive ion CI spectrum using water as a reagent. One solution to this kind of problem is to use a more acidic reagent such as methane or hydrogen, and another, which is more applicable to thermospray LC/MS, is to employ negative ions.

In the negative ion mode, OH^- ions may be produced which react with molecules more acidic than water vapor to form the deprotonated anion. If an electron beam or discharge is used to produce the initial ionization, an excess of thermal electrons is also formed in the high-pressure source. These may attach to molecules of high electron affinity to form M^- ions, and so long as M^- is stable and the electron affinity of M is greater than that of the major components present in the ion source, this process can yield high-sensitivity detection for electronegative components. Thus, both $(M - H)^-$ and M^- may be observed in the CI spectrum, depending on the relative acidity and electron affinity of the sample. If the sample has low electron affinity and is less acidic and less polar than water vapor, it will not be detected in the negative ion mode.

In most respects, the thermospray ion source behaves as a conventional CI source.[6] The only major difference appears to be caused by the relatively high gas flows involved in thermospray at conventional LC flow rates. As a result, it is necessary to locate the ionizing beam (electron beam of discharge) upstream from the ion sampling orifice rather than in the conventional position opposite the orifice; otherwise, the ions are swept downstream by the rapidly flowing gas and are sampled ineffectively. This high gas flow also limits ion residence times to values on the order or a millisecond. At first, this might seem to impose a serious limitation on sensitivity, but the flow residence time is quite similar to that determined by diffusion in the conventional CI source. In both cases, the ion–molecule reactions are sufficiently fast that most reactive molecules should be ionized efficiently, with the overall sensitivity being determined by the efficiency with which ions are extracted into the mass analyzer.

When an electron emitting filament or discharge is used to initiate ionization, the processes occurring in the thermospray ion source appear to be identical to those occurring in a conventional CI source. The only limitation in the case of thermospray is that the properties of the reagent gas are determined by the solvent used for the LC separation. More basic or more acidic reagents can be added to good effect, but it is not possible to use a more reactive reagent than the solvent vapor.

When no external source of ionization is employed, the primary ions are produced by the direct processes described above, but the gas phase ion–molecule reactions may still play a dominant role in determining the observed mass spectrum. The final set of ions observed must be essentially in equilibrium with the vapor in the ion source. In thermospray, the ions present in the charged droplet are initially formed in the gas phase clustered with solvent molecules, while in CI, molecular and fragment ions are formed initially with proton transfer and clustering reactions occurring as equilibrium is approached. While equilibrium may be approached from opposite extremes in the two cases, the final result must be the same if equilibrium is attained. All of the experimental studies of both CI and thermospray suggest that deviations from ionic equilibrium in the gas phase are insignificant under the conditions normally employed in these techniques. Some differences between the spectra obtained by thermospray without external ionization and those obtained with filament or discharge on are observed, but these can be attributed to the higher ion densities that can be obtained using the external ionizing means.

In CI, fragment ions may be observed in addition to the usual protonated and deprotonated molecular ions. Only those fragments (usually containing an even number of electrons) are observed which are unreactive with the gas in the ion source. The extent of fragmentation is determined by the internal energy of the ions at formation, which is contributed by the exoergicity of the reaction in addition to the thermal energy available. Thus, more acidic reagents and higher temperatures tend to induce more fragmentation in the positive ion mode, while less acidic reagents tend to produce more fragmentation in the negative ion mode. In addition, a thermally labile molecule may fragment prior to ionization, with the fragments then being ionized, for example, by proton transfer. It is often difficult to determine whether fragmentation or ionization occurred first under CI conditions since the products are very often the same.

Samples introduced into the ion source by the thermospray process may be ionized prior to reaching the gas phase or they may be neutral. For those ionized directly, significant fragmentation is normally not expected; however, for very labile systems, the thermal energy acquired in coming to equilibrium with the gas phase may be sufficient to cause fragmentation. Modestly volatile samples [such as those amenable to so-called direct CI (DCI)] may be vaporized efficiently by thermospray and subsequently ionized by ion–molecule reactions induced either by the ions produced directly in the thermospray process or by those formed by external means in the gas phase. Less volatile samples may not produce significant concentrations of intact molecules in the gas phase, but at sufficiently high temperatures all substances can be pyrolyzed to produce volatile products. Unvaporized material remains in the dry particles which may be produced as one of the end products of thermospray. These are swept away by the vacuum system and contain undetected material. Thus, intact molecular ions are produced from truly nonvolatile samples by direct transfer from the condensed phase to the gas phase, but molecular ions from more volatile samples and fragment ions from all samples can be produced by a variety of gas phase processes.

4.4.7. Repeller Effects

There is considerable discussion in the recent literature about the effect of repeller electrodes on the performance of thermospray [86ROO′, 86ZA3′]. Two basic configurations have been employed. One employs an electrode opposite the sampling orifice which is somewhat shielded from the upstream region of the source, such as shown in Figure 4.1, and the other places the electrode downstream from the sampling orifice. In both cases, the effects which have been observed by placing a voltage on this electrode relative to the ion source are (1) enhancement of high-mass ion intensities relative to low-mass ion intensities, and (2) with filament or discharge on, significant increase in the degree of fragmentation observed. The latter effect appears to be particularly marked for more volatile samples of low ionization potential.

The first effect may be partly due to mass discrimination caused by the high gas flows through the ion source. An ion of m/z 2000 moving with sonic velocity transverse to the ion sampling orifice has a kinetic energy of approximately 2.5 eV. Application of a positive repeller voltage will cause negative ions in the source to be collected and an excess positive space charge to be built up in the region in front of the sampling orifice. Positive ions carried along by the rapidly flowing gas will be retarded, and the probability of their exiting through the sampling orifice may be enhanced. In this model, the optimum repeller voltage should increase with increasing mass, and that effect is observed experimentally [86ROO′].

The second effect is more complex, and further studies will be required to further elucidate some of the details. It appears that the increases in the degree of fragmentation observed are much too large to be accounted for by collision-induced dissociation occurring as the result of increasing the ion kinetic energy relative to that of the gas. The nominal electrical fields employed are on the order of 100 V/cm. At these low fields, the drift velocities are small compared to thermal velocities. Thus, the kinetic energy of collisions inside the ion source is little affected by the applied voltage, and little change in fragmentation pattern would be expected. Also, the effect is not observed when no external source of ionization and no free electrons are available. It seems clear that the fragmentation is produced as the result of accelerating electrons toward the repeller. As the result of their high mobility, electrons can be readily accelerated up to the point at which they begin to lose significant energy by exciting electronic transitions of the vapor present in the ion source. These higher-energy electrons may cause increased fragmentation by ionization of neutral sample molecules followed by EI type fragmentation, by excitation of previously formed ions, or by forming more energetic reactive ions in the gas. The role of these processes has not yet been elucidated, and it may be satisfactory to say that by accelerating the electrons, the repeller has increased the effective plasma temperature in the vicinity of the sampling orifice.

4.5. OPTIMIZING THERMOSPRAY LC/MS FOR PARTICULAR APPLICATIONS

In setting up to perform an analysis by thermospray LC/MS, several choices must be made correctly for best results to be obtained. These include the following:

1. LC conditions
2. ionization technique
3. vaporizer and ion source temperatures

These choices depend on the properties of both the LC mobile phase and the sample, as well as on the kind of analytical result that is sought. While it is not yet possible to give a precise prescription that is applicable to all cases, the present level of understanding of the processes involved, as summarized in the foregoing sections, provides considerable guidance in narrowing the choices.

4.5.1. Selection and Modification of LC Conditions

Most LC procedures using standard-size LC columns can be used without modification, but in some cases changes will be required. At present, thermospray works best at flow rates between 0.5 and 1.5 ml/min. Somewhat higher and lower flow rates can be accommodated, but some loss in sensitivity generally occurs outside this range. If the separation normally uses a nonvolatile buffer or other additive such as an ion pairing reagent, it is generally necessary to switch to a volatile alternative. Phosphates and alkali salts as major components of the mobile phase must be avoided. More volatile salts such as ammonium acetate, ammonium formate, ammonium alkylsulfonates, trifluoroacetic acid, and tetrabutylammonium hydroxide are no problem and can often be substituted without severe loss in chromatographic performance, particularly if the more important parameters, such as pH, are not changed.

4.5.2. Selection of Ionization Technique

Most commercially available thermospray systems provide three alternative modes of ionization: (1) direct ion evaporation, sometimes referred to as "thermospray ionization"; (2) chemical ionization initiated by an electron beam, sometimes called "filament on" operation; and (3) chemical ionization initiated by a low-current Townsend discharge, sometimes called "discharge ionization."

In addition, the mass spectrometer is normally equipped with the capability to analyze and detect negative ions in addition to the standard positive ion capability. Both the properties of the sample and those of the mobile phase must be considered in choosing which of the six possible operating modes is likely to be best for a particular analysis. In general, for positive ion detection, samples must be more basic than the mobile phase (to form MH^+) or be sufficiently polar to form stable adducts [e.g., $(M + NH_4)^+$. For negative ion detection, samples must be more acidic than the mobile phase [to form $(M - H)^-$] or have a higher electron affinity (to form M^-). Use of either the filament or discharge is required to form M^-.

4.5.2.1. Direct Ion Evaporation

As discussed in Section 4.4, ions present in solution can be evaporated directly into the gas phase. Since this technique does not require vaporization of the sample to produce a neutral vapor, it is applicable to totally nonvolatile samples. If the sample is ionic in water or methanol, this technique can be used without the addition of a buffer, but the sensitivity is generally poor and the response is nonlinear, varying approximately as the square root of sample concentration. Addition of a volatile buffer such as ammonium acetate at a concentration in the $0.01-0.1\ M$ range generally enhances the sensitivity for these kinds of samples rather dramatically and yields a linear response over several orders of magnitude of sample concentration. In a few cases, e.g., polysulfonated azo dyes, addition of buffer decreases response. Sensitivity is best when the mobile phase is predominantly aqueous. This technique can be used with essentially 100% methanol, but the sensitivity is often marginal if the water fraction is less than ca. 20%. If a buffer is used, the sample need not be ionized in solution since often the buffer ionization can be transferred to the sample, either in the vaporizing solution or in the gas phase, to efficiently ionize the sample.

Molecules containing both acidic and basic functional groups (such as peptides) can sometimes be detected efficiently as negative ions even though they may exist in the mobile phase as positively charged species.

This mode of ionization is generally favored in the following cases:

1. The sample is ionic, polar, or nonvolatile.
2. The preferred mobile phase is water with either methanol or acetonitrile and a volatile buffer.

4.5.2.2. Filament on Operation

The use of the filament to initiate CI is most effective when the mobile phase contains a large organic fraction. It can be used with pure aqueous phases, but it may be difficult to maintain full emission and the filament lifetime may be shortened. The use of the filament is almost essential for normal phase chromatography, and it can significantly enhance the sensitivity for reversed phase separations, particularly when the samples are at least slightly volatile. Use of the filament with truly nonvolatile samples is generally not recommended, since it does not enhance sensitivity and may, in many cases, cause a decrease in the sensitivity for nonvolatile samples of interest and an increase in the solvent-related chemical noise.

4.5.2.3. Discharge Ionization

The discharge electrode provides an alternative technique for initiating chemical ionization. It is most useful when the mobile phase contains high water fractions. It can be used with organic mobile phases present, but it is not recommended for extended use with organic fractions greater than about 60% because carbon deposits on the electrode build up and short out the discharge. Carbon deposits are indicated by erratic behavior and higher than normal current readings on the front panel meter. This condition can sometimes be reversed by running the discharge for 15–30 min with pure water as the mobile phase.

4.5.2.4. Positive Ion versus Negative Ion Detection

All three of the ionization techniques discussed above produce approximately equal intensities of positive and negative ions from most common solvents and for many relatively neutral analytes. The general rule is that basic compounds give higher sensitivity in positive ion mode and acidic compounds in negative ion mode, but unfortunately the proton affinities of a great many samples are unknown. This general principle is a useful guide in deciding which mobile phases may be most suitable for analyzing particular classes of samples in either ionization mode, but in many cases the most effective method will need to be determined empirically.

4.5.3. Operating Temperatures

For optimum performance of thermospray LC/MS, two deceptively simple criteria must be met. For the best sensitivity, sufficient heat must be supplied to completely vaporize the sample, since unvaporized sample cannot contribute ions to the mass spectrum. At the same time, the sample must not be heated so much as to cause pyrolysis or other uncontrolled chemical modification. Nonvolatile samples do not produce intact molecules in the gas phase by vaporization, but they may produce ions in the gas phase by direct ion evaporation. In such cases, virtually all of the sample that is vaporized is ionized; thus, it is possible to obtain acceptable sensitivity even though the fraction of the sample vaporized may be relatively small.

Samples may be divided roughly into the following categories:

1. at least slightly volatile (can be analyzed by DCI)
2. slightly volatile, very labile (e.g., glucuronides)
3. nonvolatile, ionic
4. nonvolatile, neutral

4.5.3.1. Volatile Nonlabile Compounds

The first category encompasses a fairly large fraction of low-molecular-weight compounds. Most of these can be analyzed satisfactorily with a single set of standard operating conditions. Typical operating conditions consist of setting the vaporizer control temperature about 2–5° below takeoff, (see Figure 4.9) and the source block at ca. 250–300°C. Under these conditions, the vapor temperature in the ion source may be 25–50°C below the block temperature. Optimum conditions may vary somewhat depending on details of the ion source design, but these kinds of compounds generally give excellent results over a wide range of operating conditions.

4.5.3.2. Labile Compounds

The above conditions may also be suitable for labile compounds, but molecular ions may be weak. Molecular ions can often be increased (and fragment intensities decreased) by lowering block temperature. Block temperatures as low as 200°C with vapor temperatures of 150°C are sometimes useful for modestly volatile but thermally labile compounds. For less volatile neutral compounds, sensitivity can often be improved by increasing block temperature, but this may cause additional pyrolysis and fragmentation if the samples are thermally labile.

4.5.3.3. Nonvolatile, Ionic Compounds

The best performance for nonvolatile, ionic compounds is generally obtained under rather different conditions. To obtain intense molecular ions for these samples, it is often necessary to set the vaporizer control temperature 20–50° below the "takeoff" point and drastically increase the block temperature to as much as 450°C. Vapor temperature under these conditions may be in the 250–300°C range. Recent examples involving compounds in this class appear in Chapter 8.

4.5.3.4. Nonvolatile, Neutral Compounds

The most difficult samples for thermospray are nonvolatile neutrals, particularly those which are not water soluble. In some cases, these can be ionized by attachment of an ammonium or sodium ion, which presumably occurs prior to vaporization. Thermal conditions are generally similar to those for nonvolatile, ionic compounds when ion attachment can be made to work. Some success has been obtained using $0.1\ M$ ammonium acetate in neat methanol as a mobile phase for this class of compounds.

4.5.4. Vaporizer Tip Temperature

The vaporizer tip temperature is an important variable to monitor since it closely correlates with thermospray performance, but it is not directly controllable since it depends on the vaporizer control temperature, the flow rate, the mobile phase composition, the diameter of the vaporizer nozzle, and weakly on the temperature of the tip heater. It is important to monitor the tip temperature since if all of the above are constant, then it also should be constant. For a given set of operating conditions, changes in tip temperature normally indicate a change in the diameter of the vaporizer nozzle, although sudden changes can indicate a malfunction in the LC system. It is not unusual for the tip temperature, for a given set of operating conditions, to change by a few percent over the course of a day of running. An increase generally means that the tip nozzle is reduced in area, which may result from deposition of materials leached from the column packing. Variations in tip temperature of as much as 20% can generally be tolerated without substantially affecting performance. If the tip temperature increases more, it may be necessary to replace the vaporizer insert or to clean it using a procedure such as that described by Hsu and Edmonds [85HS14].

4.6. SUMMARY

Thermospray is now established as a practical technique for LC/MS interfacing. Its utility for a large number of applications is indicated in other sections of this volume. Despite this obvious success, there remain a number of valid criticisms which clearly indicate that additional work will be required before the development of LC/MS techniques can be considered to approach completion. Some of the deficiencies which have been noted include the following:

1. Results very often allow unambiguous determination of molecular weight, but fragmentation is either absent, insufficient, of insufficiently reproducible to allow definitive identification of known compounds or much in the way of structure elucidation of unknown compounds.

FIG. 4.27. A difficult courtship. By Patrick J. Arpino. (Reproduced from 82AR14 with permission.)

2. Sensitivity, particularly for large, nonvolatile compounds such as peptides, nucleotides, saccharides, and lipids, is disappointingly low.

3. The technique is limited to a fairly narrow range of chromatographic conditions.

4. The instruments are too expensive, too complicated, and too difficult to operate and maintain.

Considerable progress has been made on all of these problems in recent years, but a continuing research and development effort will be required to solve the remaining technical problems which cause the improbable marriage of the fish and bird (Fig. 4.27) to be less than perfect bliss.

ADDITIONAL REFERENCES

1. R. F. Browner and A. W. Boorn, "Sample Introduction: The Achilles Heel of Atomic Spectroscopy," *Anal. Chem.* **56,** 786A–798A (1984)
2. Experiments on the Atomization of Liquids in an Air Stream, E. Hope, Translator, Defense Research Board, Department of Defense, Ottawa, Canada (1950). (Cited in Ref. 1)

3. H. Ashkenas and F. S. Sherman, in *Rarefied Gas Dynamics, Vol. 2*, J. H. DeLeeuw, ed., Academic, New York (1966), p. 84.
4. S. S. Wong and F. W. Roellgen, "Field Ion Emission from a Single Organic Droplet," in *Proceedings of the 29th International Field Emission Symposium, Göteborg, 1982*, H. Nordien, and H.-O. Andrien, eds., Almqvist and Wiksell, Stockholm (1982), pp. 225–230.
5. Michael Meot-Ner (Mautner), "Heats of Hydration of Organic Ions: Predictive Relations and Analysis of Solvation Factors Based on Ion Clustering," *J. Phys. Chem.* **91**, 417–425 (1987).
6. M. S. B. Munson and F. H. Field, "Chemical Ionization Mass Spectrometry. I. General Introduction," *J. Amer. Chem. Soc.* **88**(12), 2621–2630 (1966).

Chapter 5

Particle Beam Interfaces

The major disadvantage of thermospray and other direct coupling techniques is that the ionization occurs in a bath of the solvent vapor at a relatively high source pressure of typically one torr or more. This effectively precludes the use of electron impact (EI) ionization and also limits the choice of reagents in chemical ionization (CI). Attempts to overcome this limitation have generally focused on various transport devices designed to allow removal of the solvent while transporting the sample to the ion source. The most successful of these have involved moving wires or belts, and this approach is described in Chapter 3.

An alternative transport system involving no moving parts was first described by Willoughby and Browner [84WI65]. This approach was given the acronym MAGIC, which stands for *m*onodisperse *a*erosol *g*eneration *i*nterface for *c*hromatography. A schematic diagram of this device is shown in Figure 5.1. In this device, the LC effluent is forced under pressure through a small orifice (typically 5–10 μm in diameter), and as a result of the Rayleigh instability [45RA0′], the liquid jet breaks up into a stream of relatively uniform droplets whose initial diameter is approximately 1.9 times the nozzle diameter. A short distance downstream, the stream of particles is intersected at 90° by a high-velocity gas stream (usually He) to disperse the particles and prevent coagulation. The dispersed droplets fly at relatively high velocity through the desolvation chamber where vaporization occurs at atmospheric pressure and near-ambient temperature. Heating is provided to the desolvation chamber, not to raise the aerosol temperature above ambient, but to replace the latent heat of vaporization necessary for solvent evaporation. Ideally, all of the solvent is vaporized and the sample remains as a solid particle or less volatile liquid droplet.

The monodisperse aerosol generator (MAG) is designed to handle 0.1 to 0.5 ml/min of liquid, and up to 1 liter/min of dispersion gas is required. To reach the low ion source pressures required for EI, an efficient method for separating gas from the sample particles is required. As indicated in Figure 5.1, a two-stage momentum separator is used to form a particle beam from the sample and pump away the vapor and dispersion gas. During the expansion in the first capillary

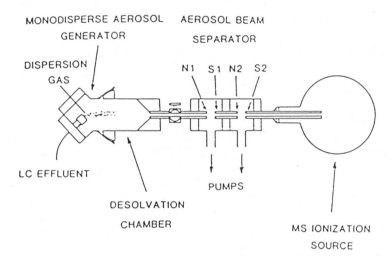

FIG. 5.1. Schematic diagram of MAGIC-LC/MS. N1, nozzle 1; N2, nozzle 2; S1, skimmer 1; S2, skimmer 2. (From 84WI65 with permission.)

nozzle, the particles are accelerated to a velocity approaching that of the gas. The high-momentum particles tend to remain on the axis of the separator while the light molecules diffuse away. The first aerosol-beam separator chamber is pumped with a 300-liters/min mechanical pump which maintains the pressure at between 2 and 10 torr. The second chamber is pumped by a 150-liters/min pump which keeps the pressure in this chamber at between 0.1 and 1 torr.

Willoughby and Poeppel [87WI9'] have described a modified version of a particle beam interface which employs a thermospray vaporizer as a nebulizer. While no details of this "Thermobeam" apparatus have yet been published, it appears that this approach may have some significant advantages. In particular, it appears that the thermospray nebulizer produces smaller initial droplets at higher temperatures, thus substantially facilitating desolvation. Furthermore, it does not require such small nebulizer orifice diameters, thus providing more immunity to plugging.

Both versions of the particle beam interface have been demonstrated to give EI spectra in good agreement with library spectra using sample injections of 100 ng or more. Generally, these spectra do not include the low-mass region where solvent interference may be expected, so it is impossible to assess the solvent removal efficiency actually achieved. These particle beam interfaces show considerable promise for future liquid chromatography/mass spectrometry (LC/MS) applications, particularly for samples with at least modest volatility, and in cases where unambiguous compound identification is required. Further development including data on sample transfer efficiency and its dependence on sample volatility, solvent removal efficiency, and reproducibility and reliability will be required before these techniques can be considered suitable for routine LC/MS applications.

Chapter 6

LC/MS of Nucleic Acid Constituents

6.1. INTRODUCTION

Nucleic acids and their constituents are of fundamental biochemical importance. Nucleotides are the biosynthetic precursors and catabolites of the oligomeric DNA and RNA functioning in the storage and transmission of genetic information and in the organization and execution of protein synthesis. The nucleotide adenosine triphosphate (ATP) is mainly involved in energy metabolism. Energy from biological oxidations is commonly converted to chemical energy in the form of ATP. Energy from the hydrolysis of the pyrophosphate bonds of ATP drives energetically unfavorable reactions. Additionally, the phosphate group may be transferred to another molecule to facilitate such a process. Nucleotides perform donor functions in many metabolic processes in which they act as carriers for the donation of groups (e.g., glycosyl, sulfate, alkyl, acyl, etc.) to the appropriate acceptor molecules. At the cellular level, adenine nucleotides are also involved in the regulation of metabolism, as mediators in hormone actions, and as intermediates in the synthesis of vitamins in which heterocyclic rings are derived from adenine.

The development of liquid chromatograpy (LC), initially ion exchange separations and, later, high-performance liquid chromatography (HPLC) operating in ion exchange, stearic exclusion, and reversed phase modes, has been highly influential in nucleic acid research. The flexibility of these methods has led to their use in a wide variety of studies including structure elucidation and investigations of the metabolism and the biosynthesis of nucleic acids. The application of liquid chromatography to nucleic acids research has been comprehensively summarized in a book edited by Brown.[1] Important areas of application have been in the separation, isolation, and purification of nucleic acids from biological samples as well as the separation and identification of constituents of hydrolyzed RNA and DNA at the nucleotide, nucleoside, and nucleobase level.[2,3] The analysis of intermediates, end products, and the enzymology of nucleic acid metabolism, including the assay of coenzymes and cofactors involved in these reactions,[4] relies extensively on liquid chromatography. Important biotechnological applications of liquid chromatography include determination of the structure and purity of synthetic oligonucleotides.[5]

Clinically, HPLC has an important role in the study of disease states of nucleic acid metabolism by the measurement of a single nucleotide, nucleoside, or nucleobase or more frequently in the profiling of a number of these compounds in complex mixtures derived from physiological fluids.[6] Similarly, the profiling of nucleic acid constituents, particularly nucleosides, has found application in the early marking of cancer.[7,8] HPLC also provides a key method for therapeutic drug monitoring in assays to determine the levels of purine and pyrimidine drugs and of chemotherapeutic agents and antimetabolites for the treatment of cancer.[9]

It is certain that liquid chromatography will continue to play a role in nucleic acid research and biotechnology. In biotechnology, the separation of intact nucleic acid from cells and tissues and the separation and purification of synthetic segments of DNA for insertion into recombinant organisms will be important.

The contribution of mass spectrometry to nucleic acid chemistry has been significant but less comprehensive than that of liquid chromatography. Historically, mass spectrometry has served principally in structural characterization, usually at the level of the nucleobase or nucleoside. Less frequently, mass spectrometry has been used in biosynthetic studies or in the measurement of purines and pyrimidines in physiological fluids. McCloskey has recently reviewed the work in this area.[10]

The ultimate significance of combined liquid chromatography/mass spectrometry (LC/MS) in nucleic acid investigations will be mixture analysis in complex matrices. In the development stages of LC/MS, interface designers have frequently evaluated their methods on "simplified" systems of nucleic acid constituents consisting of isolated single components, synthetic mixtures, etc. This might involve simple injection in solution without resort to chromatographic column. This is understandable in view of the immature technology and in the context of investigations focused principally on methods. In an early report, Vestal and co-workers commented on the analysis of nucleic acid constituents. Their report said in part:

> In most of our earlier work we used the elements of nucleic acids—bases, nucleosides and nucleotides—as test compounds which provided a convenient series of biologically important molecules of increasing difficulty. The purine and pyrimidine bases present no particular problem. . . . The nucleosides are substantially more difficult. . . . The nucleotides represent the most difficult molecules which we have studied. . . . [80BL2', 80BL26]

In this chapter, comparisons will be made of the various LC/MS methodologies, based (where possible) on the results obtained in such model experiments. This will be followed by an account of the few but increasingly frequent "real world" applications of LC/MS to nucleic acid constituents.

6.2. LC/MS OF NUCLEOBASES AND RELATED COMPOUNDS

6.2.1. Nucleobases

Purine and pyrimidine heterocyclic ring systems are the parent compounds of the two classes of nitrogenous bases found in nucleic acids. Two purine derivatives, adenine and guanine, and three pyrimidine derivatives, uracil, thymine, and cytosine, are the major bases of RNA and DNA. Additionally, a number of minor purine and pyrimidine derivatives occur in small amounts in nucleic acids. Among the most common derivatives are the pyrimidines 5-methylcytosine and 5-hydroxymethylcytosine and the purines 6-methyladenine and 2-methylguanine. Such minor bases are particularly important in transfer RNAs (tRNAs), where the level of modification may be as high as 10%.

Single nucleic acid bases or synthetic mixtures of these have been employed in the demonstration of the utility of several LC/MS interfaces. These include the particle beam precursor of the thermospray instrument [77MC6', 78BL81, 78MC8', 79BL2', 79VE97, 80BL2', 80BL26, 80PH10], the thermospray interface [84SM10', 86RU9'], the direct liquid introduction (DLI) interface [83ES07], the heated-wire concentrator/DLI device [79CH0'], and the moving-belt interface [84GA13], in combination with chemical ionization (CI) atmospheric pressure [83TH5'] and laser desorption (LD) atmospheric pressure [86HO87] interfaces.

The behavior of nucleobases in mass spectrometric experiments on solution systems which have direct or indirect connections to LC/MS interface methods has been informative. Examples include studies of adenine ions [82ZO25] and N-methylated adenine cluster ions [84SU91] from water solution under field ionization conditions. Additionally, it has been shown that the solution pH effects ion production in secondary ion mass spectrometry of adenine [84IN13].

The potential for application of the particle beam/CI methods to the examination of mixtures of nucleobases is demonstrated by the results of Blakley and co-workers [79BL2'] for the analysis of a synthetic mixture of nucleobases presented in Figure 6.1. This figure shows the UV absorbance chromatogram and reconstructed total ion current (RLC) and selected ion current chromatograms for combined LC/MS analysis using the particle beam interface device of a synthetic mixture consisting of (1) 5-hydroxymethylcytosine, (2) 5-methylcytosine, (3) uracil, (4) 7-methylguanine, and (5) N^6-methyladenine. Ions monitored are the protonated molecules, MH$^+$, which predominate under chemical ionization conditions. Under thermospray conditions, the protonated molecule is exclusively produced, and such capabilities have carried through instrument developments to the now well-developed thermospray technique.

A second set of LC/MS experiments utilizing the particle beam interface device is illustrated in Figure 6.2. Two new modified 6-aminouracil derivatives from human urine, 6-amino-3-methyl-5-(N-formylamino)uracil and 6-amino-1-methyl-5-(N formylmethylamino)uracil, have been characterized in combined chromatographic and spectroscopic studies, which included low-resolution mass

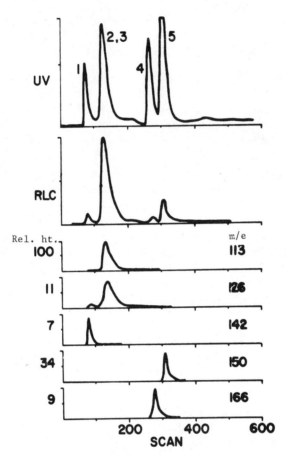

FIG. 6.1. UV absorbance chromatogram and reconstructed total ion current (RLC) and selected ion current chromatograms for the combined LC/MS Analysis using the particle beam interface device of a synthetic mixture consisting of (1) 5-hydroxymethylcytosine, (2) 5-methylcytosine, (3) uracil, (4) 7-methylguanine, and (5) N^6-methyladenine. Chromatographic conditions: Partisil PXS 10/25 ODS-2 column, 0.4 ml/min formic acid solution, MeOH gradient 0–100%, 10 min. Ionization by chemical ionization. (Reproduced from 79BL2' with permission of the authors.)

spectrometry, high-resolution mass spectrometry, and combined liquid chromatography/mass spectrometry [81DE83, 81ED9'] using the early particle beam version of the thermospray instrument operating in chemical ionization mode. In the case of these compounds, conventional mass spectrometric techniques proved to be of limited value due to the unexpected similarity of the electron impact (EI) mass spectra of the model compounds and extensive dehydration which occurred during chemical derivatization procedures. The CI mass spectra of the several candidate monomethyl and dimethyl reference compounds synthe-

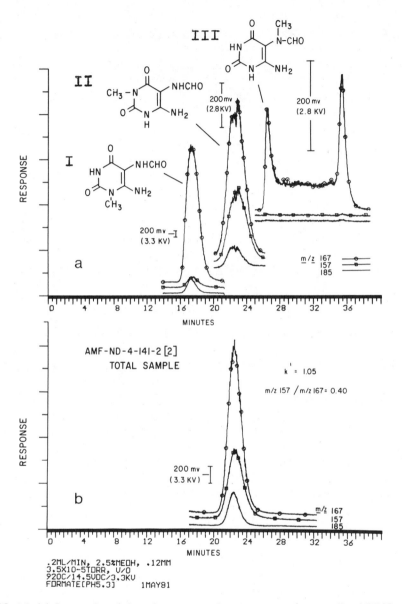

FIG. 6.2. (a) Composite of three ion current chromatograms from on-line LC/MS experiments using the particle beam interface device. Left to right: synthetic 6-amino-1-methyl-5-(*N*-formylamino)uracil (**I**), synthetic 6-amino-3-methyl-5-(*N*-formylamino)-uracil (**II**), and synthetic 6-amino-5-(*N*-formylmethylamino)uracil (**III**). (b) Ion current chromatogram from the LC/MS experiment for the base derived from urine, confirming it as the 3-methyl isomer. Chromatographic conditions: 250 × 4.6 mm Ultrasphere ODS column, 0.2 ml/min 0.01 *M* ammonium formate (pH 5.5) with 2.5% MeOH. Ionization by chemical ionization. (Reproduced from 81ED9′ with permission of the authors.)

sized for comparison with small amounts of biological compounds isolated from the urine of cancer patients were qualitatively specific. Additionally, LC retention was informative.

The upper panel of Figure 6.2 gives a composite of three ion current chromatograms from on-line LC/MS multiple ion detection (MID) experiments in which 6-amino-1-methyl-5-(N-formylamino)uracil (**I**), 6-amino-3-methyl-5-(N-formylamino)uracil (**II**) and 6-amino-5-(N-formylmethylamino)uracil (**III**) were chromatographed on a reversed phase column at 0.2 ml/min. The latter of these reference compounds shows a prominent bifurcate peak shape, which arises from the equilibrium of the open form with the closing of the 6-amine with the adjacent 5-N-formyl carbonyl to yield the closed 2-hydroxydihydroimidazole structure. The bottom panel of Figure 6.2 shows the LC/MS experiment for the isolated sample and confirms it as the 3-methyl isomer by relative abundances of the protonated molecular ion (m/z 185) and fragment ions [(MH $-$ 18)$^+$, m/z 167, and (MH $-$ 28)$^+$, m/z 157)] and by chromatographic retention. Similar experiments unambiguously characterize the dimethyl compound.

In more recent work, Rudewicz and Straub have noted the importance of ion–molecule reactions in thermospray ionization using a series of nucleobases and other compounds of known proton affinities. Data were presented supporting an ammonia chemical ionization model for thermospray ionization [86RU9']. The possibility of selective ionization of analytes through manipulation of buffer composition was demonstrated in the substitution of pyridinium acetate for ammonium acetate. With this substitution, thymine and uracil nucleobases afford relatively small responses consistent with their lower proton affinities.

As part of a study on the molecular basis of the mutagenicity of malondialdehyde in *Salmonella typhimurium,* the formation of covalent products between malondialdehyde and deoxyguanosine was assessed by thermospray LC/MS as reported by Crowley and co-workers [87CR0']. Three nucleoside adducts are observed (see Section 6.3.4), including an adduct with guanine base which is particularly sensitive in thermospray ionization and subject to degradation when examined by other methods.

6.2.2. Methylxanthine Bases

A class of purine-derived bases are the methylxanthine drugs, caffeine (1,3,7-trimethylxanthine), theophylline (1,3-dimethylxanthine), and theobromine (3,7-dimethylxanthine). These are potent stimulators of the central nervous system with particular application in respiratory disorders. These drugs and several derivatives within this class of drug are prone to interference in HPLC assays from endogenous compounds and other drugs. Extensive literature on their separation exists.[12]

The methylxanthine drugs caffeine, theophylline, and theobromine and related compounds have been frequently employed individually or in mixtures to demonstrate the utility of combined liquid chromatography/mass spectrometry systems. Direct liquid introduction has been employed in a conventional configuration [80ME87, 83CO55], with open tubular liquid chromatography, affording EI mass spectra [87NI51], and also with a sector instrument [83ST15, 85RO6']. Similar useful results have been obtained using the moving-belt interface to afford EI or NH_3 CI mass spectra [83DO65, 87KR49]. Atmospheric pressure ion sources employing nebulization with discharge ionization [83SA17] and chemical ionization [83TH2', 87SA7'] are also useful. The thermospray interface with quadrupole mass analysis [83BL50, 80CO55, 83HE2', 87LI47] and with a sector instrument [85AS79] has been used successfully. The use of supercritical fluid chromatography (SFC) with the moving-belt interface [86BE37, 86BE1'] or with direct fluid injection of SFC effluent [85CR3', 85CR71, 86LE92] has demonstrated utility.

Commonly, the mass spectra of nucleobases and related compounds in systems where CI conditions prevail are simplified and consist almost exclusively of the protonated molecule. An example of the results available from such a system is seen in the thermospray LC/MS analysis of a synthetic mixture of xanthine derivatives illustrated in Figure 6.3, taken from the work of Blakley

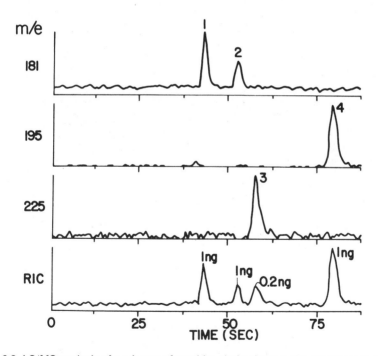

FIG. 6.3. LC/MS analysis of a mixture of xanthine derivatives: 1, theobromine; 2, theophylline; 3-β-hydroxyethyltheobromine; 4, caffeine. Chromatographic conditions: 3-μm Ultrasphere ODS column, 1.5 ml/min 0.1 M ammonium acetate with 12% acetonitrile. MS obtained with thermospray ionization. Sample quantities as indicated. (Reproduced from 83BL50 with permission of the American Chemical Society.)

and Vestal [83BL50]. This figure shows the thermospray LC/MS analysis of a
mixture of theobromine, theophylline, 3-β-hydroxyethyltheobromine, and caffeine. Here, subnanogram quantities are used in the analysis as indicated by the
amounts marked on the reconstructed ion current (RIC) chromatogram. A linear
response is obtainable over five orders of magnitude, as illustrated by Figure 6.4,
with a measurable response in the low-picogram range.

A notable LC/MS combination is direct fluid introduction of SFC effluent
into a Fourier transform mass spectrometer [86HE4', 87LE99]. In mass spectrometric systems where EI processes prevail, there is a qualitatively important
increase in structurally informative fragmentation. For example, the combination
of SFC with Fourier transform mass spectrometry [86HE4', 87LE99] affords EI
mass spectral data as shown in Figure 6.5. In the Fourier transform method,
exact mass data may be obtained. In this experiment such measurements were
made. The additional qualitative dimension provided by such data in an on-line
experiment is significant where additional confirmatory data are required in the
analysis.

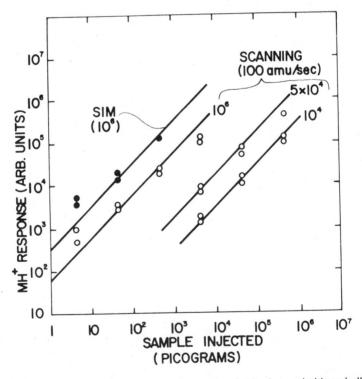

FIG. 6.4. MH+ response as a function of sample of β-hydroxyethyltheophylline injected. The parameter labeling the lines corresponding to the expected linear response
is the approximate electron multiplier gain used. (Reproduced from 83BL50 with permission of the American Chemical Society.)

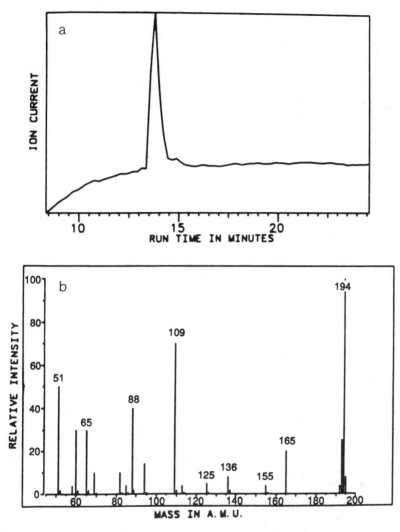

FIG. 6.5. (a) Total ion current chromatogram for the SFC/Fourier transform mass spectrometry experiment for a 25-ng sample of caffeine. (b) EI mass spectrum of caffeine. (Reproduced from 87LE99 with permission of the American Chemical Society.)

An application of thermospray LC/MS to the analysis of the methylxanthine drug caffeine is found in the recent work of Setchell and co-workers [87SE57]. In this study, the pharmacokinetics of caffeine elimination in healthy and diseased livers were assessed by the detection of caffeine in less than 1 μl of serum or saliva at a concentration of 1 μg/ml. Some of the results obtained in this study are illustrated in Figures 6.6 and 6.7. Figure 6.6 illustrates the total ion current chromatograms obtained in continuous scanning mode (m/z 110–300) from the

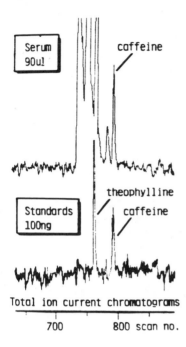

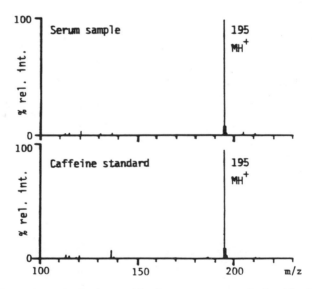

FIG. 6.6. Total ion current chromatograms obtained in continuous scanning mode (m/z 110–300) from the thermospray HPLC/MS analysis of (i) serum sample the equivalent of 90 μl injected on column (caffeine concentration, approximately 1.5 μg/ml) and (ii) a standard mixture of 199 ng of theophylline and caffeine. Vaporizer temperature, 150°C; jet block temperature, 210°C. Chromatographic conditions: 250 × 4.6 mm 5-μm Hypersil ODS, 1.5 ml/min 0.1 M ammonium acetate (pH 4.6) with 15% acetonitrile. (Reproduced from reference 87SE57 with permission of Elsevier Science Publishers B.V.)

FIG. 6.7. Thermospray ionization positive ion mass spectra obtained for the peak eluting at the HPLC retention time equivalent to that of caffeine in the serum sample shown in Figure 6.6 and for an authentic caffeine standard. Spectra are characterized by an intense protonated molecular ion (MH+) and negligible fragmentation. (Reproduced from 87SE57 with permission of Elsevier Science Publishers B.V.)

thermospray HPLC-MS analysis of (i) sample equivalent to 90 μl of serum injected on column (caffeine concentration, approximately 1.5 μg/ml) and (ii) a standard mixture of 199 ng of theophylline and caffeine. In Figure 6.7, thermospray ionization positive ion mass spectra for reference and isolated caffeine are shown. The spectrum obtained for the peak eluting at the HPLC retention time equivalent to that of caffeine in the serum sample is shown in the upper panel together with that of an authentic caffeine standard (below). Spectra are characterized by an intense protonated molecular ion (MH$^+$) and negligible fragmentation. It has been proposed that this load test is important as a dynamic assessment of liver function in a clinical context.

6.3. LC/MS OF NUCLEOSIDES

6.3.1. Liquid Chromatography of Nucleosides

The utility of chromatography in biochemical studies of nucleic acids varies with the origin of the analytical sample and the goal of the investigation. In the case of nucleosides, samples may be derived from cell extracts or physiological fluids, where they frequently occur together with nucleotides and nucleobases, or by hydrolysis from isolated nucleic acids (DNA or RNAs), themselves prepared by more or less elaborate separations and chromatographies. Gas chromatographic analysis of nucleosides is frequently constrained by the necessity to prepare derivatives of sufficient volatility and stability for the vapor phase separation.[11] Separation by column and low-pressure, ion exchange column chromatographies has a long and successful history of application. Brown and co-workers have been prominent in the development of applications of bonded anion exchange and reversed phase packings for modern high-performance separations of nucleosides (and their bases and nucleotides) derived from tissue and physiological fluids. Highly efficient, selective, and stable reversed phase separations of complex nucleoside mixtures have been developed.[2,3,7,13,14] These have found their principal application in the examination of DNA and RNA hydrolysates and also in the analysis of nucleosides in physiological fluids. The method of Buck and co-workers[14] is particularly distinguished by its practicality and routine applicability.

6.3.2. Methods Employed for Nucleosides

Several interface methods have demonstrated applicability to nucleosides. These include DLI, in this instance including the several gas and vacuum nebulizing variations; DLI on quadrupole [79ME3', 80ME87, 81OK97, 82ES8', 82KE5', 82TS1', 82YO14, 83GO1', 84DO67, 45ES5', 84MI256, 85ES8', 85ES21, 85HI39, 86ES5'] and sector instruments [83CH4', 83CH61, 83ST15]; and DLI with supercritical fluids [84SM95, 85SM31]. The heated-wire concentrator DLI device [80CH1RE'] has also been employed.

The moving-belt methods on quadrupole [80GA8', 80GA73, 81BR0', 81EC79, 81GA31, 81GA84, 84GA17] and sector [80QU2', 81BR0'] instruments have been employed. Special ionization methods including fast atom bombardment (FAB) [83LE0', 84FA60], secondary ion mass spectrometry (SIMS) [84FA60], and laser desorption ionization [80HA6', 81HA9', 81HA32, 83HA5'] have been explored.

As the Vestal quotation suggests, nucleosides have figured prominently in the development of the thermospray technology and its application beginning with the particle beam precursor of thermospray (EI/CI) [77MC6', 78BL81, 79BL2', 80BL2', 80BL26, 80BL86, 80PH10] and continuing with the thermospray method with quadrupole [80BL0', 80BL21, 81BL5', 82ED6', 83ED6', 83ED9', 83ES07, 83HE2', 84HE3', 84SM10', 84ST97, 84VO7', 84VO78', 85ED37, 85ED45', 85GA34, 85GO5', 85VE73, 85VO77, 85VO33, 85VO0', 86BE5', 86DE23, 86ED6', 86PH0', 87DA0', 87ED79, 87HA28, 87MC53, 87PH22] and sector [81BL5', 85CH47, 85CH33] instruments. Some of the most significant applications of LC/MS methods to nucleic acid investigation have involved this method. The thermospray system with ancillary discharge or filament ionization [85VE1'] has also been successfully used with model nucleosides.

Other interface methods which have been employed with nucleoside test compounds include atmospheric pressure ionization (API) with corona discharge [82KA43, 83SA17, 87SA7'] and ion evaporation [81IR9', 82TH9', 82TH49, 82ZO25, 83IR01], liquid ionization [83TS65, 86OT12], and electrospray [85WH75].

6.3.3. Performance Comparisons for Nucleosides

Nucleosides, as a class of compounds, are suited to testing the critical aspects of the performance of an LC/MS interfacing method. The principal ions in the chemical ionization[15] and thermospray [80BL26, 83BL50, 85ED37] mass spectra are the nucleobase fragment, BH_2^+, and the protonated molecule, MH^+, ions. Additional ions derived from the sugar moiety are also observed. Most other ionization modes of interest in the present context afford the same ions or ions of related structure. The relative abundance of these ions may be adopted as a broadly useful index of LC/MS utility for nucleosides across a wide range of interface regimes.

An example of the utility of performance testing with nucleoside models is illustrated in Figure 6.8 from the thermospray development work of Vestal and co-workers [85VE73]. This illustrates an experiment which formed part of an argument on the important advantages of direct joule heating of an interface capillary as a thermospray vaporizer device. With the ion source maintained at 250°C and a constant 60 W power dissipated in a 30-cm length of vaporizer capillary, a test solution of adenosine (A) and guanosine (G) is delivered at flow rates between 1.2 and 2.0 ml/min. Ion intensities are measured for the MH^+ and

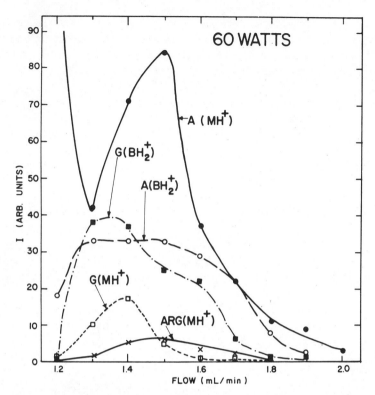

FIG. 6.8. Thermospray ion intensities versus flow at constant vaporizer power with downstream heating of jet to 250°C. Test solution injected contains 10^{-4} M adenosine (A), arginine (Arg), and guanosine (G) in 0.1 M aqueous ammonium acetate. (Reproduced from 85VE73 with permission of the American Chemical Society.)

BH_2^+ ions which the thermospray ionization affords as the fraction of the solution vaporized increases. These pass through distinct maxima corresponding to the optimum desolvation of analyte-bearing charged droplets from which these ions are derived. It is seen that optimum conditions for operation, i.e., those that afford maximum MH^+ or fragment ion intensity or the maximization of the response ratio MH^+/BH_2^+, vary with the properties of the analytes. That is, for relatively apolar analytes (e.g., adenosine), optimum conditions are found at higher fraction vaporized than for relatively polar nucleosides (e.g., guanosine). Accordingly, the conditions chosen for an analysis may differ depending on the nature of the compounds of interest and the information sought.

In other studies of LC/MS performance for nucleosides, the collision-induced decomposition effect of a repelling electrode in the thermospray ion source has been tested by Bencsath and Field [86BE5']. As this potential on an electrode located within the thermospray ion source near the point of ion sampling is increased from 0 V to a limit of approximately 500 V (at which voltage discharge is initiated), the ratio MH^+/BH_2^+ for adenosine decreases from 4 to 0.1 with an increase in the total ionization of more than 30-fold. Additionally, in related mass spectrometric investigations, it has been shown that the solution pH of the sample has an important effect on which ions produced in SIMS of adenosine [84IN13].

Sensitivity is the most significant parameter for the evaluation of perfor-mance of the combination of a liquid chromatograph with a mass spectrometer. A system's sensitivity may be measured by the introduction of a known quantity of a test substance into the mass spectrometer via the chromatographic inlet. With particular respect to nucleosides, a number of semiquantitative observa-tions on LC/MS system performance are worth comparing:

(a) For DLI systems, Melera reports [79ME3'] full-scan spectra for test compounds (including adenosine) obtained with 50–500 ng of the compound and detection in selected ion monitoring mode of 2–20 ng. Henion and co-workers [83CO55, 83HE2'] use 10-ppm adenosine solution at 300 µl/min for DLI and thermospray operation affording strong signals for performance optimization. Direct injection of 20 ng of 2-deoxyadenosine in a stream of supercritical am-monia in direct fluid injection gave a good mass spectrum [84SM95].

(b) Laser desorption ionization from a moving metal belt [81HA32, 83HA5'] requires samples (including nucleotides and nucleosides) in the range of 0.01–1 µg. By this method, detection limits for nucleosides [84FA60] are "a few nanograms" and about 100 times higher than with the UV absorbance moni-tor [84HA62].

(c) In thermospray operation for a variety of analytes including adenosine, Yergey and co-workers [83YE0'] have reported full scans on 1–10 pmol and selected ion sensitivities of 1–10 pg.

(d) The ion evaporation API performance [82TH49] is obtained on solutions of 10^{-4} M adenosine and guanosine in 10^{-4} M HCl. By this method, most test compounds (including nucleosides and nucleotides) are detectable in the low parts per billion (w/w) range [81IR9'].

(e) The electrospray ionization mass spectra [85WH75] of adenosine (AMP and ADP) were obtained from solutions of 01.-µg/µl concentration delivered to the electrospray source at 6 µl/min, and the mass spectrum was recorded scan-ning 0–600 amu in 5 min.

(f) Measurable peak areas for m/z 268 of adenosine are obtained in liquid ionization mass spectrometry [86OT12] down to concentrations on the order of 10^{-1} ng/µl.

Several more rigorous discussions of performance have also been reported. These are confined to the thermospray device and related designs and do not form a satisfactory basis for critical comparison of LC/MS system performance for nucleosides. Vestal and co-workers have discussed sensitivity for the particle beam/CI precursor of the thermospray device [80BL26]. The smallest amount of sample detectable was 1.4 ng (norleucine), corresponding to a minimum detect-able input rate of 60 pg/s. Using the same instrument design, McCloskey and co-workers have reported [82ED6'] a measured system sensitivity for the MH^+ ion of adenosine (m/z 268) of $3.3 \times 10^{-13} C/\mu g$. With the same instrument operated in thermospray mode [80BL21, 80BL0'], sample input rates of 1–10 ng/s afford clean full-scan mass spectra. In the case of adenosine, 200 pg gave an integrated ion current profile (m/z 268) five times the noise level. In this case, the measured sensitivity figure is $2.5 \times 10^{-11} C/\mu g$ [82ED6'], with performance of the same order reported in a later study [83ED9']. Measured sensitivities as much as a factor of 10 higher may be observed (C. G. Edmonds, unpublished experiments). Vestal and co-workers, using the direct joule-heated thermospray vaporizer and carefully optimized conditions, report [85VE73] a limit of detection for adenosine (S/N = 2) of about 20 pg. Similar results are reported by McCloskey and co-workers [85ED37]. With such a system, under favorable conditions, detection limits for most nucleosides are in the range of 0.1–10 ng using selected ion monitoring, with requirements for full-scan experiments being 10–50 ng [85ED37, 86ED07].

The foregoing has focused on the quantitative aspects of LC/MS performance with emphasis on the abundance of the molecular ion. An alternative perspective might be that the relatively more informative mass spectrum will contain the more abundant molecular ion and would consequently be least affected in a practical way by interfering chemical and instrument noise. There exists a much more extensive collection of qualitative data, namely, mass spectra of reference nucleosides, which form a basis for the comparison of the performance of LC/MS interface designs. These are summarized in Table 6.1, in which the MH^+/BH_2^+ ratios for adenosine and guanosine are compiled for all interface devices for which these data are presently available. This is an expanded version of such a comparative table presented by Mizuno and co-workers which uses the MH^+/BH_2^+ ratio for adenosine as a figure of merit for performance of a DLI LC/MS interface [84MI256].

The data presented in Table 6.1 fail to present an uncomplicated case for the superiority of any given LC/MS interface method on the basis of MH^+/BH_2^+ ratios for nucleosides. Best values for adenosine are found for atmospheric pressure ionization with corona discharge and liquid ionization. In the former case, the adaptability of the API technique to a variety of chromatographic systems is an additional point in its favor. However, the result for liquid ionization is for an off-line experiment, and its utility must be considered only potential until on-line developments of the method [83TS65, 86OT12] reach a more mature stage. Thermospray LC/MS seems to provide the most consistently high values of the MH^+/BH_2^+ ratio from a variety of laboratories, but examples of equivalent performance have been reported for both moving-belt and DLI interface methods. Perhaps the most striking feature of the data presented is the high variability of the MH^+/BH_2^+ ratio for nucleosides among measurements made with a given interface design. This is a consequence of the sensitivity of nucleosides to the variations in design details and other parameters within a given interfacing method. Accordingly, in the absence of clear mass spectrometric

TABLE 6.1. MH$^+$/BH$_2^+$ Ratios for Reference Nucleosides in Various LC/MS Interface Methods

Method	Citation	Conditions	Adenosine MH$^+$/BH$_2^+$	Guanosine MH$^+$/BH$_2^+$
Direct liquid	81OK97	CI (suppl. CH$_4$)	0.10	—
introduction		CI (suppl. NH$_3$)	0.70	—
	82YO14	CI	0.13	—
	83CH61	CI	0.58	—
	83CO55	CI	0.15	—
	83ES07	CI	1.23	—
	83ST15	CI	0.56	—
	84MI256	CI	1.1	—
	85HI39	CI	15.0	—
		CI (neg. ion)	0.84[a]	—
Moving belt	80GA8'	CI	0.20	—
	80GA73	CI	0.20	—
	81EC79	CI (NH$_3$)	0.20	—
		CI (CH$_4$)	0.13	—
	81GA31	CI (CH$_3$Cl)	0.78[b]	—
	81GA84	CI (CH$_4$)	0.21	—
		CI (CH$_3$Cl)	0.48[b]	—
	84GA17	CI (NH$_3$)	0.12–3.84[c]	—
	84DO67	CI (i-C$_4$H$_{10}$)	0.58 [d]	—
	80HA6'	Laser	—	0.59[e]
	84HA62	Laser	—	MH$^+$ only[e]
		Laser (neg. ion)	—	0.90[f]
	83LE0'	FAB	0.27	—
Particle beam	78BL81	EI	No M$^+$	—
	78MC8'	CI	0.11	0.32
	79BL2'	CI	—	0.15
	79BL2'	EI	0.007	—
	80BL2'	CI	1.52	—
	80BL26	CI	1.56	—
	80BL86	CI	0.11	—
	82ED6'	CI	1.0	—
Thermospray	82ED6'	Thermospray	1.74	—
	83CO55	Thermospray	2.2	—
	84VO7'	Thermospray	1.94[g]	—
	85CH33	Thermospray	10.2	—
	85VE73	Thermospray	2.5	2.12
	86BE5'	Thermospray	4.1	—
		Thermospray	0.21[h]	—
	85VE1'	Filament CI	0.91[i]	—
		Discharge CI	2.06[i]	—
Atmospheric pressure ionization	82KA43	Corona discharge	6.0	0.90

TABLE 6.1. *(Continued)*

Method	Citation	Conditions	Adenosine MH^+/BH_2^+	Guanosine MH^+/BH_2^+
Ion evaporation	81IR9'	Ion evaporation	—	0.90
	82TH49	Ion evaporation	—	0.97
Liquid ionization	83TS65	Metastable Ar	5.0	—
	86OT12	Metastable Ar	4.0^j	—
Electrospray	85WH75	Electrospray	2.2	—

[a] $\dfrac{m/z\ 266}{m/z\ 134}$.

[b] Negative ion chloride attachment, $M \cdot Cl^-/B^-$.

[c] Comparison among several inlet designs.

[d] $\dfrac{m/z\ 267}{m/z\ 136}$; very noisy.

[e] Abundant (3–5 times) cationized adduct ions of BH and M.

[f] $\dfrac{m/z\ 282}{m/z\ 151}$.

[g] Collisionally induced dissociation of cluster ion after ion source.

[h] +500-V repeller CID.

[i] Average of 4 noisy measurements.

[j] Also M_2H^+, 7.6%.

advantages, collateral considerations of simplicity, practicality, and chromatographic compatibility figure prominently in the selection of LC/MS interface methods most suitable for nucleoside analysis.

6.3.4. Application of LC/MS to Nucleoside Analysis

The moving-belt interface has been employed for the analysis of simple mixtures of the common ribo- and deoxyribonucleosides [80GA73]. In a similar context, the DLI interface has proven useful [82ES8', 83ES07]. Of more practical significance is the work of Esmans and co-workers in the analysis of nucleotides isolated from urine [84ES5', 85ES21, 85AL45]. In these studies, microcolumn methods were employed with flow rates of 70 μl/min in the detection of nucleobases, common deoxyribonucleosides, and the tRNA degradation products 5,6-dihydrouridine and pseudouridine.

The analysis of nucleoside antimetabolite drugs and antiviral agents in the support of synthetic programs and in the verification of purity for biological testing has been significantly facilitated by LC/MS methods. In model experiments, 5-fluorouracil and its derivatives 5-fluoro-2'-deoxyuridine 5'-monophosphate, 5-flourouridine, 5-fluoro-2'-deoxyuridine and ftorafur [(1-tetrahydrofuranyl)-5-fluorruracil] may be simultaneously analyzed by LC/MS using the particle beam interface device with negative ion detection [81ED9',

82MC7AL′]. Practical applications include the analysis of substituted pyridinium C-nucleosides by microbore DLI LC/MS [85ES8′, 86ES5′] and the utilization of thermospray LC/MS for the cancer chemotherapeutic agents tetrahydrouridine, triazone [84VO78′, 85VO33], 3-deazauridine [85VO0′, 85VO77], 3′-azido derivatives of 2′, 3′-dideoxythymidine and 4-thio-2′,3′-dideoxythimidine, and related compounds [87KE6′].

Chemical assessment of oxidative and radiation damage to DNA has been investigated relying principally on LC/MS analysis. In a thermospray LC/MS experiment, Davidson and co-workers [87DA0′] investigated radiation-induced damage to DNA. The base thymine 5,6-glycol, derived from the oxidation of thymine, the corresponding deoxyribonucleoside, and several other products were observed by thermospray LC/MS carried out on a sector instrument. High-resolution and MS/MS experiments were included in this study. Experiments modeling radiation damage with polyadenylic acid [87AL6′, 87AL94] have been reported. The principal damage products after enzymatic hydrolysis of the irradiated nucleic acid were identified in the liquid chromatogram, shown in Figure 6.9. The presence of 4-amino-5-formylamino-6-(ribosyl)aminopyridine (peak 1), R- and S-8, 5-cycloadenosine (peaks 4 and 13), adenine (peak 6), 8-hydroxyadenine (peak 7), α-adenosine (peak 8), and 8-hydroxyadenosine (peak 18) was demonstrated. Experiments were carried out on an atmospheric pressure chemical ionization interface with a triple-quadrupole mass spectrometer. Identifications were by liquid chromatographic and mass spectrometric comparison with reference materials and by the "fingerprinting" of the daughter ion spectra of the closely related purine derivatives.

In a second study, the chemical basis of the effect of the anticancer drugs mitomycin C and porfiromycin was studied by the reaction of the enzymatically reduced drugs with calf thymus DNA, enzymatic digestion of the adducted nucleic acid to the nucleoside level, and examination of the resulting mixture by thermospray LC/MS. Figure 6.10 shows the thermospray mass spectra of the principal adducts for the two drugs together with their structures. Protonated molecular ions are observed (m/z 570 and m/z 584 for the mitomycin and porfiromycin adducts, respectively) together with structurally informative fragments including loss of the deoxyribosyl related (m/z 454 and m/z 468, respectively) and drug related (m/z 244 and m/z 258, respectively) portions.

Other studies of nucleosides include the analysis of dimeric nucleosides derivatized as their O-tert-butyldimethylsilyl ethers and using a moving-belt interface on a sector mass spectrometer [80QU2′]. The utility of deuterium exchange of active hydrogens in structure determination is suggested by experiments of Tsuchiya and co-workers [83TS3′] employing a liquid ionization mass spectrometer using adenosine as a model compound. In this work, experiments were conducted off-line, but recent development of this ionization technique for on-line LC/MS experiments [85TS8′, 87TS20′] makes it potentially a particularly useful mode of experimentation.

Naturally modified nucleosides exist in great structural diversity in both

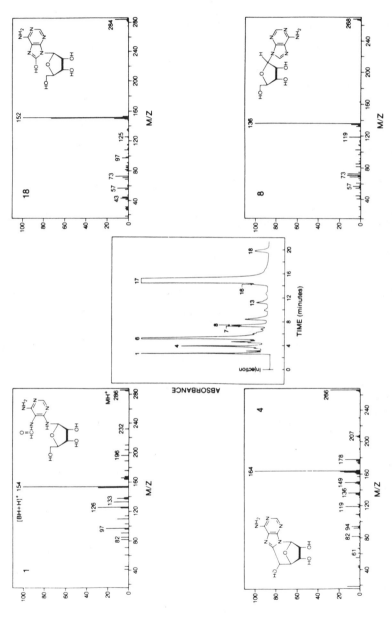

FIG. 6.9. UV absorbance chromatogram and mass spectra obtained by LC combined with atmospheric pressure chemical ionization MS and MS/MS analysis of the products of irradiated polyadenylic acid. Peaks shown are 4-amino-5-formylamino-6-(ribosyl)aminopyridine (peak 1), R-8,5-cyclo-adenosine (peak 4), α-adenosine (peak 8), and 8-hydroxyadenosine (peak 18). Identification is by comparison with authentic compounds and MS/MS daughter ion spectra. (Reproduced from 87AL5' with permission of the authors.)

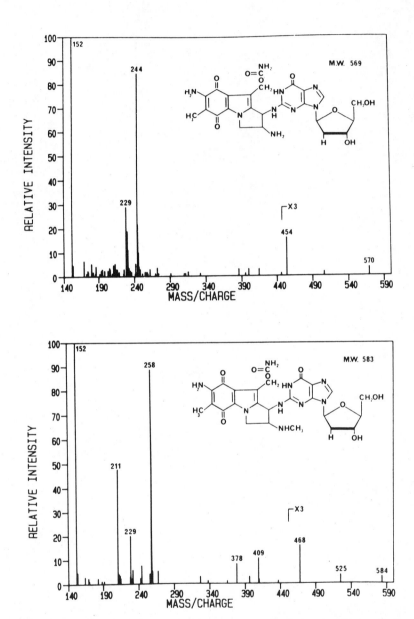

FIG. 6.10. Thermospray mass spectra of mitomycin C (upper) and porfiromycin (lower) adducts of deoxyguanosine obtained by enzymatic digestion after incubation of DNA and the drug with reducing enzymes. Separation was on a C_{18} reversed phase column with 0.1 M ammonium formate buffer. (Reproduced from 87MU6′ with permission of the authors.)

DNA and RNA. More than 80 are now recognized, of which 63 are known in transfer RNAs.[16-18] These compounds are produced by post-translational enzymatic transformations, the simplest of which is a simple base or sugar ($2'$-O) methylation. A number of the so-called "hypermodified" structures arise by much more complicated processes. In this context, the identification of known nucleosides and the recognition and characterization of unknown structures must be performed in a mixture of considerable complexity. In a given isoaccepting tRNA molecule, a modified residue may occur one time and there may be as many as 15 such residues. Thus, the analysis might proceed on a multicomponent mixture where these analytes represent 1% of the sample.

McCloskey has reviewed the role of mass spectrometric techniques in the characterization of modified nucleosides derived from transfer RNAs [82MC7AL', 86MC5GA']. Such investigations generally involve the combined application of several methods including preparation of volatile derivatives and their examination by EI mass spectrometry at low and high resolution. This may be supplemented by FAB mass spectrometry and the count of active hydrogens in the molecule of interest by exchange with D_2O and mass spectrometric evaluation of the resulting mass increment. In this context, application of combined LC/MS methods extends the range of information that can be obtained by application of mass spectrometry to these investigations. The combined method offers significant advantages over chromatographic methods with nonselective detection which rely solely on chromatographic mobility for identification, or over MS and GC/MS procedures which require preparation of volatile derivatives of frequently sensitive chemical entities.

Thermospray LC/MS has been applied to the examination of nucleoside mixtures derived from isoaccepting tRNAs [83ED9', 83ED6', 85ED37, 86PH0', 87PH22]. Figure 6.11 demonstrates the simultaneous detection of all modified species occurring in 0.05 $A_{260\ nm}$ units (2.5 μg) of rabbit liver tRNA^{Val}. This experiment followed a full-scan run and was an MID confirmation run monitoring the channels of MH$^+$ and BH$_2$$^+$ suggested by the scanning experiment and the published sequence of the tRNA. The modified nucleosides detected (abbreviation, BH$_2$$^+$ and MH$^+$) included dihydrouridine (D, m/z 247), pseudouridine (U, m/z 245), 1-methyladenosine (m^{1}A, m/z 150 and m/z 282), 5-methylcytidine (m^{5}C, m/z 126 and m/z 258), 7-methylguanosine (m^{7}G, m/z 166), inosine (I, m/z 137), 1-methylguanosine (m^{1}G, m/z 166), and N^2-methylguanosine (m^{2}G, m/z 166 and m/z 298).

The presence and initial characterization of a new nucleoside, 7-{5-(2,3-epoxy-4,5-dihydroxycyclopent-1-yl)amino]methyl}-7-deazaguanosine (**IV**), in tRNA^{Tyr} from *E. coli* MRE 600 was recognized in a thermospray LC/MS experiment [86PH0', 87PH22]. Full structure proof of this compound, an epoxide derivative of the "hypermodified" nucleoside queuosine and the first epoxide structure recognized in nucleic acids, required a considerable combined study involving derivatization and EI mass spectrometry at low and high resolution and a scaled-up isolation of the compound in quantities sufficient for ^{1}H nuclear

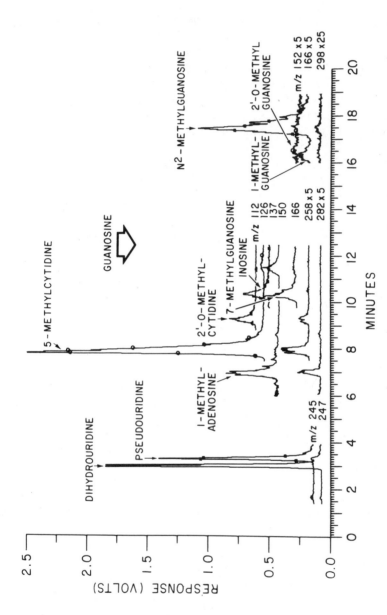

FIG. 6.11. Thermospray LC/MS multiple ion detection (MID) analysis of 0.05 $A_{260 \, nm}$ units (2.5 μg) of a hydrolysate of rabbit liver $tRNA^{Val}$. BH_2^+ and MH^+ ions monitored: D (247), U (245), m^1A (150, 282), m^5C (126, 258), m^7G (166), I (137), m^1G (166), m^2G (166, 298). Chromatographic conditions: 4.0 × 300 mm Waters μBondapak C-18, 1.5 ml/min 0.1 M ammonium acetate (pH 5.6), linear gradient 0–15% MeOH over 30 min. (Reproduced from 85ED37 with permission of IRL Press Limited.)

IV

magnetic resonance spectroscopy. At points during the course of the investigation, thermospray LC/MS was used to track the compound of interest through
multiple chromatographies. Additionally, thermospray LC/MS was employed to
facilitate the verification of microchemical transformations important in the
structure proof. For example, in Figure 6.12 the isopropylidene derivative of the
epoxy-modified queuosine is represented. In this case, the vicinal diols can be
identified on the basis of the reactivity with 2,2-dimethoxypropane, and thermospray LC/MS analysis verifies the presence of two such moieties, one in the
side chain portion of the molecule (viz., an increment of 40 amu for the putative
aminoepoxycyclopropanediol fragment ion, m/z 132 to m/z 172) and one in the
ribose moiety (viz., 2 × 40 amu for the protonated molecule, m/z 426 to m/z
506).

In investigations of tRNAs, the quantity and purity of the nucleic acid
available is frequently an analytically pivotal question. This, in turn, is critically
affected by the organism being studied and the effort required for its culture and
the subsequent isolation of the nucleic acid. The preparation of isoaccepting
tRNAs in high purity generally requires arduous multiple chromatographies as
well as considerable sample. Except in a few straightforward (and frequently
commercially available) cases, these are rarely obtained in quantities exceeding a
few tens of micrograms. On the other hand, depending only on the difficulty of
growth of the organism, bulk unfractionated tRNA may be isolated in multi-
milligram quantities. Thus, in the progression from isoacceptor to partially pu-
rified to bulk tRNA, there is a decreasing difficulty of preparation (and thus an
increasing availability of sample) accompanied by an increasing difficulty of
analysis as the abundance of a single modified residue decreases from the 1%
level in a mixture of increasing complexity. Thermospray LC/MS has been

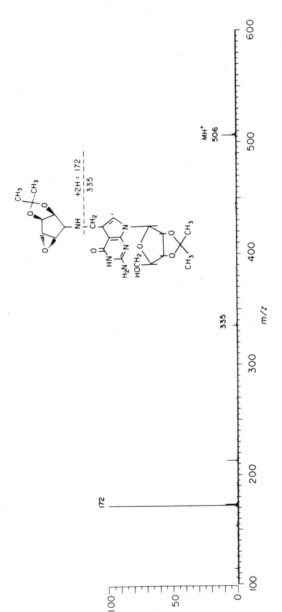

FIG. 6.12. Thermospray mass spectrum of the di-*O*-isopropylidene derivative of the epoxy-modified queuosine, oQ. Chromatographic conditions: 250 × 4.6 mm Supelcosil LC-18 DB; 2 ml/min 0.25 *M* ammonium acetate (pH 6) with multilinear gradient to 40% acetonitrile according to Reference 14. (Reproduced from 87Ph22 with permission of the American Society for Biochemistry and Molecular Biology, Inc.)

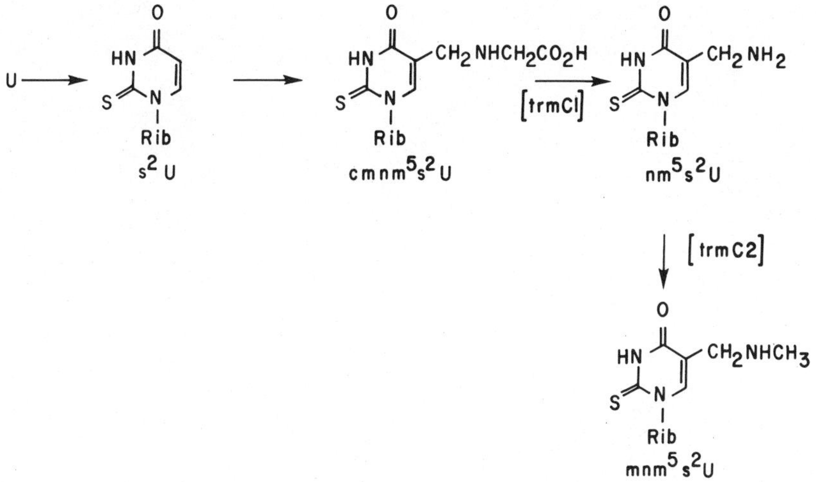

$U \longrightarrow$

$s^2 U$ Rib (HN, O, S, N)

$cmnm^5s^2U$ $CH_2NHCH_2CO_2H$, Rib

$[trmCl] \longrightarrow$

nm^5s^2U CH_2NH_2, Rib

$\downarrow [trmC2]$

mnm^5s^2U CH_2NHCH_3, Rib

SCHEME 6.1

useful in the examination of nucleoside mixtures derived from unfractionated tRNAs [85ED37, 85ED45', 86ED6', 87ED79, 87HA28, 87MC53, 87PH22].

As part of the study of the epoxy-modified queuosine nucleoside mentioned above [87PH22], the presence and relative abundance of queuosine and its epoxy derivative were assessed in unfractionated tRNAs from strains of *E. coli* in a target compound analysis by multiple-ion detection using thermospray LC/MS. In a second set of studies [87HA28], the pathway for the post-transcriptional modification of uridine to the 5-methylaminomethyl-2-thiouridine in the glutamine, glutamic acid, and lysine isoaccepting tRNAs of *E. coli* was investigated in mutant strains which afford undermodified derivatives of the biosynthetic product. In the biosynthetic scheme (see Scheme 6.1 above) the putative sequence is outlined.

This scheme is unusual in that the elaboration of 5-methylaminomethyl-2-thiouridine proceeds by the synthesis of the carboxymethylaminomethyl side chain followed by the degradation to the 5-aminomethyl compound and subsequent reextension to the final 5-methylaminomethyl-2-thiouridine. Unfractionated tRNAs from two mutant strains, which interrupt modification at the points indicated (strains *trmC1* and *trmC2*) were examined, along with the tRNA of the genetically wild type. Figure 6.13 is a composite of ion current profiles for the biosynthetic intermediates observed in two chromatographic systems (A, C, and E: cyanodecyl; B, D, and F: octadecyl) with different selectivities for these analyses. Arrows mark the points of elution of the available standards. Satisfactory consistency is obtained between chromatographic systems, including a response attributable to 5-aminomethyl-2-thiouridine. Higher chemical noise levels in the less efficient cyanodecyl system obscure weak protonated molecular ions for 5-methylaminomethyl-2-thiouridine. The putative 5-aminomethyl-2-

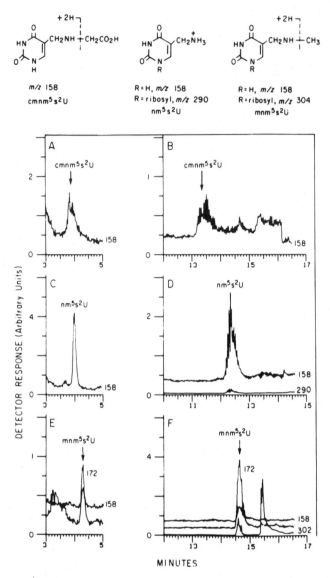

FIG. 6.13. Selected ion chromatograms from the combined liquid chromatographic/ thermospray mass spectrometric analysis of enzymatic hydrolysates of *E. coli* tRNA. A and B, strain *trmC1*; C and D, strain *trmC2*; E and F, strain TH10 (wild type). Chromatographic conditions: A, C, and E, 250 × 4.6 mm Supelcosil LC-18 DB, 2 ml/min 0.25 M ammonium acetate (pH 6) with multilinear gradient to 40% acetonitrile according to Reference 14; B, D, and F, series connection of two 250 × 4.6 mm Supelcosil CN10 (prototype packing), 2 ml/min 0.25 M ammonium acetate (pH 6). Elution times of authentic nucleosides determined from separate experiments are indicated by arrows. Ions monitored are derived according to the insert at the top of the figure. (Reproduced from 87HA28 with permission of the American Society for Biochemistry and Molecular Biology, Inc.)

thiouridine is observable in the cyanodecyl system. These results together with other biochemical and analytical data support the proposed biosynthetic scheme.

Thermospray LC/MS has been used as an important analytical method in the examination of unfractionated tRNAs from archaebacteria [85ED37, 85ED45', 86ED6', 87ED79, 87MC53]. Archaebacteria are a recently recognized third primary kingdom, derived from the division of the prokaryotes into the eubacteria and the archaebacteria. Aspects of their metabolism and growth optima at the extremes of salt strength, pH, and temperature, similar to the conditions which are believed to have prevailed during the early history of life on earth, suggest that they may be the most ancient kingdom.[19]

For example, representatives of the archaebacteria are the only organisms capable of growth above 90°C. The investigation of the identities of the modifications occurring in tRNAs is the first step in a broader consideration of their relationship to the structure and function of tRNAs. In the case of archaebacteria, this is motivated by the insight which may be gained into a metabolism which functions under conditions that would render more familiar biochemical machinery useless. McCloskey has reviewed the current understanding of nucleoside modification in archaebacterial tRNA and commented on the utility of LC/MS in

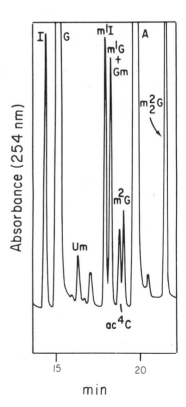

FIG. 6.14. Portion of reversed phase chromatographic separation of nucleosides from 0.6 A_{260} nm units of unfractionated *H. volcanni,* with UV detection at 254 n. Chromatographic conditions: 250 × 4.6 mm Supelcosil LC-18 DB; 2 ml/min 0.25 M ammonium acetate (pH 6) with multilinear gradient to 40% acetonitrile according to Reference 14. (Reproduced from 85ED37 with permission of IRL Press Limited.)

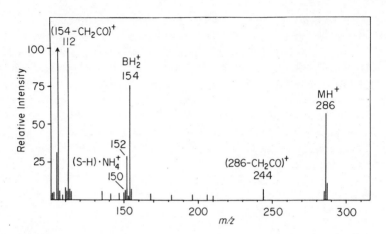

FIG. 6.15. Thermospray mass spectrum of N^4-acetylcytidine directly acquired from an HPLC separation of nucleosides from 0.6 $A_{260 \ mm}$ units of unfractionated *H. volcanii* tRNA. Chromatographic conditions as for Figure 6.14. (Reproduced from 85ED37 with permission of IRL Press Limited.)

V VI VII

VIII IX

advancing this understanding.[20] In the examination of nucleoside mixtures derived from archaebacterial tRNAs, the LC/MS experiment facilitates the rapid recognition of known nucleosides based on chromatographic retention and characteristic ions. An example is shown in Figures 6.14 and 6.15. In Figure 6.14, a portion of the UV absorbance chromatogram for the separation of nucleosides from 0.6 $A_{260\,nm}$ units (30 μg) of unfractionated tRNA from *Halobacterium volcanii* obtained during an LC/MS experiment is shown. The mass spectrum afforded by the minor known nucleoside N^4-acetylcytidine is shown in Figure 6.15. With the recognition of known structures, there follows the important function of obtaining initial structural information on nucleosides of unknown structure directly on the crude hydrolysate, not necessarily with chromatographic resolution of the components from the mixture. In the case of archaebacteria, several such new structures have been reported. These include 5,2'-*O*-dimethylcytidine (**V**), N^4-acetyl-2'-*O*-methylcytidine (**VI**), 2-thio-2'-*O*-methyluridine (**VII**), $N^2,N^2,2'$-*O*-trimethylguanosine (**VIII**) [85ED44', 87ED79], and a new, highly fluorescent member of the tricyclic Y nucleoside family, 3-(β-D-ribofuranosyl)-4,9-dihydro-4,6,7-trimethyl-9-oxoimidazo[1,2-*a*]purine (**IX**) [87MC53].

6.4. LC/MS OF NUCLEOTIDES

6.4.1. Methods Employed for Nucleotides and Dinucleotides

At the ultimate level of mass spectrometric difficulty are the nucleotides. The principal modes of chromatographic separation of nucleotides are reversed phase and ion exchange. These separations have been reviewed by McKeag.[21] LC/MS interface methods and related mass spectrometric techniques which have been reported useful cover the full range and history of application. These include DLI [79ME3', 83GO1', 82KE5', 83VO6'], moving belt with EI and DCI [81BR0', 83GA93] and with laser desorption ionization [80HA6', 81HA9', 81HA32, 83HA5'], the particle beam precursor to thermospray with EI and CI [77MC6', 78BL81, 78BL81, 79VE97, 80BL67, 80BL26] and thermospray itself [80PH10, 80BL2', 80BL0', 80BL21, 81BL5', 85ED45', 85ED6', 86VO8', 87VO29], atmospheric pressure ionization methods employing ion evaporation [81IR9', 82TH9', 82TH49] and corona discharge ionization [87SA7'], and electrospray ionization [84YA93, 85WH75].

6.4.2. Applications of LC/MS to Nucleotides and Dinucleotides

Kenyon and Goodley [82KE5'] have shown a CI mass spectra of adenosine 5'-monophosphate recorded in both positive and negative ion modes which show rich fragmentation and molecular ions (MH$^+$ and *sic.* M$^-$, respectively) of 20% relative intensity. Brunnee and co-workers [81BR0'] and Games [83GA93] have

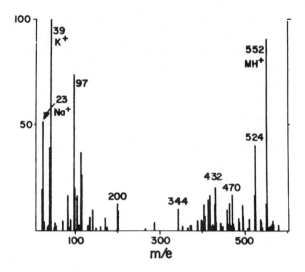

FIG. 6.16. Laser desorption mass spectrum of the disodium salt of adenosine 5-triphosphate (Na$_2$ATP) continuously electrosprayed on a rotating stainless steel belt. Sample density, 1 μg/cm². (Reproduced from 80HA6' with permission of the authors.)

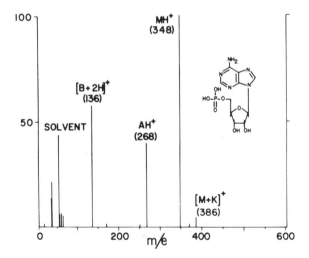

FIG. 6.17. Positive ion mass spectrum of AMP obtained by the thermospray technique. The sample of AMP (20μl of 0.1 mg/ml solution) was injected using the LC without column as inlet system. The liquid phase was 0.2 M formic acid at a flow rate of 0.5 ml/min, and the spectrum was scanned from 10 to 600 amu in 4 s. (Reproduced from 80BL21 with permission of the American Chemical Society.)

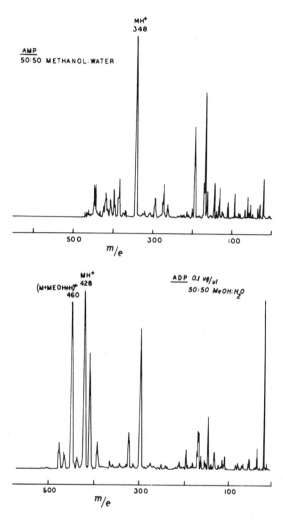

FIG. 6.18. Electrospray mass spectra of adenosine monophosphate (upper) and adenosine diphosphate (lower). Samples were present at concentrations of 0.1 μg/ul in 50 : 50 methanol/water. The liquid flow rate was 6 μl/min. The full mass scan took 5 min. (Reproduced from 85WH75 with permission of the American Chemical Society.)

employed moving-belt methods and reported mass spectra of 5′-AMP and 5′-dAMP containing positive protonated molecular ions in the range of 5–25% relative abundance. The laser desorption mass spectrum for the disodium salt of adenosine triphosphate [80HA6′] presented to the mass spectrometer on a moving stainless steel belt is shown in Figure 6.16. The relative abundance of the protonated molecular ion is 92% in this spectrum with abundant alkali metal ions and complex fragmentation.

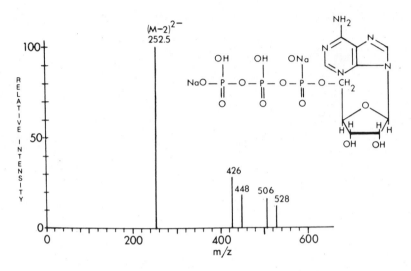

FIG. 6.19. Negative ion evaporation spectrum of 10^{-4} M disodium salt of adenosine triphosphate. (Reproduced from 82TH49 with permission of the American Chemical Society.)

In their initial reports of the application of the particle beam LC/MS interface, Vestal and co-workers [78BL81] presented EI mass spectra for the series of compounds uracil, uridine, 5′-UMP, 5′-UDP, and 5′-UTP. In each of these, the nucleobase ion predominates, with sugar and weak phosphate fragment ions observed in the spectra of the latter members of the series. Subsequent developments [80BL2′, 80BL26] afforded the CI mass spectra with weak but measurable responses [2.7% MH$^+$ and 1.5% (M − H)$^-$ in positive and negative ion detection modes, respectively] for the molecular ion. A significant advance was the introduction of the thermospray technique, which afforded the mass spectrum of adenosine 5′-monophosphate shown in Figure 6.17. Here the molecular ion is observed as base peak in the mass spectrum, with other prominent ions corresponding to the protonated nucleoside and nucleobase.

With other methods of potential or actual applicability for the combined LC/MS experiment, interesting advances on this performance seem to be possible. Fenn and co-workers [85WH75] have made use of the electrospray concept to obtain the mass spectra of adenosine 5′-monophosphate and adenosine 5′-diphosphate (Figure 6.18). In this case, a solution of nucleotide (0.1 μg/μl, 50:50 MeOH/H$_2$O) was delivered to the electrospray ion source at 6 μl/min to afford mass spectra in which the protonated molecule is the most intense ion. The ion evaporation spectrum of the disodium salt of adenosine 5′-triphosphate is shown in Figure 6.19. In this case, an aqueous solution (10^{-3} to 10^{-4} mol/liter) was delivered to the electrospray API ion source at 1 ml/min to afford a spectrum containing the doubly charged molecular naion (m/z 252.5) as the base peak

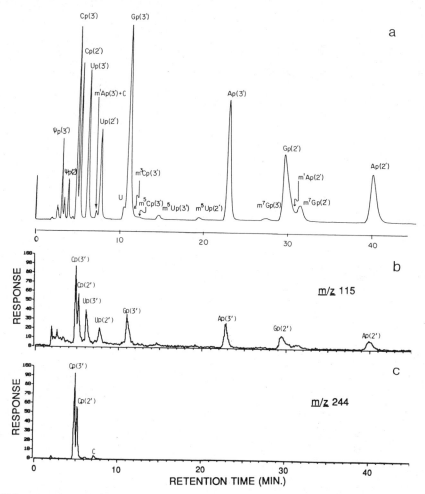

FIG. 6.20. UV absorbance (254 nm) chromatogram (A) and reconstructed ion current chromatograms for the common sugar fragment, m/z 115 (B) and cytidine nucleoside fragment (NH_2^+), m/z 244 (C) for the thermospray LC/MS analysis of an alkaline hydrolysate of baker's yeast tRNA. Chromatographic conditions: 250 × 4.6mm Supelcosil LC-18 DB, 2 ml/min mobile phase [A = 0.1 M ammonium formate (pH 5), B = 0.1 M ammonium formate (pH 3)], linear gradient 0–100% B over 20 min. Vapor temperature, 240–250°C. (Reproduced from 85ED6′ with permission of the authors.)

together with m/z 528 (MNa⁻), m/z 506 (MH⁻), and m/z 448 and 426, the corresponding ions for adenosine diphosphate.

The examination of the nucleotide mixtures by on-line LC/MS has been reported by McCloskey and co-workers [85ED6′, 85ED45′]. The thermospray method was employed, and positive ion mass spectra of reference ribonucleotides and deoxyribonucleotides contained prominent ions for the proto-

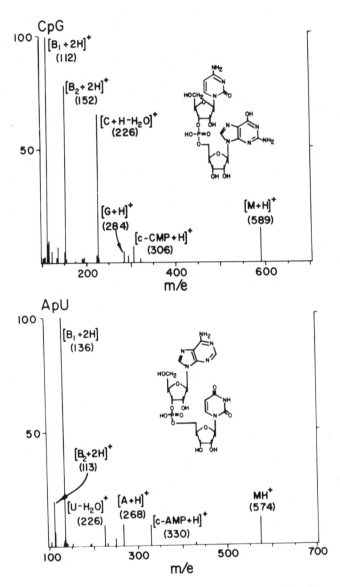

FIG. 6.21. Positive ion mass spectra of the nucleotide 3',5'-monophosphates CpG (upper) and ApU (lower) obtained under thermospray conditions. The sample of dinucleotide (20μl of 0.1 mg/ml solution) was injected using the LC without column as inlet system. The liquid phase was 0.2 *M* formic acid at a flow rate of 0.5 ml/min, and the spectrum was scanned from 100 to 700 amu in 4 s. (Reproduced from 80BL21 with permission of the American Chemical Society.)

nated nucleoside and nucleobase together with ions from the sugar (ribose, $2'$-O-methylribose, and deoxyribose = S) corresponding to (S − H)NH$_4{}^+$. Intact nucleoside ions are variable in intensity and occasionally absent (e.g., pG, pm^{7}G, pdC, and pdG), and intact molecular ions are of generally low abundance (0.3–1.9%) or absent (e.g., pG, pGm, pm^{6}A, pm^{5}C, pm^{7}G, pdC, and pdG). Figure 6.20 shows the application of this method to an alkaline hydrolysate of the bulk tRNA derived from Baker's yeast. The UV absorbance chromatogram and ion current profiles for common (S − H)NH$_4{}^+$ and cytidine nucleobase fragment ions are shown, illustrating the utility of the method for mixture analysis and confirmation of peak identity.

The thermospray method has been applied to the analysis of dinucleotide monophosphates by Vestal and co-workers [80BL0′, 80BL21] and Voyksner [87VO29]. Figure 6.21 illustrates such results for the $3',5'$-phosphodiester-linked compounds CpG and ApU. Prominent protonated molecular and nucleobase fragment ions are seen. Additionally, nucleoside and sequence-specific ions of composition corresponding to the $3',5'$-cyclophosphate structure of the $5'$-nucleoside are observed. The data presented by Voyksner [86VO8′, 87VO29] are significant for the analysis of nucleotides and dinucleotide monophosphates. Thermospray mass spectra are obtained in the absence of the ammonium acetate buffer conventionally employed to afford symmetric charging of thermospray droplets and thus charge-assisted evaporation of analyte ions. With residual ions of pure water and microgram level injections of ionic compounds (approximately 10^{-7} M), a pure ion evaporation phenomenon is observable, uncomplicated by the ion–molecule chemistries which arise from the presence of abundant ammonium ions in conventional thermospray ionization. In this simplified regime, the predicted relationship of the intensity of MH$^+$ to solution pH and also the relative affinities of a series of metal ions for nucleotides are observable.

6.5. FUTURE PROSPECTS

Because of their sensitivity to operating conditions and their topicality, nucleic acid constituents (nucleosides, nucleotides, and higher oligomers) are likely to remain an important standard of performance evaluation for LC/MS interface innovators. Expansion of combined liquid chromatography/mass spectrometry into applications in nucleic acid investigations will ultimately rely on the straightforward applicability of the interface technique to conventional chromatographies, with particular emphasis on complex mixtures and sample matrices. The broad community of liquid chromatography users is being led by chromatography innovators and manufacturers to "modern" techniques, some of which have important advantages in the engineering of the interface (e.g., micro-column and open tubular liquid chromatography columns). In this, the biochemical fraternity will be the slowest to follow. The reader is reminded that among biochemists the application of such veteran techniques as paper chromatography and two-dimensional thin layer chromatography continues in the hands of highly pragmatic investigators.

So far, the LC/MS interface method which has had the most extensive "field test" in nucleic acid investigations is the thermospray method. Here, McCloskey and co-workers have emphasized the complementary aspects of thermospray methods and other important analytical methods. For example, derivatization/EI and FAB mass spectrometric methods are frequently encumbered by sensitivity to sample matrix and, in particular, to interference from salts inevitably present in biochemical samples. Thermospray LC/MS is, in general, not susceptible to these interferences. Thermospray is most frequently limited by the vigor of the vaporization process and the thermal and hydrolytic degradation of sensitive analytes. It is here in particular that FAB methods frequently reinforce thermospray LC/MS. Such thermal and hydrolytic degradations will form the ultimate barrier to the extension of thermospray methods within any class of involatile and unstable analytes.

In the analysis of nucleotide and oligonucleotide mixtures, two existing methods seem to have a potential not exploited so far. The use of the moving-belt interface device with modern "desorption ionization" such as laser desorption or FAB offers an important prospect for the union of conventional chromatographic methods with modern mass spectrometric capabilities. Other interface techniques which do not involve significant heating as an integral feature of the method include atmospheric pressure ionization techniques which employ ion evaporation or electrospray ionization. These methods may minimize the problems associated with thermal and hydrolytic degradation noted for thermospray. Similarly, with the effective adaptation of micro liquid chromatography to accommodate flow rate limitations, the flowing FAB probe might provide important advantages. The combination of capillary electrophoresis with electrospray ionization mass spectrometry provides the exciting prospect of very high efficiency separation combined with very gentle ionization of charged analytes without the application of heat.

ADDITIONAL REFERENCES

1. P. R. Brown (ed.), *HPLC in Nucleic Acids Research, Chromatogr. Sci. Series,* Vol. 28, Dekker, New York (1984).
2. C. W. Gehrke, K. C. Kuo, and R. W. Zumwalt, in *Modified Nucleosides in Transfer RNA,* Vol. 2, P. F. Agris, ed., Alan R. Liss, New York (1982), pp. 59–91.
3. T. Kaneko, K. Katoh, M. Fugimoto, M. Kumagai, J. Tanaoka, and Y. Katayama-Fujimura, *J. Microbiol. Methods* **4,** 229–240 (1986).
4. A. P. Halfpenny, in *HPLC in Nucleic Acids Research: Methods and Applications,* P. R. Brown, ed., *Chromatogr. Sci. Series,* Vol. 28, Marcel Dekker, New York (1984), pp. 285–302.
5. (a) L. W. McLaughlin, *Trends Anal. Chem.* **5,** 215–219 (1986); (b) K. Ashman, A. Bosrehoft, and R. Frank, *J. Chromatogr.* **397,** 137–140 (1987).
6. K. Nakano, in *HPLC in Nucleic Acids Research: Methods and Applications,* P. R. Brown, ed., *Chromatogr. Sci. Series,* Vol. 28, Marcel Dekker, New York (1984), pp. 247–266.

7. (a) C. W. Gehrke, K. C. Kuo, and R. W. Zumwalt, *J. Chromatogr.* **188**, 129–147 (1980); (b) C. W. Gehrke, R. W. Zumwalt, R. A. McCune, and K. C. Kuo, in: *Recent Results in Cancer Research,* Vol. 84, G. Nass, ed., Springer, Berlin (1983), pp. 344–359.
8. M. Uziel. *J. Chromatogr.* **377**, 175–182 (1986).
9. J. R. Miksic, in *HPLC in Nucleic Acids Research: Methods and Applications,* P. R. Brown, ed., *Chromatogr. Sci. Series,* Vol. 28, Marcel Dekker, New York (1984), pp. 317–338.
10. (a) J. A. McCloskey, in *Proc. 4th International Roundtable on Nucleosides, Nucleotides and Their Biological Applications,* F. C. Alderweireldt and E. L. Esmans, eds., University of Antwerp, Antwerp (1982), pp. 47–67, and references therein; (b) J. A. McCloskey, in *Mass Spectrometry in the Health and Life Sciences,* A. L. Burlingame and N. Castagnoli, Jr., eds., Elsevier, Amsterdam (1985), pp. 521–546, and references therein; (c) J. A. McCloskey, in *Mass Spectrometry in Biomedical Research,* S. J. Gaskell, ed., Wiley, New York (1986), pp. 75–95, and references therein.
11. K. H. Schram and J. A. McCloskey, in *GLC and HPLC Determination of Therapeutic Agents, Part 3,* K. Tsuji, ed., Marcel Dekker, New York (1979), pp. 1149–1190.
12. K. Nakano, in *HPLC in Nucleic Acids Research: Methods and Applications,* P. R. Brown, ed., *Chromatogr. Sci. Series,* Vol. 28, Marcel Dekker, New York (1984), pp. 339–364.
13. (a) G. E. Davis, C. W. Gehrke, K. C. Kuo, and P. F. Agris, *J. Chromatogr.* **173**, 281–298 (1979); (b) C. W. Gehrke, K. C. Kuo, R. A. McCune, K. O. Gehrhardt, and P. F. Agris. *J. Chromatogr.* **230**, 297–308 (1982); (c) R. W. Zumwalt, K. C. T. Kuo, P. F. Agris, M. Ehrlich, and C. W. Gehrke, *J. Liq. Chromatogr.* **5**, 2041–2060 (1982).
14. M. Buck, M. Connick, and B. Ames, *Anal. Biochem.* **129**, 1–13 (1983).
15. M. S. Wilson and J. A. McCloskey. *J. Am. Chem. Soc.* **97**, 3436–3444, (1975).
16. R. A. J. Warren. *Ann. Rev. Microbiol.* **34**, 137–157 (1980).
17. S. Hattman, in: *The Enzymes, Vol. 14,* P. D. Bayer, ed., Academic Press, New York (1981), Part A, pp. 517–548.
18. M. Sprinzl, J. Moll, F. Meissner, and T. Hartmann, *Nucleic Acids Res.* **13**, r1–149 (1985).
19. C. W. Woese, L. J. Magrum and G. E. Fox. *J. Mol. Evol. 11,* 245–252, (1978).
20. J. A. McCloskey. *System. Appl. Microbiol.* **7**, 246–252 (1986).
21. M. McKeag, in *HPLC in Nucleic Acids Research: Methods and Applications,* P. R. Brown, ed., *Chromatogr. Sci. Series,* Vol. 28, Dekker, New York (1984), pp. 215–246.

Chapter 7

LC/MS of Conjugated Molecules

7.1. INTRODUCTION

Acid-catalyzed nucleophilic substitution of lipophilic substances is one of the major metabolic pathways for internal transport or elimination in urine or stools of such substances. The highly polar glucuronic and sulfuric acid conjugates of steroids, bile salts, and other endogenous and exogenous compounds in urine are among the most common products of such reactions. Other important members of this class of materials include acylcarnitines, encountered in fatty acid transport and excretion, and glycosides, conjugates of an aglycon and at least one saccharide molecule, that are encountered in plant metabolism.

Identification and determination of these complex materials as intact molecules is of great interest and importance in studies of drug metabolism, disease diagnosis and treatment, and natural product chemistry, as well as other areas of biological research. All of these materials are relatively labile thermally and very polar, and they are generally found in complex mixtures, such as urine or plant extracts. In real systems, these materials are found in complex mixtures of similar and dissimilar materials, and therefore chromatographic separation is necessary before any mass spectrometric measurements are made on them. While all of these molecules, except the acylcarnitines, contain chromophores that enable them to be detected by UV irradiation after liquid chromatographic separation, UV detectors lack the specificity, and frequently the sensitivity, of a mass spectrometric detector. Gas chromatographic separation of these substances, as intact molecules without hydrolysis, prior to mass spectrometry has generally not been successful due to high operating temperatures and the need for extensive derivatization, both of which tend to degrade the conjugates into their components. It is a major success for liquid chromatography/mass spectrometry (LC/MS) systems to be able to both separate and generate mass spectra of these materials. This success may be one of the most important single justifications for the development of LC/MS interfaces.

7.2. GLUCURONIDES

Games and Lewis [80GA73] were the first to report spectra of a glucuronide introduced through an LC/MS interface. A moving-belt apparatus allowed them to generate an ammonia chemical ionization (CI) mass spectrum of 2-naphthyl-β-D-glucuronide from a chromatographic system of 90 : 10 acetonitrile/water at 0.8 ml/min on a 10 cm $\times$ 5 mm Hypersil ODS(C_{18}). ODS is an octadecylsilyl (i.e., C_{18}-silyl) residue attached to the column packing. The spectrum showed molecular ions at m/z 338 and m/z 320 at 50% and 20%, respectively, and corresponding to ammoniation and (M + NH_4 − $H_2O)^+$. The base peaks of the spectrum were an ion at m/z 144, corresponding to the aglycon, and an ion at m/z 194. The latter corresponds to a fragment that is characteristic of many glucuronide mass spectra where ammonium ions are present. In a series of experiments using ammonia and perdeutero ammonia, Cairns and Siegmund [82CA46] showed this sugar fragment ion to be formed in the process shown in Scheme 7.1.

While Games and Lewis obtained spectra from a compound injected on column, they performed no separation of it from other standard materials. The first demonstration of a separation of glucuronide standards was published by Kenyon et al. [83KE58] using two steroid glucuronides. The Hewlett-Packard diaphragm direct liquid introduction (DLI) interface operating at 250°C using a 10 cm $\times$ 3mm ODS column at 0.6 ml/min with an acetonitrile/water mobile phase in a gradient from 10–25% H_2O over 3 min showed a separation of estriol-17-β-D-glucuronide from 5β-pregnene-3α,20β-D-glucuronide. The negative ion spectra of both of these molecules showed 25% (M − H)⁻ molecular ions as well as single and double water molecule losses, with the latter being the base peak in both.

Watson et al. [86WA35], in addition to reporting the negative ion spectra of 18 steroid glucuronides, showed the separation of 100 ng each of 12 of them using a thermospray source maintained at 200°C with a 5-μm Novapak C_{18} HPLC column, 15 cm $\times$ 4.6 mm, and a mobile phase gradient from 100% water to 20 : 80 water/acetonitrile over 15 min flowing at 1.5 ml/min. Although a thermospray source was employed, no buffer was present nor was any external means of ionization used. The spectra reported in their paper consisted entirely of protonated molecules showing optimal sensitivity response at a vaporizer tem-

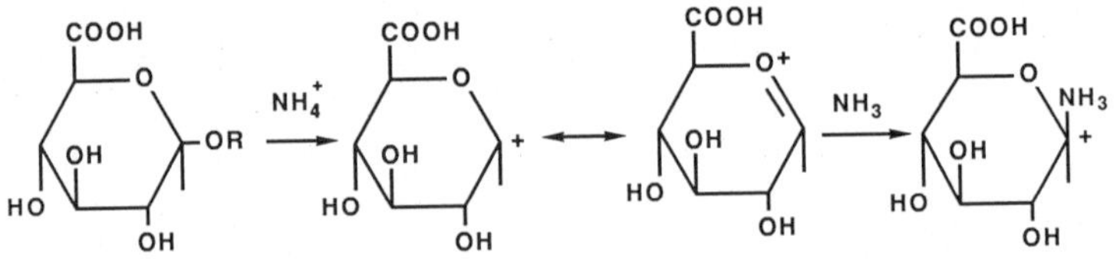

SCHEME 7.1

perature of 150°C. The separations obtained, while not always complete, were usable since those compounds that coeluted could be distinguished by their specific ion chromatograms.

The first reported LC/MS separation of glucuronides derived from a biological source rather than from standards was made by Liberato et al. [83LI51]. These workers showed the separation of a mixture of two diastereomers of propranolol glucuronide from a liver microsome preparation using a thermospray system operating at a jet temperature of 215°C and a vaporizer temperature of 230°C with a mobile phase of 47.5% methanol/52.5% aqueous ammonium acetate (0.05 M) at 1 ml/min on an Altex Utrasphere 5-μm ODS column. The base peak of the propranolol glucuronide spectrum was the ammoniated molecular ion at m/z 436 with small peaks, both less than 20%, corresponding to the aglycon, m/z 260, and glucuronic acid fragments, m/z 194. The separation achieved is shown in Figure 7.1. The figure shows the UV detector trace along with the reconstructed ion chromatograms for the protonated aglycon and the ammoniated molecule.

Comparison of the UV and m/z 436 traces of Figure 7.1 shows excellent chromatographic fidelity through the LC/MS interface. Calculation of band broadening due to the interface, using the method of Kenyon et al. [81KE56], shows that there is zero broadening. While there are very few reports of glucuronide separations involving comparisons of UV and LC/MS bandwidths, both the comparison in the paper by Liberato et al. [83LI51] and a report on the

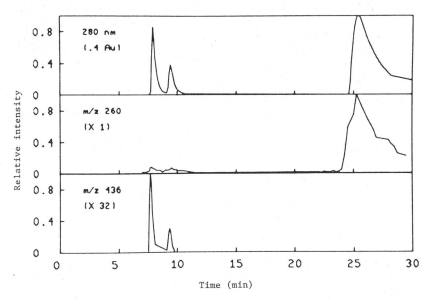

FIG. 7.1. Separation of diastereomeric glucuronides of racemic propranolol glucuronide and propranolol: 47.5% MeOH/52.5% (0.05 M NH$_4$OAc), 1 ml/min. (Reproduced from 83LI51 with permission.)

TABLE 7.1a. *p*-Nitrophenyl-β-D-glucuronide: Positive Ions[a]

Glucuronide ions					
m/z	315	357	316	333	316
Origin	$(M + NH_4 - H_2O)^+$	$(MH + MeCN)^+$	$(MH)^+$	$(M + NH_4)^+$	$(MH)^+$
RI	5	5	30	69	30
Aglycon ions					
m/z	157	181		157	
Origin	$(A + NH_4)^+$	$(A + MeCN)^+$		$(A + NH_4)^+$	
RI	100	100		20	
Sugar fragment ions					
m/z	194	218	200	194	177
Origin	$(G + NH_4 - H_2O)^+$	$(G + MeCN)^+$	$(218 - H_2O)^+$	$(G + NH_4 - H_2O)^+$	$(G + H)^+$
RI	100	100	25	100	3
Other ions					
m/z	110	124		149	
Origin	—	—		—	
RI	40	30		4	
Solvent system	NH_3 CI	MeCN/H_2O		0.05 M NH_4Ac	
Interface	Moving belt	Diaphragm DLI		Thermospray	
Reference	82CA46	81DI5'		83LI51	

[a]Abbreviations: RI = relative intensity, M = intact molecule, A = aglycon, G = glucuronic acid.

TABLE 7.1b. p-Nitrophenyl-β-D-glucuronide:
Negative Ions[a]

Glucuronide ions			
m/z	315	298	314
Origin	M^-	$(M - OH)^-$	$(M - H)^-$
RI	25	40	100
Aglycon ions			
m/z	139	138	
Origin	A^-	$(A - H)^-$	
RI	100	17	
Sugar fragment ions			
m/z	—	175	
Origin		$(G - H)^-$	
RI		5	
Solvent system	MeCN/H_2O	0.05 M NH$_4$Ac	
Interface	Diaphragm DLI	Thermospray	
Reference	81DI5'	83LI51	

[a]Abbreviations: RI = relative intensity, M = intact molecule, A = aglycon, G = glucuronic acid.

identification and confirmation of furosemide 1-O-acyl glucuronides by Rachmel and co-workers [85RA35] using a thermospray interface showed zero band broadening by the interfaces.

The spectra of p-nitrophenyl-β-D-glucuronide and estriol-17β-glucuronide are given in Tables 7.1 and 7.2 as a basis for comparing results from the several interface types. Full spectra for a given compound using a particular interface appear in the columns of the tables while the grouping of the rows allows comparison between interface types for particular types of ions in each spectra.

TABLE 7.2a. Estriol-17β-glucuronide: Positive Ions[a]

Glucuronide ions				
m/z	—			
Aglycon ions				
m/z	306	289	287	271
Origin	$(A + NH_4)^+$	$(AH)^+$	$(A - H)^+$	$(AH - H_2O)^+$
RI	100	10	10	5
Sugar fragment ions				
m/z	194			
Origin	$(G + NH_4 - H_2O)^+$			
RI	60			
Solvent system		NH$_3$ CI		
Interface		Moving belt		
Reference		82CA46		

[a]Abbreviations: RI = relative intensity, A = aglycon, G = glucuronic acid.

TABLE 7.2b. Estriol-17β-glucuronide: Negative Ions[a]

Glucuronide ions				
m/z	463	445	428	463
Origin	$(M - H)^-$	$(M - H - H_2O)^-$	$(M - H - 2H_2O)^-$	$(M - H)^-$
RI	25	5	100	100
Aglycon ions				
m/z	—			—
Sugar fragment ions				
m/z	—			—
Solvent system		MeCN/H$_2$O		MeCN/H$_2$O
Interface		Diaphragm DLI		Thermospray (no buffer, filament or discharge)
Reference		83KE58		85WA35

[a]Abbreviations: RI = relative intensity, M = intact molecule.

The positive ion spectra of p-nitrophenyl-β-D-glucuronide in Table 7.1a all show molecular ions that vary in composition with the solvent system. The molecular ion intensity variations are most probably due to differing degrees of thermal degradation, with thermospray being the most gentle ionization technique. Interestingly, the relative intensities in the thermospray spectrum are in excellent agreement with those in a FAB spectrum of this compound published in a comparative study of the two techniques [84FE69]. All three interface techniques show sugar fragment ions that could potentially be used to signal the presence of a glucuronide in a chromatograph of unknown materials. The ammoniated sugar fragment at m/z 194 seen in both thermospray and moving-belt spectra (Table 7.1a) has been seen for all glucuronides when ammonium ions are available in the ion source. The negative ion spectra of p-nitrophenyl-β-D-glucuronide show less fragmentation than the positive ion spectra, but there is a sugar fragment ion in the thermospray spectrum that appears to be characteristic when ammonium ions are available for ionization.

The only published report of a positive ion spectrum for steroid glucuronides is Cairn and Siegmund's work using a moving-belt interface [82CA46]. Their spectrum for estriol-17β-glucuronide is given in Table 7.2a. The spectrum shows a series of ions related to the parent ion and differing by ionizing agent and water losses. The m/z 194 characteristic sugar fragment ion is also present. The negative ion spectra of this compound given in Table 7.2b show glucuronide ions only. The loss of two moles of water from the hydride loss parent ion accounts for the base peak of the spectrum produced in a diaphragm DLI source, while the buffer-free thermospray spectrum shows only the hydride loss parent.

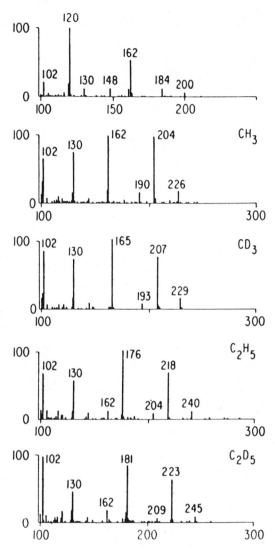

FIG. 7.2. Mass spectra of L-carnitine and its acetyl, [²H₃]acetyl, propionyl, and [²H₅]pro-pionyl derivatives. (Reproduced from 84YE98 with permission.)

7.3. ACYLCARNITINES

Yergey and his colleagues [84YE98, 85LI43, 85MI59] have published the only work to date on the separation and mass spectrometric detection of acylcar-nitines. Since neither carnitine nor the fatty acids commonly encountered in

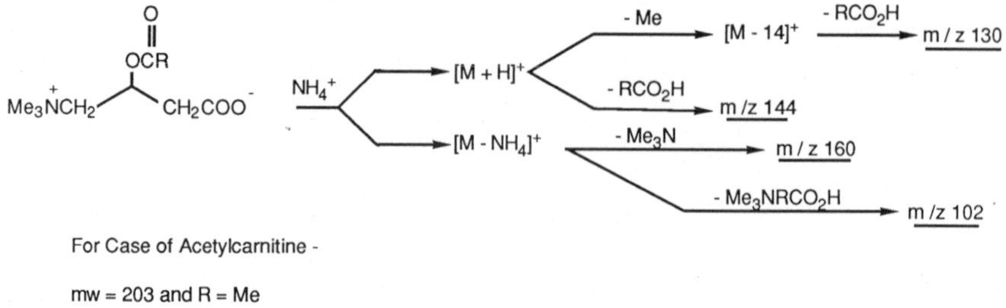

For Case of Acetylcarnitine -

mw = 203 and R = Me

SCHEME 7.2

mitochondrial metabolism have chromophores that would make UV detection possible, and since the intact molecules cannot be separated by gas chromatography, LC/MS is the only analytical technique capable of determining intact acylcarnitines in mixtures. Chromatographic separations of acylcarnitine standards and urine extracts were performed using an Altex Ultrasphere 5-µm ODS column (4.6 mm × 15 cm) with isocratic methanol/0.1 M ammonium acetate solutions at pH 6.85, the methanol fraction varying from 5–50% depending on the particular acylcarnitines being separated, and at solvent flow rates of 1 ml/min [85LI43].

The thermospray mass spectra of carnitine and a series of acylcarnitines and perdeutero acylcarnitines are reproduced in Figure 7.2. The fragmentation pathways that were deduced from these spectra are illustrated in Scheme 7.2 for the case of acetylcarnitine.

A number of consistent features are seen. First, protonated, and in some cases natriated and kalinated, molecular ions are seen in good abundance. Ions at $(M - 14)^+$ and $(M - 42)^+$, resulting from demethylation and loss of trimethylamine from the protonated and ammoniated molecule and characteristic of particular acyl groups, are always observed. The demethlyation has been seen in the thermospray spectrum of acetylcholine [86LI31], and the trimethylamine loss has been seen in the spectra of phosphatidylcholines by ammonia CI.[1] Ions that are characteristic of all acylcarnitines are those that result from losses of the acyl group, either alone (m/z 144) or in combination with other losses (m/z 130 and 102).

7.4. SULFATES

The only spectra reported for intact conjugates of sulfuric acid are the steroid sulfate spectra of Watson et al. [85WA20, 85WA34]. In addition to giving the thermospray mass spectra of 26 steroid monosulfates and 6 steroid disulfates, these authors report the spectra of two bile acid disulfates [85WA20] in this same paper. The thermospray mass spectra were obtained in a buffer-free system in the absence of any external ionization sources, either filament or discharge. These authors further demonstrated the ability to separate chromatographically 12 steroid monosulfate standards using a 5-μm Lichrosorb-RP2 HPLC column, 25 cm × 1.5 mm, with a 1.5-ml/min flow of deionized water. Optimal response was obtained at a source temperature of 150°C. It should be noted that in the Finnigan MAT version of the thermospray source, vaporizer and source temperatures are the same.

The negative ion mass spectra of the steroid monosulfates are extremely simple, consisting solely of the intact molecular anion. The disulfate spectra are considerably more interesting. All eight compounds give doubly charged anions as the base peak in their spectra at source temperatures below 150°C, but the two bile acid disulfates and 17β-estriol have the doubled charged anion as the base peak even at that temperature. With one exception, however, at the lower temperatures the total response is less than 50% of the response at 150°C. At 150°C, the base peak in the spectra is typically the protonated monosulfate ion, $(M - SO_3 + H)^+$, with the protonated, dehydrated monosulfate ion $(M - SO_3 - H_2O + H)^+$ and protonated and otherwise cationized ions all being at intensities of 50% or less of the base peak.

ADDITIONAL REFERENCES

1. C. G. Crawford and R. D. Plattner, "Ammonia Chemical Ionization Mass Spectrometry of Intact Diacyl Phosphatidylcholine," *J. Lipid Res.* **24**(4), 456–460 (1983).

Applications of LC/MS to Amino Acids, Peptides, and Proteins

8.1. BACKGROUND

The potential for application of liquid chromatography/mass spectrometry (LC/MS) to sequencing of peptides and proteins has long been recognized. This was among the earliest applications of the direct liquid introduction technique (DLI) pioneered by McLafferty and co-workers [75MC36, 76AR61]. Despite some early success, LC/MS was not applicable to larger peptides until the observation was made that thermospray could provide a "soft ionization" technique [80BL21] which yielded protonated molecules from peptides and other nonvolatile molecules without employing any external source of ionization. This new direct thermospray ionization process produced spectra very similar to those obtained by field desorption (FD), and subsequent work [83VE27] showed that a direct ion evaporation mechanism similar to that discussed earlier by Iribarne and Thomson [76IR47] was involved.

The first attempt to employ thermospray LC/MS for peptide sequencing by Pilosof et al. [84PI13] involved direct detection of the amino acids released by on-line enzymatic digestion. The underivatized peptide solutions were injected through a column containing immobilized carboxypeptidase Y (CPY), and the amino acids released from the C-terminus were directly transported by a continuous flowing aqueous buffer into a thermospray mass spectrometer where the amino acids were detected and quantitated as the protonated molecular ions. Comparison of integrated protonated molecule intensities obtained by digestion of a peptide or protein with those obtained from injection of a standard mixture of amino acids allowed the sequence of several known peptides and small proteins to be correctly determined based on relative yields of the released amino acids.

For example, the mass chromatograms for MH^+ ions from several amino acids present in the C-terminal portion of bovine ribonuclease are shown in Figure 8.1. These results were obtained using a 2.1 mm i.d. $\times$ 20 cm CPY

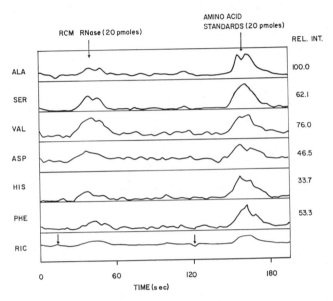

FIG. 8.1. Mass chromatograms obtained for a 20 pM injection of RCM-RNase onto a CPY column compared to an amino-acid standard. The arrows on the reconstructed total ion trade (RIC) at the bottom indicate the injection times.

column with a total enzyme content of 8.6 mg using ammonium acetate buffer at a flow of 0.7 ml/min. The first peak shows the response to injection of 20 pmol of the carboxymethylated ribonuclease, and the second peak to injection of a standard containing 20 pmol of each of the common amino acids. In Table 8.1, the integrated response from the substrate for the MH^+ ions for the amino acids observed is compared to that obtained for a similar injection of the standard.

TABLE 8.1. Integrated Responses and Yields Obtained from Data Shown in Figure 8.1

Amino acid	Peak area (arbitrary units)		Yield[a]
	Cleaved amino acid	Standard amino acid	
Val	7.39×10^4	8.80×10^4	0.84
Ser	3.33×10^4	6.54×10^4	0.51
Ala	4.26×10^4	9.61×10^4	0.44
Asp	2.06×10^4	4.99×10^4	0.41
Phe	1.78×10^4	5.32×10^4	0.33
His	1.02×10^4	3.60×10^4	0.28

[a]Yield is expressed as the number of moles of amino acid/number of moles of peptide substrate injected.

From the observed yields the correct sequence (Phe-Asp-Ala-Ser-Val) can be determined. The sensitivity of this method is excellent because only the low-mass portion of the spectrum (ca. 70–250 amu) is scanned, and the first few residues from the C-terminus can usually be determined using only a few pmol of result in similar yields and, thus, the order of cleavage that is used to elucidate sequence is ambiguous.

More recent work has focused on direct detection of the peptides using combinations of on-line enzyme columns with an HPLC column [84PI66, 84KI64, 88ST08, 88ST7MC']. As discussed in detail in Chapter 4, the thermospray interface can be optimized to provide reliable molecular weight information on peptides. The thermospray vaporizer can also be adjusted so that extensive fragmentation is observed. Unfortunately, the observed fragmentation, at least in the case of larger peptides, seems to be generally indicative of pyrolysis rather than unimolecular fragmentation and thus is not useful for structural elucidation. As has been discussed previously, the control temperature on the thermospray vaporizer can be calibrated to correspond to the fraction of the mobile phase vaporized within the thermospray capillary. By plotting the exit temperature as a function of the control temperature, the observed breaks can be correlated with the onset and the completion of vaporization, and in between these two points the fraction vaporized is directly proportional to the control temperature.

The intensity of the ions produced by direct ion evaporation in the thermospray is a strong function of the fraction of the effluent vaporized within the capillary. Also, for many nonvolatile and thermally labile compounds, such as peptides, extensive nonspecific fragmentation is often observed at higher fractions vaporized. This is presumably the result of pyrolyzing the molecule with subsequent ionization of the pyrolytic fragments by ion–molecule reactions in the gas phase. An example of the optimization of thermospray for peptides is

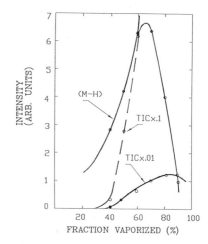

FIG. 8.2. Total ion current and deprotonated molecular ion (M-H⁻) for brandykinen as a function of fraction of the LC effluent vaporized in the thermospray capillary. 5 nanomoles of peptide in 0.1 M ammonium acetate at 1 ml/min.

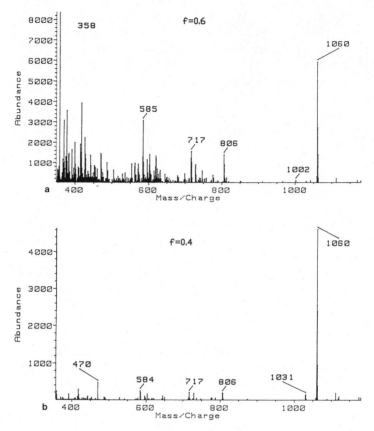

FIG. 8.3. Thermospray negative ion spectra from 5 nanomoles of bradykinen, (a) at 60% (an average of 9.15 to 9.25 min) and (b) at 40% (an average of 2.13 to 2.30 min) vaporization in the capillary.

illustrated in Figure 8.2, where both the total negative ion intensity and the intensity of the deprotonated molecular ion are plotted as functions of fraction vaporized. Spectra corresponding to 40 and 60% vaporization are shown in Figure 8.3. Clearly, the optimum operating point for obtaining molecular weight without cluttering the spectrum with nonspecific fragmentation occurs at about 40% vaporization even though the absolute intensity of the molecular ion is somewhat higher when a larger fraction is vaporized within the capillary. To obtain useful sensitivity at these low fractions vaporized, it is necessary to move the droplets and particles toward complete vaporization downstream by operating the ion source quite hot (ca. 450–500°C). Interestingly, operating the ion source at such high temperatures does not appear to introduce additional pyrolysis of samples.

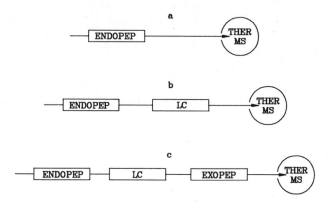

FIG. 8.4. Possible configurations of the enzymic–thermospray LC/MS protein sequencing system.

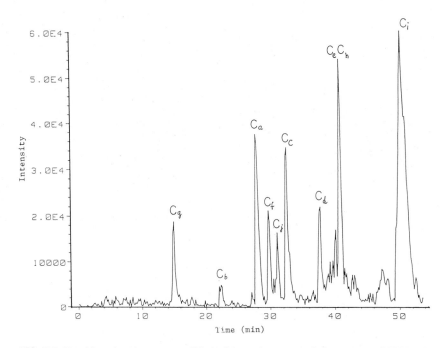

FIG. 8.5. Total ion chromatogram of the primary chymotryptic fragments of CM-bovine pancreatic trypsin inhibitor. The instrument is configured as in Figure 8.4b, with chymotrypsin as the endopeptidase and RP-304 as the chromatographic support. A gradient of 0 to 15% propanol in 0.1 M ammonium acetate was used as the eluent.

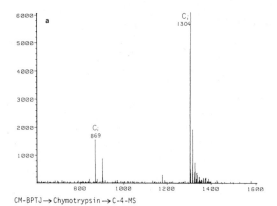

CM-BPTJ→Chymotrypsin→C-4-MS

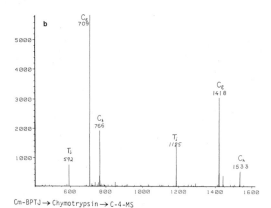

Cm-BPTJ→Chymotrypsin→C-4-MS

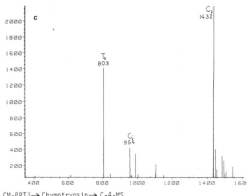

CM-BPTJ→Chymotrypsin→C-4-MS

FIG. 8.6. (a) Mass spectrum of peak C_i from the chymotryptic digest shown in Figure 8.5. Major peaks at 1304 and 869 correspond to the doubly and triply deprotonated molecular ions from the indicated peptide with molecular weight 2609. (b) Mass spectrum of peak at 41 min in Figure 8.5, showing singly and doubly deprotonated molecular ions from two chymotryptic fragments plus a tryptic fragment due to trypsin impurity in the enzyme. (c) Mass spectrum of small peak at 31 min in Figure 8.5, showing singly and doubly deprotonated molecular ions from the highest-molecular-weight fragment in this digest (MW 2866) and a small tryptic fragment. [Averages of (a) 46.650 to 47.596 min; (b) 40.619 to 41.312 min; and (c) 31.085 to 31.417 min. All data from [88ST08].

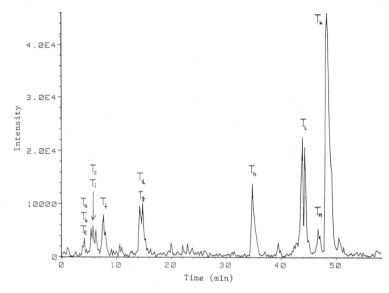

FIG. 8.7. Total ion chromatogram of primary tryptic fragments of CM-bovine pancreatic trysin inhibitor. Conditions are the same as in Figure 8.5.

8.2. APPLICATIONS USING IMMOBILIZED ENZYME DIGESTIONS

To provide structural information, the use of thermospray LC/MS with on-line immobilized enzymes has been developed [84PI66, 84KI64, 88ST08]. This continuous flow system consists of a series of analytical columns connected on-line and terminating with the thermospray LC/MS as illustrated in Figure 8.4. First, a column of immobilized trypsin or chymostrypsin is used to cleave the peptide or denatured protein being analyzed into primary fragments; next, an HPLC column separates these primary fragments; a second enzyme column containing an immobilized carboxy- or aminopeptidase then digests each fragment as it elutes from the HPLC column; and finally, the thermospray mass spectrometer detects the peptides produced and determines their molecular weights. Several HPLC columns have been evaluated for this work including reversed phase amino, C_{18}, C_8, and C_4 columns and ion exchange on a polyethyleneimine (PEI)-coated silica gel. By far the best results have been obtained using the C_4 reversed phase column, with the only problem being that the organic modifier required to elute the more strongly retained peptides partially inhibits to some extent the activity of the exopeptidases in the subsequent step.

In recent work, a Hewlett-Packard model 5988 mass spectrometer equipped with the 2000-amu mass range and negative ion options has been employed, together with a Vestec model 360Q thermospray system. With this system and

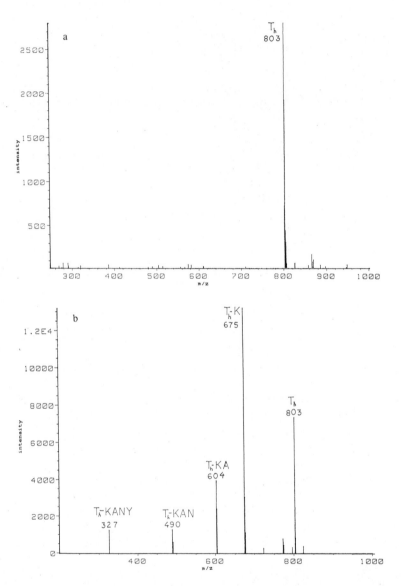

FIG. 8.8. Mass spectrum of 35 min component in Figure 8.7. (a) Spectrum of the peptide (T_h) without further proteolysis. (b) Spectrum of T_h after digestion after passage through columns of immobilized CPY and CPB in the exopeptidase position in Figure 8.4c. Peaks are identified by indicating the amino acids cleaved, using the single-letter code.

immobilized enzyme columns recently developed [88ST08], results are now obtained relatively routinely on 1–10 nanomol of sample injected.

As an example of the performance of the thermospray LC/MS protein sequencing system, the chymotryptic LC peptide map obtained from 10 nanomol of bovine pancreatic trypsin inhibitor (MW 6510) is shown in Figure 8.5. Examples of spectra obtained for particular peaks in the chromatogram are shown in Figure 8.6. In addition to the expected fragments resulting from cleavage adjacent to phenylalanine and tyrosine, some tryptic peptides are also observed at somewhat lower levels. These apparently resulted from contamination of the enzyme with trypsin. With the additional purification of the trypsin used, these peaks are not seen in the most recent results. The LC/MS chromatogram for the on-line tryptic digest of the same substrate is shown in Figure 8.7. An example of results obtained from a triple-column experiment in which the HPLC column is followed by a column containing an immobilized mixture of carboxypeptidases B and Y is shown in Figure 8.8. This result allows the C-terminus of the peptide fragments to be read directly from the spectrum except for mass degeneracies such as in the case of Leu and Ile. A total of six such experiments employing either trypsin or chymotrypsin with reversed phase HPLC on a C_4 column and with and without endopeptidase columns containing either an amino- or carboxypeptidase have provided information sufficient to determine about one-half of the total sequence of bovine pancreatic trypsin inhibitor.

8.3. LIMITATIONS AND FUTURE WORK

Two problems inhibit the immediate application of this technique to unknowns. First, the sensitivity of the present thermospray system for peptides is such that about 5–10 nanomol of sample are required to obtain high-quality full-scan spectra for all of the peptides produced. Second, the conditions required for LC separation of the peptide fragments are only marginally compatible with the exopeptidases. As a result, the sequence information obtained is significantly inferior to that obtained by using the exopeptidase alone on-line with the thermospray without the organic modifier required for the LC separation. It appears that the present strategy whereby as many as three columns on-line are employed with the thermospray may not be the most fruitful approach, since it requires satisfying four masters simultaneously. This has been shown to be possible by the work described above, but some serious compromises are required, and under the conditions presently used, the performance of the exopeptidases suffers most severely. A more practical approach may be one such as that depicted in Figure 8.9. In this approach, immobilized trypsin or other endopeptidase is used in series with the C_4 reversed phase column. A fraction of the flow is directed to the thermospray LC/MS interface, and the other portion is diverted through a capillary delay to a fraction collector controlled by the MS data system. When the MS signal exceeds a preset threshold, fractions are collected at

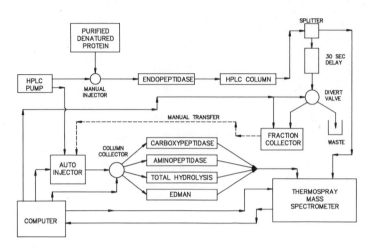

FIG. 8.9. Block diagram of the proposed semiautomatic method for protein sequencing using thermospray LC/MS.

appropriate intervals, and when no peaks are present, the effluent is discarded to waste. Under ideal conditions, the number of fractions should not be much larger than the number of peaks in the chromatogram. Since the sensitivity of the MS will normally be larger for the lower-molecular weight-peptides and amino acids, these fractions can be divided into several aliquots for further analysis. For example, one aliquot might be injected on a carboxypeptidase column, a second injected on an aminopeptidase column, a third subjected to total hydrolysis and injected on an ion exchange column for amino acid analysis, and a fourth subjected to Edman degradation. Full implementation of this scheme should allow complete, or nearly complete, sequences to be determined in a very short time, perhaps as short as one day per sequence.

The major factor impeding the wider use of this technique for many applications is the lack of sufficient sensitivity. For applications which are not sample limited, results are obtained very rapidly. An excellent sample is the recent results on identification of hemoglobin variants (K. Stachowiak and D. F. Dyckes, unpublished results). In most cases, only one or two injections were required to find and unambiguously identify the modification. Attempts to improve the sensitivity of the technique are presently focusing on first achieving a better fundamental understanding of the processes involved in producing gas phase ions directly from the condensed phase and then implementing design changes which will lead to more efficient ionization and detection. These processes are thought to be important not only in thermospray, but also in other desorption ionization techniques.

From these fundamental studies, it is apparent that the sensitivity of thermospray can be substantially improved by changes in several parts of the apparatus. First, it is important to achieve more efficient nebulization of the liquid into more, and smaller, droplets. Preliminary results suggest that this can be achieved by using smaller vaporizer nozzle diameters. Second, it is clear that if additional net charge could be induced on the initially formed droplets, without necessarily adding more electrolyte, then enhanced sensitivity for difficult samples such as peptides would result. Third, less volatile ions are vaporized at lower rates; thus, more nearly complete vaporization of the droplets and particles as they fly through the ion source leads to improved ionization efficiency for difficult samples. Finally, the ion sampling efficiency from the present thermospray ion sources is very low (ca. 1 per 100,000 formed), and the detection efficiency for high-mass ions with the low acceleration voltages used in quadrupole analyzers may also be quite low. It should be noted that improvements in any of these areas translates directly into improved sensitivity because, unlike particle impact desorption techniques, thermospray is not presently limited by chemical or ion noise.

Until recently, application of thermospray to higher-mass peptides has been severely limited by the mass range of the mass analyzers used, since most of the work has been done using relatively low-mass-range quadrupole analyzers. Recently, thermospray has been increasingly available on extended-mass-range quadrupoles and high-performance magnetic instruments. The current molecular weight record for thermospray on peptides appears to be *N*-acetylated eglin C, MW 8133, reported by Jones and Krolik [87JO17] and obtained using a VG 70 SE magnetic sector mass spectrometer. These workers found that the absolute sensitivity for thermospray in their instrument was about an order of magnitude less than for fast atom bombardment (FAB) on high-molecular-weight peptides but that interference from fragment ions and solvent background was significantly less with thermospray. For example, the FAB and thermospray spectra for glucagon, MW 3483, are compared in Figure 8.10. These results provide additional support for the view that, with improved sensitivity, thermospray may provide the preferred technique for determining molecular weights of unknown peptides.

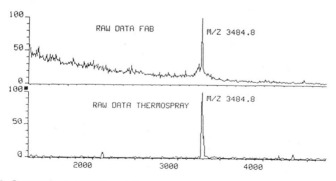

FIG. 8.10. Comparison of FAB and thermospray mass spectra for glucagon, MW 3483. (From 87JO17 with permission.)

Thermospray LC/MS is a useful technique for rapidly obtaining sequence information on peptides and proteins. Improvements in the sensitivity of thermospray LC/MS are needed before the technique can be extended to the many applications where sample quantities are severely limited, but recent fundamental studies provide strong indications that substantial improvements will be realized shortly. Such improvements, together with the development of automated, less expensive instrumentation, may make thermospray LC/MS a valuable practical tool for the protein chemist in the not-too-distant future.

Chapter 9

A Bibliography of Combined Liquid Chromatography/Mass Spectrometry

9.1. INTRODUCTION

The development of directly combined liquid chromatography/mass spectrometry (LC/MS) from its nascence in the work of McLafferty and co-workers and of the Tal'roze group in the USSR to May 1988 is presented in the following compilation of citations. Evidence of the approaching maturity of the technique may be found in the increasing rate of appearance of papers in which combined LC/MS appears as only one component of a multi-method analytical investigation. As a consequence, it is no longer possible to follow developments in combined liquid chromatographic/mass spectrometric methods and their application simply by the appearance of the named method or its acronym in the title of scientific reports.

9.1.1. Methods

This bibliographic listing represents a continuation of the listing presented in *Biomedical Mass Spectrometry* in 1983.[1] These references are presented along with additional citations gathered from ongoing bibliographic efforts including the use of conventional abstracting and current awareness publications in their printed and machine-accessible forms. These include *Chemical Abstracts,*[2] *Science Citation Index,*[3] *Mass Spectrometry Bulletin,*[4] and the now defunct *Liquid Chromatography Mass Spectrometry Abstracts.*[5] For the purposes of this listing, liquid chromatography/mass spectrometry is defined as the on-line, dynamic combination of the two techniques. The use of liquid chromatography in its diverse forms for the analysis and preparation of analytes for subsequent off-line mass spectrometric examination is specifically excluded. Direct analysis of thin-layer chromatographic media by mass spectrometry without intermediate manipulation is included as a conceptually allied topic. Additionally, a selection of citations on instrumental methods and applications dealing with related aspects of physics or mass spectrometry are included. Reviews of the principal topic are included as well as references to review papers in which combined liquid chromatography/mass spectrometry is included as a subsidiary topic. Patents for interface devices are also compiled. Contributions to symposia and meetings without formal published proceedings have generally not been included. Exceptions are the Annual Conference on Mass Spectrometry and Allied Topics of the American Society for Mass Spectrometry and meetings of the American Chemical Society and the American Oil Chemist Society, whose abstracts are generally available. Additionally, occasional or periodically issued unrefereed reports from instrument manufacturers are included where the content is of sufficient substance to warrant inclusion.

Author and subject indices have been compiled. In the subject index, the organizing principle has been the classification of research reports by the method of liquid chromatographic/mass spectrometric combination and the compound or compound class for which this combination has a demonstrated utility.

9.1.2. Organization of the Bibliography

The format conforms to that already employed,[1] which is in turn modeled on the very useful bibliographic contributions of Klein and Klein.[6] References are arranged in two dimensions: first, by year of publication; and second, by author. Each reference is assigned a coden constructed in the following manner.

87ED79   87 = year of publication (e.g., 1987);
          ED = first two letters of the author's surname;
          7 = last digit of volume number;
          9 = last digit of initial page.

Thus, in the arrangement of codens in sequential order by year and author the last two digits do not necessarily follow in strict numerical order. Where a complete correspondence in coden occurs, distinction is obtained by the inclusion of a third digit corresponding to the last digit of the final page of the article.

Where a reference appears in a proceedings for which no editor is listed, or where the reference is to a patent, the coden becomes:

87LE0'   0 = last digit of first page or patent number;
          ' = unspecified editor of a non-journal article or patent.

For articles or chapters in edited books or symposium proceedings:

87YE7VE' 7 = last digit of first page;
          VE = first two letters of first editor's name;
          ' = non-journal article, i.e., book.

ADDITIONAL REFERENCES

1. C. G. Edmonds, J. A. McCloskey and V. A. Edmonds. *Biomed. Mass Spectrom.* **10** *(4)*, 237–252 (1983).
2. *Chemical Abstracts,* American Chemical Society, Columbus, Ohio 43202, U.S.A.
3. *Science Citation Index,* Institute for Scientific Information, Philadelphia 19106, U.S.A.
4. *Mass Spectrometry Bulletin,* The Royal Society of Chemistry, The University, Nottingham NG7 2RD, U.K.
5. *Liquid Chromatography Mass Spectrometry Abstracts* (I. C. Aylott, ed.), PRM Science and Technology Agency Ltd., 261 Finchley Road, London NW3 6LU, U.K.
6. E. R. Klein and P. D. Klein, *Biomed. Mass Spectrom.* **6**, 515–545 (1979) and earlier bibliographies cited therein.

9.2. BIBLIOGRAPHY

45RA0'
The Theory of Sound. J.W.S. Rayleigh. Macmillan, New York, (1945).
53DO43
The Statistics of Liquid Spray and Dust Electrification by the Hopper and Laby Method. E.R. Dodd. *J. Appl. Phys. 24*, 73–80, (1953).
58LO0'
Static Electrification. L. Loeb. Springer-Verlag, Berlin, (1958).
68DO16
Gas Phase Macroions. M. Dole, R.L. Hines, L.L. Mack, R.C. Mobley, L.D. Ferguson and M.B. Alice. *Macromolecules 1*, 96–99, (1968).
68DO90
Molecular Beams of Macroions. M. Dole, L.L. Mack, R.L. Hines, R.C. Mobley, L.D. Ferguson and M.B. Alice. *J. Chem. Phys. 49*, 2240–2249, (1968).
68TA28
Capillary System for the Introduction of Liquid Mixtures into an Analytical Mass Spectrometer. V.L. Tal'roze, G.V. Karpov, I.G. Gordetskii and V.E. Skurat. *Russ. J. Phys. Chem. 42*, 1658–1664, (1968).
68TA5'
Mass Spectrometric Analysis of Liquid Mixtures. V.L. Tal'roze, G.V. Karpov, I.G. Gordetskii and V.E. Skurat. U.S.S.R. Patent 226,235 (Cl. G01N), 5 Sept. 1968, Appl. 5 July 1965; *Izobret. Prom. Obraztsy, Tovarnye Znaki 45*, 95, (1968).
68TA22
Preparation of Inlet Devices for a Capillary System for Introducing Liquid Mixtures into a Mass Spectrometer and Measurement of Their Parameters. V.L. Tal'roze, G.V. Karpov, I.G. Gordetskii and V.E. Skurat. *Kh. Fiz. Khim. 42*, 3112–3117, (1968).
69TA38
Analysis of Mixtures of Organic Substances on a Mass Spectrometer with a Capillary System for the Introduction of Liquid Specimens. V.L. Tal'roze, G.V. Karpov, I.G. Gordetskii and V.E. Skurat. *Russ. J. Phys. Chem. 43*, 198–201, (1969).
69TA37
Analysis of Mixtures of Organic Substances in a Mass Spectrometer with a Capillary System for Introducing Liquid Samples. V.L. Tal'roze, G.V. Karpov, I.G. Gordetskii and V.E. Skurat. *Zh. Fiz. Khim. 43*, 367–372, (1969).
69TA43
Analysis of Mixtures of Polar Organic Substances on a Mass Spectrometer with a Capillary System

for Introducing Liquids. V.L. Tal'roze, G.V. Karpov and V.E. Skurat. *Zh. Anal. Khim. 24*, 1603–1604, (1969).

70GI49
Problems in the Gas-Phase Isolation of Nonvolatile Molecules by High-Pressure Jets for Mass Spectrometry. J.C. Giddings, M.N. Myers and A.L. Wahrhaftig. *Int. J. Mass Spectrom. Ion Phys. 4*, 9–20, (1970).

70MA1′
Molecular Beams of Macroions. L.L. Mack. Thesis, 261p. Univ. Microfilms Int., Ann Arbor, MI, U.S.A. Order no. 70–6493. *Diss. Abstr. Int. B 30*, 4584–4585, (1970).

70MA27
Molecular Beams of Macroions. II. L.L. Mack, P. Kralik, A. Rheude and M. Dole. *J. Chem. Phys. 52*, 4977–4986, (1970).

70TA8OG′
New Techniques in Chromato-Mass Spectrometry. V.L. Tal'roze, V.D. Grishin, V.E. Skurat and G.D. Tantsyrev. In: *Recent Developments in Mass Spectrometry 1969* (K. Ogata and T. Hagakawa, eds.), p. 1218–1225, Univ. Park Publ., Baltimore, (1970).

71CL01
Molecular Beams of Macroions. III. Zein and Polyvinylpyrrolidone. G.A. Clegg and M. Dole. *Biopolymers 10*, 821–826, (1971).

72GI6′
Application of the Ion-Drift Spectrometer to Macromass Spectroscopy. J. Gieniec, H.L. Cox Jr., D. Teer and M. Dole. Presented at 20th Ann. Conf. on Mass Spectrom. and Allied Topics, Dallas, TX, U.S.A. June 4–9, 1972. Abstr. O2, p. 276–280.

72SK60
Equation for the Capillary Flow of a Volatile Liquid from a Region of Atmospheric or High Pressure into a Vacuum. V.E. Skurat. *Russ. J. Phys. Chem. 46*, 570–571, (1972).

73BA71
Liquid Chromatography-Mass Spectrometry Interface. I. The Direct Introduction of Liquid Solutions into a Chemical Ionization Mass Spectrometer. M.A. Baldwin and F.W. McLafferty. *Org. Mass Spectrom. 7*, 1111–1112, (1973).

73GI2′
Mobility Spectra of Some Electrohydrodynamically Atomized Solutions. J. Gieniec, R.P. Blickensderfer, D. Teer and M. Dole. Presented at 21st Ann. Conf. on Mass Spectrom. and Allied Topics, San Francisco, CA, U.S.A. May 20–25, 1973. Abstr. R11, p. 362.

73KA63
Evaluation of the Plasma Chromatograph as a Qualitative Detector for Liquid Chromatography. F.W. Karasek and D.W. Denney. *Anal. Lett. 6*, 993–1004, (1973).

73LO53
Liquid Chromatography-Mass Spectrometry: Coupling of a Liquid Chromatograph to a Mass Spectrometer. R.E. Lovins, S.R. Ellis, G.D. Tolbert and C.R. McKinney. *Anal. Chem. 45*, 1553–1556, (1973).

74AR10
Liquid Chromatography-Mass Spectrometry II. Continuous Monitoring. P.J. Arpino, M.A. Baldwin and F.W. McLafferty. *Biomed. Mass Spectrom. 1*, 80–82, (1974).

74AR24
A Liquid Chromatography/Mass Spectrometry System Providing Continuous Monitoring with Nanogram Sensitivity. P.J. Arpino, B.G. Dawkins and F.W. McLafferty. *J. Chromatogr. Sci. 12*, 574–578, (1974).

74CA3′
A Liquid Chromatograph-Mass Spectrometer System Using Atmospheric Pressure Ionization (API) Techniques. D.I. Carroll, I. Dzidic, R.N. Stillwell, M.G. Horning and E.C. Horning. Presented at

22nd Ann. Conf. on Mass Spectrom. and Allied Topics, Philadelphia, PA, U.S.A. May 19–24, 1974. Abstr. L4, p. 243–244.
74CA66
Subpicogram Detection System for Gas Phase Analysis Based Upon Atmospheric Pressure Ionization (API) Mass Spectrometry. D.I. Carroll, I. Dzidic, R.N. Stillwell, M.G. Horning and E.C. Horning. *Anal. Chem. 46*, 706–710, (1974).
74CO65
Electrohydrodynamic Ionization Mass Spectrometry. B.N. Colby and C.A. Evans. *Adv. Mass Spectrom. 6*, 565–570, (1974).
74HO25
Atmospheric Pressure Ionization (API) Mass Spectrometry. Solvent-Mediated Ionization of Samples Introduced in Solution and in a Liquid Chromatograph Effluent Stream. E.C.Horning, D.I. Carroll, I. Dzidic, K.D. Haegele, M.G. Horning and R.N. Stillwell. *J. Chromatogr. Sci. 12*, 725–729, (1974).
74HO93
Liquid Chromatograph-Mass Spectrometer Computer Analytical Systems. Continuous Flow System Based on Atmospheric Pressure Ionization Mass Spectrometry. E.C. Horning, D.I. Carroll, I. Dzidic, K.D. Haegele, M.G. Horning and R.N. Stillwell. *J. Chromatogr. 99*, 13–21, (1974).
74LO67
Coupling of a Liquid Chromatograph to a Mass Spectrometer. R.E. Lovins, S.R. Ellis, G.D. Tolbert and C.R. McKinney. *Adv. Mass Spectrom. 6*, 457–462, (1974).
74SC95
Interface for On-Line Liquid Chromatography-Mass Spectroscopy Analysis. R.P.W. Scott, C.G. Scott, M. Munroe and J. Hess, Jr. *J. Chromatogr. 99*, 395–405, (1974).
74SC65
On-Line Liquid Chromatography-Mass Spectrometry System. R.P.W. Scott, C.G. Scott, M. Munroe and J. Hess, Jr. *Ciba Found. Symp. 26*, 155–169, (1974).
74SI51
Electrohydrodynamic Ionization Mass Spectrometry. Ionization of Liquid Glycerol and Nonvolatile Organic Solutes. D.S. Simons, B.N. Colby and C.A. Evans, Jr. *Int. J. Mass Spectrom. Ion Phys. 15*, 291–302, (1974).
75CA79
Atmospheric Pressure Ionization Mass Spectrometry: Corona Discharge Ion Source for Use in Liquid Chromatograph-Mass Spectrometer-Computer Analytical System. D.I. Carroll, I. Dzidic, R.N. Stillwell, K.D. Haegele and E.C. Horning. *Anal. Chem. 47*, 2369–2373, (1975).
75DO04'
Application of the Ion-Drift Spectrometer to Macromass Spectroscopy. II. M. Dole and J. Gieniec. Presented at 23rd Ann. Conf. on Mass Spectrom. and Allied Topics, Houston, TX, U.S.A. May 25–30, 1975. Abstr. C–8, p. 84–86.
75FI5'
Detector for Liquid Chromatography. D. Fischer and E.G. Kohl. Ger. Offen. 2,424,985 (Cl. GO1N), 4 Dec. 1975, Appl. P 24 24 985.2, 22 May 1974, 14p.
75GR9'
Particulate Impact Mass Spectrometry. F.T. Greene. Presented at 23rd Ann. Conf. on Mass Spectrom. and Allied Topics, Houston, TX, U.S.A. May 25–30, 1975. Abstr. Z–2, p. 659.
75JO70
A Liquid Chromatograph/Mass Spectrometer Interface. P.R. Jones and S.K. Yang. *Anal. Chem. 47*, 1000–1008, (1975).
75MC73
Continuous Mass Spectrometric Monitoring of a Liquid Chromatograph with Subnanogram Sensitivity Using an On-Line Computer. F.W. McLafferty, R. Knutti, R. Venkatarghavan, P.J. Arpino and B.G. Dawkins. *Anal. Chem. 47*, 1503–1505, (1975).

75MC36
Continuous Monitoring of Liquid Chromatography by Mass Spectrometry: Applications on Polypeptide Sequencing. F.W. McLafferty and B.G. Dawkins. In: *High Efficiency Liquid Chromatography and Mass Spectrometry* (J.H. Knox and E. Bayer, eds.). *Biochem. Soc. Trans. 3*, 856–858, (1975).

76AR61
Amino Acid Sequencing of Oligopeptides by Mass Spectrometry. P.J. Arpino and F.W. McLafferty. In: *Determination of Organic Structures by Physical Methods* (F.C. Nachod, J.J. Zuckerman and E.W. Randall, eds.) Vol. 6, pp. 1–89, Academic Press, New York, (1976).

76GR2'
Mass Spectrometry of Nonvolatile Materials and Solutions by the Particulate Impact Technique. F.T. Greene. Presented at 24th Ann. Conf. on Mass Spectrom. and Allied Topics, San Diego, CA, U.S.A. May 9–13, 1976. Abstr. Z–6, p. 552.

76HE4'
LC/MS/COM Using an Unchanged Commercially Available Quadrupole Mass Spectrometer. J.D. Henion. Presented at 24th Ann. Conf. on Mass Spectrom. and Allied Topics, San Diego, CA, U.S.A., May 9–13, 1976. Abstr. S–6, p. 414–415.

76HO95
Analytical Systems and Analytical Methods Based upon Mass Spectrometry and Liquid Chromatography or Gas Chromatography. E.C. Horning, D.I. Carroll, I. Dzidic, K.D. Haegele, M.G. Horning and R.N. Stillwell. *Kagaku No Ryoiki Zokan 109*, 85–95, (1976).

76IR47
On the Evaporation of Small Ions from Charged Droplets. J.V. Iribarne and B.A. Thomson. *J. Chem. Phys. 64*, 2287–2294, (1976).

76MC29
Direct Analysis of Liquid Chromatographic Effluents. W.H. McFadden, H.L. Schwartz and D.C. Bradford. *J. Chromatogr. 122*, 389–396, (1976).

76MC9'
Liquid Chromatography-Mass Spectrometry System and Method. F.W. McLafferty and M.A. Baldwin. US 3,997,289 (CI. 23–253R; GO1N31/08), 14 Dec. 1976, Appl. 553,519, 27 Feb. 1975, 14p.

76MC63
Mass Spectrometer, Minicomputer, Liquid Chromatograph: An Ideal Combination. F.W. McLafferty. *Chem. Weekbl. Mag. 6*, 333,335,337, (1976).

76NI95
Recent Development in Mass Spectrometry. T. Nishishita. *Sekiyu Gakkaishi 19*, 555–559, (1976).

76TA91
Development of GC/MS and LC/MS Combination Analysis. A. Tatematsu, H. Miyazaki and M. Suzuki. *Kagaku to Kogyo (Tokyo) 29*, 891–896, (1976).

77AR9'
On-Line Liquid Chromatography-Mass Spectrometry. P.J. Arpino, H. Colin and G. Guiochon. Presented at 25th Ann. Conf. on Mass Spectrom. and Allied Topics, Washington D.C. U.S.A. May 29–June 3, 1977. Abstr. A 8, p. 189–190.

77BL8'
A New Crossed-Beam LC-MS for Involatile Samples. C.R. Blakley and M.L. Vestal. Presented at 25th Ann. Conf. on Mass Spectrom. and Allied Topics, Washington, D.C., U.S.A., May 29–June 3, 1977. Abstr. FP 12, p. 718–720.

77DA54
Fractionation of Coal Liquids by HPLC with Structural Characterization by LCMS. W.A. Dark, W.H. McFadden and D.C. Bradford. *J. Chromatogr. Sci. 15*, 454–464, (1977).

77DA87
Polypeptide Sequencing Utilizing a Liquid Chromatograph/Mass Spectrometer. B.G. Dawkins.

Thesis, 115p. Univ. Microfilms Int., Ann Arbor, MI 48106, U.S.A. Order no. 7806791. *Diss. Abstr. Int.B 38*, 5317, (1977).
77ER27
A New System for Lipid Analysis by Liquid Chromatography-Mass Spectrometry. W.L. Erdahl and O.S. Privett. *Lipids 12*, 797–803, (1977).
77GA5'
Applications of Mass Spectral Methods in Studies of Drugs and Their Metabolites. D.E. Games, J.L. Gower, M.G. Lee, I.A.S. Lewis, M.E. Pugh and M. Rossiter. Presented at 25th Ann. Conf. on Mass Spectrom. and Allied Topics, Washington, D.C., U.S.A. May 29–June 3, 1977. Abstr. FB 11, p. 535–536.
77HE7'
The Application of LC/MS/COM to the Separation and Identification of Biologically Important Substances. J.D. Henion. Presented at 25th Ann. Conf. on Mass Spectrom. and Allied Topics, Washington, D.C., U.S.A. May 29–June 3, 1977. Abstr. WD 8, p. 377–379.
77HO33
Development and Use of Bioanalytical Systems Based on Mass Spectrometry. E.C. Horning, D.I. Carroll, I. Dzidic, K.D. Haegele, S.N. Lin, C.U. Oertli and R.N. Stillwell. *Clin. Chem. 23*, 13–21, (1977).
77MC6'
Applications of Crossed-Beam LC-MS to Involatile Biological Samples. M.J. McAdams, C.R. Blakley and M.L. Vestal. Presented at 25th Ann. Conf. on Mass Spectrom. and Allied Topics, Washington D.C., U.S.A. May 29–June 3, 1977. Abstr. WD–7, p. 376.
77MC56
Applications of Combined Liquid Chromatography/Mass Spectrometry. W.H. McFadden, D.C. Bradford, D.E. Games and J.L. Gower. *Am. Lab.* 55–56, 58–60, 62–64, (1977).
77MC7'
Applications of Combined Liquid Chromatography/Mass Spectrometry. (LCMS). W.H. McFadden and D.C. Bradford. Presented at 25th Ann. Conf. on Mass Spectrom. and Allied Topics, Washington D.C., U.S.A. May 29–June 3, 1977. Abstr. FR 11, p. 717.
77MC87'
Liquid Chromatograph/Mass Spectrometer Interface. W.H. McFadden. US 4,055,987 (Cl 73–61.1C; GO1N31/08), 1 Nov. 1977, Appl. 664,058, 4 March 1976, 4p.
77SC13
Spectroscopic Detectors. R.P.W. Scott. *J. Chromatogr. Libr. 11*, 223–245, (1977).
78AR6'
Enrichment of the Sample through a Direct Liquid Inlet (DLI) Interface for OnLine LC/MS. P.J. Arpino and P. Krien. Presented at 26th Ann. Conf. on Mass Spectrom. and Allied Topics, St. Louis, MO, U.S.A. May 28–June 2, 1978. Abstr. RE 6, p. 426.
78BL81
Crossed-Beam Liquid Chromatograph-Mass Spectrometer Combination. C.R. Blakley, M.J. McAdams and M.L. Vestal. *J. Chromatogr. 158*, 261–276, (1978).
78BR7'
Ionization of Organic Substances in Liquid Chromatography-Mass Spectrometric Analysis Apparatus. C. Brunnee, J. Franzen and S. Meier. Ger. Offen. 2,654,057 (CI. GO1N31/08), 27 April 1978, Appl. 29 Nov. 1976, 5p.
78DA69
The Role of HPLC and LC/MS in the Separation and Characterization of Coal Liquefaction Products. W.A. Dark and W.H. McFadden. *J. Chromatogr. Sci. 16*, 289–293, (1978).
78DA9TS'
The Mass Spectrometer as a Detector for High-Pressure Liquid Chromatography. B.G. Dawkins and F.W. McLafferty. In: *GLC and HPLC Determination of Therapeutic Agents, Part 1*. (K. Tsuji and W. Morozowich, eds.) *Chromatogr. Sci. 9*, 259–275, (1978).

78DA51
Mass Spectrometric Studies of Peptides. 9. Polypeptide Sequencing by Liquid Chromatography Mass
Spectrometry. B.G. Dawkins, P.J. Arpino and F.W. McLafferty. *Biomed. Mass Spectrom.* 5, 1–
6, (1978).

78DO53
Electrospray Mass Spectrometry. M. Dole, H.L. Cox, Jr. and J. Gieniec. *Adv. Chem. Ser.* 125, 73–
84, (1978).

78DY9'
Application of HPLC/Positive-Negative CI/MS to the Analysis of PCB Mixtures and Individual PCB
Metabolites. P. Dymerski, D. Barnard and L. Kaminsky. Presented at 26th Ann. Conf. on Mass
Spectrom. and Allied Topics, St. Louis, MO, U.S.A. May 28–June 2, 1978. Abstr. RE 7, p. 429.

78GA6'
Combined Liquid Chromatography-Mass Spectrometry. D.E. Games, M.L. Games, E. Lewis, M.
Rossiter and N.C. Weerasinghe. Presented at 26th Ann. Conf. on Mass Spectrom. and Allied
Topics, St. Louis, MO, U.S.A. May 28–June 2, 1978. Abstr. RP 3, p. 656–658.

78GA5RE'
Scope of HPLC-MS and of Soft Ionization MS in Quantitation. D.E. Games, J.L. Gower, M.G. Lee,
I.A.S. Lewis, M.E. Pugh and M. Rossiter. In: *Blood Drugs and Other Analytical Challenges.* (E.
Reid and E. Horwood, eds.), pp. 185–194, Chichester, London (1978).

78GA19
Soft Ionization Mass Spectral Methods for Lipid Analysis: Appendix—Comments upon an Alternate
Sample Inlet System for LC/MS. D.E. Games. *Chem. Phys. Lipids 21*, 389–402, (1978).

78GA51
Some Applications of Newer Mass Spectral Techniques in the Analysis of Organic Compounds. D.E.
Games, J.L. Gower, M.G. Lee, I.A.S. Lewis, M.E. Pugh and M. Rossiter. *Anal. Proc. (London)*
15, 101–104, (1978).

78HA38
Development of Liquid Chromatography-Mass Spectrometry LC-MS. H. Hatano. *Kagaku (Kyoto)*
33, 738–741, (1978).

78HE75
The Application of LC/MS/COM to the Separation and Identification of Biologically Important
Substances. J.D. Henion. *Adv. Mass Spectrom. 7B*, 865–877, (1978).

78HE0'
Continuous Monitoring of Total LC Eluant by Direct Injection LC/MS/COM. J.D. Henion. Pre-
sented at 26th Ann. Conf. on Mass Spectrom. and Allied Topics, St. Louis, MO, U.S.A. May 28–
June 3, 1978. Abstr. E 2, p. 420–422.

78HE07
Drug Analysis by Continuously Monitored Liquid Chromatography/Mass Spectrometry with a Quadrupole Mass Spectrometer. J.D. Henion. *Anal. Chem. 50*, 1687–1693, (1978).

78HO03
Development and Use of Bioanalytical Systems Based on Mass Spectrometry with Ionization at
Atmospheric Pressure. E.C. Horning, D.I. Carroll, I. Dzidic and R.N. Stillwell. *Pure Applied
Chem. 50*, 113–127, (1978).

78KE33
Combination of Liquid Chromatography and Mass Spectrometry. E. Kenndler and E.R. Schmid. *J.
Chromatogr. Libr. 13*, 163–177, (1978).

78LO3'
Laser-Enhanced Ionization of Low Volatility Samples in Solution. E.R. Lory, B.G. Dawkins, P.J.
Arpino, P.A. Hoffman and F.W. McLafferty. Presented at 26th Ann. Conf. on Mass Spectrom.
and Allied Topics, St. Louis, MO, U.S.A. May 28–June 2, 1978. Abstr. RE 4, p. 423.

78MC8'
Applications of Crossed-Beam LC-MS to Involatile Biological Samples. M.J. McAdams, C.R.

Blakley and M.L. Vestal. Presented at 26th Ann. Conf. on Mass Spectrom. and Allied Topics, St. Louis, MO, U.S.A. May 28–June 2, 1978. Abstr. RE 1, p. 418–419.
78MC4'

Mixture Analysis Using Collision Induced Dissociation of Mass Selected Ions as an Alternative to HPLC/MS Analysis. C.N. McEwan. Presented at 26th Ann. Conf. on Mass Spectrom. and Allied Topics, St. Louis, MO, U.S.A. May 28–June 2, 1978. Abstr. RE 5, p. 424–425.
78MC2'

Applications of Combined Liquid Chromatography/Mass Spectrometry. W.H. McFadden, D.C. Bradford, G. Eglinton, S. Hajibrahim, W.A. Dark and N. Nicolaides. Presented at Conf. Mass Spectrom. and Allied Topics, St. Louis, MO, U.S.A. May 28–June 2, 1978. Abstr. RE 3, p. 422.
78ME77

Design, Operation and Application of a Novel LC/MS CI Interface. A. Melera. *Adv. Mass Spectrom. 7*, 1597–1615, (1978).
78MI4'

Intermediate System for Use in a Combined Liquid Chromatography-Mass Spectrometry System. H. Miyagi, F. Nakajima and Y. Arikawa. Ger. Offen. 2,728,944 (CI. G01N27/62), 12 Jan. 1978, Japan Appl. 76/76,362, 30 June 1976, 19p.
78PR11

Practical Aspects of Liquid Chromatography-Mass Spectrometry (LC-MS) of Lipids. O.S. Privett and W.L. Erdahl. *Chem. Phys. Lipids 21*, 361–387, (1978).
78RA03

Dense Gas Chromatograph/Mass Spectrometer Interface. L.G. Randall and A.L. Wahrhaftig. *Anal. Chem. 50*, 1703–1705, (1978).
78RY0'

Analyzing a Liquid Sample Using a Liquid Chromatograph and a Mass Spectrometer or a Mass Spectrograph. R. Ryhage. Ger. Offen. 2,811,300 (CI. G01N31/08), 5 Oct. 1978, Swed. Appl. 77/2,900, 15 March 1977, 19p.
78SE5'

A New LC/MS System. J.W. Serum and A. Melera. Presented at 26th Ann. Conf. on Mass Spectrom. and Allied Topics, St. Louis, MO, U.S.A. May 28–June 2, 1978. Abstr. RP 2, p. 655.
78ST52

Electrohydrodynamic Ionization Mass Spectrometry of Biochemical Materials. B.P. Stimpson and C.A. Evans, Jr. *Biomed. Mass Spectrom. 5*, 52–63, (1978).
78ST20

Mass Spectrometry of Solvated Ions Generated Directly from the Liquid Phase by Electrohydrodynamic Ionization. B.P. Stimpson, D.S. Simons and C.A. Evans, Jr. *J. Phys. Chem. 82*, 660–670, (1978).
78SU15

Development of Combined Liquid Chromatography/Mass Spectrometry. M. Suzuki. *Kagaku No Ryoiki Zokan 121*, 105–118, (1978).
78TA09

On-Line Coupling of a Micro Liquid Chromatograph and Mass Spectrometer Through a Jet Separator. T. Takeuchi, Y. Hirata and Y. Okumura. *Anal. Chem. 50*, 659–660, (1978).
78TA79

Round Table on High-Pressure Liquid Chromatography/Mass Spectrometry. V.L. Tal'roze, F.W. McLafferty, M. Story, J. Henion and P.J. Arpino. *Adv. Mass Spectrom. 7B*, 949–961, (1978).
78UD9'

An LC/MS Interface. H.R. Udseth, R.G. Orth and J.H. Futrell. Presented at 26th Ann. Conf. on Mass Spectrom. and Allied Topics, St. Louis, MO, U.S.A. May 28–June 2, 1978. Abstr. RP 4, p. 659.
78WR7'

Liquid Chromatography/Mass Spectrometric Methods of Analysis for Carbamate Pesticides. L.H.

Wright and E.O. Oswald. Presented at 26th Ann. Conf. on Mass Spectrom. and Allied Topics, St. Louis, MO, U.S.A. May 28–June 2, 1978. Abstr. MA 3, p. 47–48.

79AR12
LC/MS Coupling: Report. P.J. Arpino and G. Guiochon. *Anal. Chem. 51*, 682A–701A, (1979).

79AR1KL'
On-Line Liquid Chromatography Mass Spectrometry: The Monitoring of HPLC Effluents by a Quadrupole Mass Spectrometer and a Direct Liquid Interface. P.J. Arpino. In: *Proceedings of International Symposium on Instrumental Applications in Forensic Drug Chemistry*, 1978. (M. Klein, A.V. Kruegel and S.P. Sobol, eds.) pp. 151–156, Govt. Printing Office, Washington, D.C. (1979).

79AR59
Optimization of the Instrumental Parameters of a Combined LC-MS Coupled by an Interface for Direct Liquid Introduction. 1. Performance of the Vacuum Equipment. P.J. Arpino, G. Guiochon, P. Krien and G. Devant. *J. Chromatogr. 185*, 529–547, (1979).

79AR6'
Progress in LC/MS Interfacing Based on the DLI Concept. P.J. Arpino, G. Guiochon, G. Devant and P. Krien. Presented at 27th Ann. Conf. on Mass Spectrom. and Allied Topics, Seattle, WA, U.S.A. June 3–8, 1979. Abstr. TPMP 20, p. 366–367.

79BA9'
LC-MS Analysis of Drugs. J.D. Baty, D.A. Yorke and B.N. Green. Presented at 27th Ann. Conf. on Mass Spectrom. and Allied Topics, Seattle, WA, U.S.A. June 3–8, 1979. Abstr. TPMP 22, p. 369.

79BE6'
Method and Apparatus for Separating and Detecting Substances. A. Benninghoven. Ger. Offen. 2,732,746 (Cl. G01N27/62), 01 Mar. 1979, Appl. 20 July 1977, 14p.

79BL2'
LC/MS Interface Using Molecular Beam Techniques. C.R. Blakley, M.J. McAdams and M.L. Vestal. Presented at 27th Ann. Conf. on Mass Spectrom. and Allied Topics, Seattle, WA, U.S.A. June 3–8, 1979. Abstr. RPMOC 10, p. 622–623.

79CH0'
LC/MS Using Continuous Sample Pre-Concentration. R.G. Christensen, H.S. Hertz, S. Meiselman and E. White V. Presented at 27th Ann. Conf. on Mass Spectrom. and Allied Topics, Seattle, WA, U.S.A. June 3–8, 1979. Abstr. RPMOC 9 p. 620–621.

79DE8'
An HPLC-MS Study of Some Polycyclic Aromatic Hydrocarbons, Carbamates and Hydrazines. F.L. De Roos and R.L. Foltz. Presented at 27th Ann. Conf. on Mass Spectrom. and Allied Topics, Seattle, WA, U.S.A. June 3–8, 1979. Abstr. TPMP 15, p. 358–359.

79DY95
Analysis of Individual Polychlorinated Biphenyls (PCBs) and Their Hepatic Microsomal Metabolites by HPLC-MS. P. Dymerski, M. Kennedy and L. Kaminsky. NBS Spec. Publ. 519, p. 685–690, (1979).

79EV59
Directly Coupled Liquid Chromatography-Mass Spectrometry and Its Use in the Analysis of Biological Material. S. Evans. *Acta Pharm. Suec. 15*, 479, (1979).

79EV64
Electron Impact, Chemical Ionization and Field Desorption Mass Spectra of Some Anthraquinone and Anthrone Derivatives of Plant Origin. F.J. Evans, M.G. Lee and D.E. Games. *Biomed. Mass Spectrom. 6*, 374–380, (1979).

79GA7'
Combined LC/MS in Studies of Natural Products. D.E. Games, C. Eckers, P. Hirter, E. Lewis and K.R.N. Rao. Presented at 27th Ann. Conf. on Mass Spectrom. and Allied Topics, Seattle, WA, U.S.A. June 3–8, 1979. Abstr. RPMOC 13, p. 627–628.

79GA6'
* The Interface of Desorption Chemical Ionization (DCI) in Mass Spectral Ionization Methods. D.E. Games, J.L. Gower, E. Lewis, U. Rapp and G. Dielmann. Presented at 27th Ann. Conf. on Mass Spectrom. and Allied Topics, Seattle, WA, U.S.A. June 3–8, 1979. Abstr. TPMOC 11, p. 336–337.
79GA0'
Studies of Herbicides and Drugs by Combined LC/MS. D.E. Games, M.E. Knight and E. Lewis. Presented at Ann. Conf. on Mass Spectrom. and Allied Topics, Seattle, WA, U.S.A. June 3–8, 1979. Abstr. RPMOC 13, p. 350.
79HA26
Development of Gas Chromatography (GC)-Plasma Chromatography (PC)-Mass Spectrometry (MS)-Computer (COM)-High Performance Liquid Chromatography(HPLC) PC-MS-COM Systems. H. Hatano. *Bunseki 2*, 166–169, (1979).
79HI5'
High Efficiency Radially Compressed Liquid Chromatography/Mass Spectrometry (RCLC/MS)—Is It the Answer? D. Hilker, P.P. Dymerski and P. Champlin. Presented at 27th Ann. Conf. on Mass Spectrom. and Allied Topics, Seattle, WA, U.S.A. June 3–8, 1979. Abstr. RPMOC 12, p. 625–626.
79HI46
The Application of a New Sampling Technique Using an Atomizer for Chemical Ionization Mass Spectrometry of Free Amino Acid, Drug Components, High Phthalates and Oligomers of Styrene and Ethyleneglycol. Y. Hirata, T. Takeuchi, S. Tsuge and Y. Yoshida. *Org. Mass Spectrom. 14*, 126–128, (1979).
79HO1'
Liquid Chromatograph/Mass Spectrometer Interface. R.L. Horton. U.S. Patent US 4,160,161 (Cl. B01D 59/44), 3 July 1979, Appl. 30 May 1978, 5p.
79KA14
On-Line Reversed Phase Liquid Chromatography-Mass Spectrometry. B.L. Karger, D.P. Kirby, P. Vouros, R.L. Foltz and B. Hidy. *Anal. Chem. 51*, 2324–2328, (1979).
79KA6'
An On-Line Segmented Flow LC/MS Combination. B.L. Karger, D.P. Kirby, P. Vouros, B. Hidy and R.L. Foltz. Presented at 27th Ann. Conf. on Mass Spectrom. and Allied Topics, Seattle, WA, U.S.A. June 3–8, 1979. Abstr. RPMOC 6, p. 616.
79KE1'
LC/MS Analysis of Thermally Labile Compounds: LSD and Psilocybin. P.E. Kelley, R.F. Skinner and C.R. Philips. Finnigan Corp. Application Report No. 9. LC/MS PPINICI 4021 GC/MS 1979, Sept., 5p.
79LO2'
Laser-Enhanced Ionization of Involatile Samples in Solution. E.R. Lory and F.W. McLafferty. Presented at 27th Ann. Conf. on Mass Spectrom. and Allied Topics, Seattle, WA, U.S.A. June 3–8, 1979. Abstr. TPMOC 8, p. 332.
79MC88
Application of Crossed-Beam LC-MS to Polar and Volatile Compounds in Water. M.J. McAdams, C.R. Blakley and M.L. Vestal. *Am. Chem. Soc. Abstr. 178*, 58, (1979).
79MC8'
Environmental Applications of Crossed-Beam LC/MS. M.J. McAdams, C.R. Blakley and M.L. Vestal. Presented at 27th Ann. Conf. on Mass Spectrom. and Allied Topics, Seattle, WA, U.S.A. June 3–8, 1979. Abstr. RAMP 11, p. 548–549.
79MC78
Applications of Combined Liquid Chromatography/Mass Spectrometry (LC/MS): Analysis of Petroporphyrins and Meibomian Gland Waxes. W.H. McFadden, D.C. Bradford, G. Eglinton, S.K. Hajibrahim and N. Nicolaides. *J. Chromatogr. Sci. 17*, 518–520, (1979).

79MC72
Interfacing Chromatography and Mass Spectrometry. W.H. McFadden. *J. Chromatogr. Sci. 17*, 2–16, (1979).

79MC95
Gradient Pressure Chemical Ionization. F. McLafferty. *Govt. Rep. Announce. Index (U.S.) 79*, 75, (1979).

79ME3'
Chemical and Analytical Applications of an LC/MS Interface. A. Melera. Presented at 27th Ann. Conf. on Mass Spectrom. and Allied Topics, Seattle, WA, U.S.A. June 3–8, 1979. Abstr. TPMP 19, p. 363–365.

79SC95
The Wire Transport LC/MS System. R.P.W. Scott. In: *Trace Organic Analysis: A New Frontier in Anal. Chem. Proceedings of 9th Materials Res. Symp. Apr. 10–13, 1978, NBS, Gaithersburg, MD, U.S.A.* Govt. Printing Office, NBS Spec. Publ. 519, p. 635–645, (1979).

79SK1'
The Determination of Herbicide Photolysis Products by LC/MS: Application Report. R.F. Skinner, Q. Thomas, J. Giles and D.G. Crosby. Finnigan Appl. Report No. 8 LC/MS PPINICI 4021 GC/MS Aug. 1979, 4p.

79TS4'
Liquid Ionization at Atmospheric Pressure: Cluster Ions from Liquid Produced by Charge Transfer. M. Tsuchiya, T. Taira and K. Hiroaka. Presented at 27th Ann. Conf. on Mass Spectrom. and Allied Topics, Seattle, WA, U.S.A. June 3–8, 1979. Abstr. MAMP 13, p. 94–95.

79TS165
Analysis by Liquid Chromatography-Mass Spectrometry Combined System. S. Tsuge. *Bunseki Kagaku Kashukai Tekisuto 21*, 6–15, (1979).

79TS87
Analysis of Polar and/or Large Molecules by Directly Coupled LC/MS Using a New Vacuum Nebulizing Interface. S. Tsuge, Y. Yoshida, T. Takeuchi and Y. Hirata. *Am. Chem. Soc. Abstr. 178*, 57, (1979).

79TS14
Prospect of Liquid Chromatography-Mass Spectrometry Combined System. S. Tsuge. *Bunseki Kagaku Kohukai Tekisuto 21*, 14–19, (1979).

79TS16
Vacuum Nebulizing Interface for Direct Coupling of Micro-Liquid Chromatograph and Mass Spectrometer. S. Tsuge, Y. Hirata and T. Takeuchi. *Anal. Chem. 51*, 166–169, (1979).

79VE97
Techniques for Combined Liquid Chromatography-Mass Spectrometry. M.L. Vestal. *NBS Spec. Publ. 519*, 647–654, (1979).

79WE08
Use of Variable pH Interface to a Mass Spectrometer for the Measurement of Dissolved Volatile Compounds. J.C. Weaver and J.H. Abrams. *Rev. Sci. Instrum. 50*, 478–481, (1979).

79WR2'
Characterization of Underivatized Phenols in Human Urine by HPLC/MS. L.H. Wright and T.R. Edgerton. Presented at 27th Ann. Conf. on Mass Spectrom. and Allied Topics, Seattle, WA, U.S.A. June 3–8, 1979. Abstr. FAMP 15, p. 742.

79YO4'
A New HPLC/MS Interface. D.A. Yorke, P. Burns and D.S. Millington. Presented at 27th Ann. Conf. on Mass Spectrom. and Allied Topics, Seattle, WA, U.S.A. June 3–8, 1979. Abstr. RPMOC 11, p. 624.

79ZA38
Determination of the Chemical Composition of Petroleum Oils by Continuous Two-Stage Liquid Chromatographic and Mass Spectrometric Methods. V.A. Zakupra, V. Kozak, E.V. Kolosova and N.I. Vykhrestyuk. *Khim. Tekhnol. Topl. Masel 3*, 58–63, (1979).

79ZE19
The Combination Liquid Chromatography Mass Spectrometry. L.F. Zerilli. *Chromatogr. Symp. Ser.* *1*, 59–71, (1979).
80AH02
Ultratrace Analyses. S. Ahuja. *Chemtech. 10*, 702–707, (1980). ·
80AR84
LC/MS Analysis in Forensic Studies. Analysis of Extract of Cannabis Leaves. P.J. Arpino and P. Krien. *J. Chromatogr. Sci. 18*, 104, 112–115, (1980).
80BE59
Application of a Secondary Ion Mass Spectrometer as a Detector in Liquid Chromatography. A. Benninghoven, A. Eicke, M. Junack, W. Sichtermann, J. Krizek and H. Peters. *Org. Mass Spectrom. 15*, 459–462, (1980).
80BL2'
Combined Liquid Chromatograph-Mass Spectrometer for Involatile Biological Samples. C.R. Blakley, J.J. Carmody and M.L. Vestal. Presented at 28th Ann. Conf. on Mass Spectrom. and Allied Topics, New York, NY, U.S.A. May 25–30, 1980. Abstr. TPMOC 9, p. 312–313.
80BL67
Combined Liquid Chromatograph-Mass Spectrometer for Involatile Biological Samples. C.R. Blakley, J.J. Carmody and M.L. Vestal. *Clin. Chem. 26*, 1467–1473, (1980).
80BL26
Liquid Chromatograph-Mass Spectrometer for Analysis of Nonvolatile Samples. C.R. Blakley, J.J. Carmody and M.L. Vestal. *Anal. Chem. 52*, 1636–1641, (1980).
80BL86
A New Liquid Chromatograph/Mass Spectrometer Interface Using Crossed-Beam Techniques. C.R. Blakley, M.J. McAdams and M.L. Vestal. *Adv. Mass Spectrom. 8B*, 1616–1623, (1980).
80BL0'
A New Soft Ionization Technique for Complex Involatile Molecules. C.R. Blakley, J.C. Carmody and M.L. Vestal. Presented at 28th Ann. Conf. on Mass Spectrom. and Allied Topics, New York, NY, U.S.A. May 25–30, 1980. Abstr. TPMOC 13, p. 320–322.
80BL21
A New Soft Ionization Technique for Mass Spectrometry of Complex Molecules. C.R. Blakley, J.J. Carmody and M.L. Vestal. *J. Am. Chem. Soc. 102*, 5931–5933, (1980).
80BR65
A Packed Microbore Liquid Chromatography Column Used as a Direct Probe Inlet for a Chemical Ionization Mass Spectrometer. J.J. Brophy, D. Nelson and M.K. Withers. *Int. J. Mass Spectrom. Ion Phys. 36*, 205–212, (1980).
80BU24
Mass Spectrometry. A.L. Burlingame, T.A. Baillie, P.J. Derrick and O.S. Chizhov. *Anal. Chem. 52*, 214R–258R, (1980).
80CH8RE'
Combined LC/MS Techniques for Direct Quantitative Analysis of Individual Organic Compounds in Complex Mixtures. R.G. Christensen, H.S. Hertz, S. Meiselman and E. White V. In: *Technical Activities 1980, Center for Analytical Chemistry* (C.W. Reimann, R.A. Velapoldi, L.B. Hagen and J.K. Taylor, eds.), pp. 148–149, Govt. Printing Office, Washington D.C., NBS 80–2164, (1980).
80CH1RE'
Liquid Chromatography/Mass Spectrometry. R.G. Christensen, H.S. Hertz, W.E. May, S. Meiselman and E. White V. In: *Technical Activities 1979, Center for Analytical Chemistry* (C.W. Reimann and L. Hagen, eds.), p. 151, 157–158, Govt. Printing Office, Washington D.C., NBS 80–1995, (1980).
80DY4'
The First Viable EI/CI LC/MS Interface. P.P. Dymerski. Presented at 28th Ann. Conf. on Mass Spectrom. and Allied Topics, New York, NY, U.S.A. May 25–30, 1980. Abstr. RPMP 21, p. 624.

80EC86

Studies of Natural Products and Pesticides and Their Metabolites by LCMS and Other Mass Spectral Methods. C. Eckers, D.E. Games, E. Lewis, K.R.N. Rao, M. Rossiter and N.C.A. Weerasinghe. *Adv. Mass Spectrom. 8B*, 1396–1404, (1980).

80EG23

Petroporphyrins: Structural Elucidation and the Application of HPLC Fingerprintng to Geochemical Problems. G. Eglinton, S.K. Hajibrahim, J.R. Maxwell, J. Quirke and E. Martin. *Phys. Chem. Earth. 12* (Adv. Org. Geochem. 1979), 193–203, (1980).

80ER78

Analysis of Lipids by a New Liquid Chromatography-Mass Spectrometry Computer System. W.L. Erdahl, W.R. Beck, D.E. Jarvis and O.S. Privett. *J. Am. Oil Chem. Soc. 57*, A118, (1980).

80FI6'

Methods and Apparatus for Mass Spectrometric Analysis of Constituents in Liquids. W.L. Fite. U.S. 4,209,696 (CI. 250–281; BO1D59/44), 24 June 1980, Appl. 835,160, 21 Sept. 1977, 10p.

80GA86

Analysis of the Metabolites of CIPC (Chloropropham). D.E. Games and N.C.A. Weerasinghe. *J. Chromatogr. Sci. 18*, 106–107, 112–115, (1980).

80GA72

Applications of Combined High-Performance Liquid Chromatography Mass Spectrometry. D.E. Games. *Anal Proc. (London) 17*, 322–326, (1980).

80GA17

Combined HPLC-MS and Its Potential in Clinical Studies. D.E. Games, C. Eckers, J.L. Gower, P. Hirter, M.E. Knight, E. Lewis, K.R.N. Rao and N.C. Weerasinghe. *Clin. Res. Cent. Symp. (Harrow, Engl.) 1*, 97–118, (1980).

80GA70

Combined High-Performance Liquid Chromatography-Mass Spectrometry. D.E. Games. *Anal. Proc. (London) 17*, 110–116, (1980).

80GA8'

Combined LC/MS in Studies of Natural Products. D.E. Games, M.L. Games, C. Eckers, W. Kuhnz and E. Lewis. Presented at 28th Ann. Conf. on Mass Spectrom. and Allied Topics, New York, NY, U.S.A. May 25–30, 1980. Abstr. TPMOC 12, p. 318–319.

80GA73

Combined Liquid Chromatography Mass Spectrometry of Glycosides, Glucuronides, Sugars and Nucleosides. D.E. Games and E. Lewis. *Biomed. Mass Spectrom. 7*, 433–436, (1980).

80GA83

Structural and Quantitative Studies of Drugs and Their Metabolites by Combined LCMS. D.E. Games, E. Lewis, N.J. Haskins and K.A. Waddell. *Adv. Mass Spectrom. 8B*, 1223–1240, (1980).

80GA6'

Studies of Pesticides by Combined LC/MS. D.E. Games, N.C.A. Weerasinghe and S.A. Westwood. Presented at Ann. Conf. on Mass Spectrom. and Allied Topics, New York, NY, U.S.A. May 25–30, 1980. Abstr. TPMOC 11, p. 316–317.

80GR6'

Further Development of Particulate Impact Mass Spectrometry. F.T. Greene. Presented at 28th Ann. Conf. on Mass Spectrom. and Allied Topics, New York, NY, U.S.A. May 25–30, 1980. Abstr. TAMOA 10, p. 206.

80HA84

Factor Analysis (Principal Components and Matrix Rand Analyses) Applied to Some Arrays of Repetitively Scanned GCMS, LCMS and Direct Insertion Probe Mass Spectra. J.M. Halket. *Adv. Mass Spectrom. 8B*, 1554–1563, (1980).

80HA6'

Laser Desorption Mass Spectrometry of Involatile Samples. E.D. Hardin and M.L. Vestal. Presented

at 28th Ann. Conf. on Mass Spectrom. and Allied Topics, New York, NY, U.S.A. May 25–30, 1980. Abstr. RPMP 16, p. 616–617.
80HE815

A Comparison of Direct Liquid Introduction LC/MS Techniques Employing Microbore and Conventional Packed Columns. J.D. Henion. *J. Chromatogr. Sci. 18*, 101–102, 112–115, (1980).
80HE81

Direct Injection LCMS/COM of Total LC Eluants Applied to Drugs and Metabolism Studies. J.D. Henion. *Adv. Mass Spectrom. 8B*, 1241–1250, (1980).
80HE3TO′

Direct Injection Micro-Liquid Chromatography/Mass Spectrometry Applied to Equine Drug Testing. J.D. Henion. In: *Proc. 3rd. Int. Symp. Equine Med. Control, 1979*. (T. Tobin, J.W. Blake and W.E. Woods, eds.), pp. 133–140, Cornell Univ., New York, (1980).
80HE75

Drug Analysis by Direct Liquid Introduction Micro Liquid Chromatography-Mass Spectrometry. J.D. Henion and G.A. Maylin. *Biomed. Mass Spectrom. 7*, 115–121, (1980).
80HE08

Developments in the Analysis of Priority Pollutants by Liquid Chromatography Mass Spectrometry and by Immunoassay Procedures. H.S. Hertz, R.G. Christensen, D.J. Reeder and E. White V. *Am. Chem. Soc. Abstr. 180*, 88, (1980).
80HI27

The Hy-phen-ated Methods. T. Hirschfeld. *Anal. Chem. 52*, 297A–312A, (1980).
80HU0′

Laser Desorption CI Mass Spectrometry. D.F. Hunt, W.M. Bone and J. Shabanowitz. Presented at 28th Ann. Conf. on Mass Spectrom. and Allied Topics, New York, NY. May 25–30, 1980. Abstr. RPMP 18, p. 620.
80KA81

Reversed Phase LC/MS Using a Continuous Extraction Interface. B.L. Karger, D.P. Kirby and P. Vouros. *J. Chromatogr. 18*, 111–115, (1980).
80KE8′

Applications of the Baldwin-McLafferty LC/MS Interface to the Analysis of Molecules of Biochemical Importance. C.N. Kenyon. Presented at 28th Ann. Conf. on Mass Spectrom and Allied Topics, New York, NY, U.S.A. May 25–30, 1980. Abstr. RPMP 11, p. 608–609.
80KE83

Use of the Direct Liquid Inlet LC/MS System for the Analysis of Complex Mixtures of Biological Origin. C.N. Kenyon, A. Melera and F. Erni. *J. Chromatogr. Sci. 18*, 103–104, 112–115, (1980).
80KI4′

Ion Pair Reversed Phase Liquid Chromatography-Mass Spectrometry. D.P. Kirby, B.L. Karger, P. Vouros, B. Petersen and B. Hidy. Presented at 28th Ann. Conf. on Mass Spectrom. and Allied Topics, New York, NY, U.S.A. May 25–30, 1980. Abstr. TPMOC 10, p. 314–315.
80KI95

Ion Pairing Techniques: Compatibility with On-Line Liquid Chromatography Mass Spectrometry. D.P. Kirby, P. Vouros and B.L. Karger. *Science 209*, 495–497, (1980).
80LE87

Chemical Ionization, Field Ionization and Field Desorption. K. Levsen. *Adv. Mass Spectrom. 8A*, 897–917, (1980).
80LO4′

Ionization of Involatile Compounds by Solution Droplet Vaporization. E.R. Lory and F.W. McLafferty. Presented at 28th Ann. Conf. on Mass Spectrom. and Allied Topics, New York, NY, U.S.A. May 25–30, 1980. Abstr. FAMOB 2, p. 644–645.
80LO84

Ionization of Involatile Compounds by Submicrosecond Solution Vaporization. E.R. Lory and F.W. McLafferty. *Adv. Mass Spectrom. 8A*, 954–960, (1980).

80MA95
Analysis of Multicomponent Liquid Crystal Mixtures Using LC-MS-DS Techniques. T.I. Martin and W.E. Hass. *Am. Chem. Soc. Abstr. 179*, 55, (1980).

80MA84
Applications of LC/MS Techniques to the Analysis of Liquid Crystal Mixtures. T.I. Martin. *J. Chromatogr. Sci. 18*, 104–106, 112–115, (1980).

80MA9'
Mass Spectrometer Ion Source. T. Matsuo, I. Katakuse and H. Matsuda. Ger. Offen. 2,906,359 (CI. HO1J39/35), 14 Feb. 1980, JP. Appl. 78/98,575, 12 Aug 1978, 14p.

80MC80
Analysis of Priority Pollutants by Cross-Beam LC/MS. M.J. McAdams and M.L. Vestal. *J. Chromatogr. Sci. 18*, 110–111, 112–115, (1980).

80MC87
Liquid Chromatography/Mass Spectrometry Systems and Applications. W.H. McFadden. *J. Chromatogr. Sci. 18*, 97–115, (1980).

80MC19
Separation/Identification Systems Applicable to Complex Mixtures. F.W. McLafferty. In: *Biochem. Appl. Mass Spectrom.* (G.R. Waller and O.C. Dermer, eds.), pp. 1159–1168. (1st Suppl. Vol.), Wiley, New York. (1980).

80ME0'
Sample Insertion Device for a Mass Spectrometer. A. Melera and A. Neukermans. Ger. Offen. 3,013,620 (CI. GO1N27/62), 4 Dec. 1980, US Appl. 42,477, 25 May 1979, 11p.

80ME87
Design, Operation and Applications of a Novel LC/MS CI Interface. A. Melera and H. Weaver. *Adv. Mass Spectrom. 8B*, 1597–1615, (1980).

80MI89
A New Liquid Chromatography-Mass Spectrometry Interface. D.S. Millington, D.A. Yorke and P. Burns. *Adv. Mass Spectrom. 8B*, 1819–1825, (1980).

80NO9'
Method and Apparatus for the Analysis of Fluids Flowing Out of a Chromatograph. H.G. Noeller, H.D. Polaschegg and R. Wechsung. Ger. Offen. 2,837,799 (Cl. G0IN27/62), March 1980; *Chem. Abstr. 93*, 36487y (1980).

80PH10
Crossed Beam LC/MS Using Laser Vaporization for Samples of Environmental Interest. M.J. Phinney. Ph.D. Thesis. 134p. Univ. Microfilms Int., Ann Arbor, MI 48106, U.S.A. Order no. 8025017. *Diss. Abstr. Int. 41*, 1760–1765, (1980).

80PR0'
Simple Interface Probe for LC/MS. P.C. Price and S.L. Wellons. Presented at 28th Ann. Conf. on Mass Spectrom. and Allied Topics, New York, NY, U.S.A. May 25–30, 1980. Abstr. RPMP 12, p. 610–611.

80QU2'
Chemical Derivatization for Combined Liquid Chromatography-Mass Spectrometry. M.A. Quilliam and E.Y. Oseitwum. Presented at Ann. Conf. on Mass Spectrom. and Allied Topics, New York, NY, U.S.A. May 25–30, 1980. Abstr. RPMP 13, p. 612–613.

80RA4'
TLC/CIMS: A Thin-Layer Chromatogram Scanner Directly Coupled for Chemical Ionization Mass Spectrometry. L. Ramaley, W.D. Jamieson and R.G. Ackman. Presented at 28th Ann. Conf. on Mass Spectrom. and Allied Topics, New York, NY, U.S.A. May 25–30, 1980. Abstr. TPMP 3, p. 324.

80RA80
The Interface of Desorption Chemical Ionization (DCI) in Mass Spectral Ionization Methods. U. Rapp, G. Dielmann, D.E. Games, J.L. Gower and E. Lewis. *Adv. Mass Spectrom. 8B*, 1660–1668, (1980).

80RE2'
Method and Appartus for Analyzing Fluid. G. Renner, R. Wechsung and E. Unsoeld. Ger. Offen.
2,844,002 (Cl. G0IN27/62) May 1980; *Chem. Abstr. 93*, 197174j (1980).

80RO5'
Application of Combined LC/MS to the Analysis of Selected Cephalosporins and Penicillins. T.A.
Roy, F.L. DeRoos, B.J. Hidy and C.C. Howard. Presented at Ann. Conf. on Mass Spectrom. and
Allied Topics, New York, NY, U.S.A. May 25–30, 1980. Abstr. RPMP 15, p. 615.

80SC39
Separation and Positive Identification of Compounds in Complex Sample Mixtures Using On-Line
LC-UV/VIS and LC-MS Direct Coupling Techniques. R. Schuster. *Chromatographia 13*, 379–
385, (1980).

80SK88
The Determination of Herbicide Photolysis Products by LC/MS. R.F. Skinner, Q. Thomas, J. Giles
and D.G. Crosby. *J. Chromatogr. Sci. 18*, 108–109, 112–115, (1980).

80SK84
Mass Spectrometry of Ions Field Evaporated from Glycerolic and Aqueous Solutions. V.E. Skurat,
N.B. Zolotoi, G.V. Karpov, V.L. Tal'roze, Yu. V. Vasyuta and G.I. Ramendik. *Adv. Mass
Spectrom. 8A*, 1054–1060, (1980).

80SM97
Liquid Chromatography Mass Spectrometry of Thermally Unstable Compounds Using Secondary Ion
Emission. R.D. Smith and J.E. Burger. *Am. Chem. Soc. Abstr. 179*, 167, (1980).

80SM0'
An LC-MS Using Ion Impact. R.D. Smith and J.E. Burger. Presented at 28th Ann. Conf. on Mass
Spectrom. and Allied Topics, New York, NY, U.S.A. May 25–30, 1980. Abstr. TPMOC 8, p.
310–311.

80ST87
Application of LC/MS to Chromatographic Separation of Aromatics Using Carbon as a Stationary
Phase. D.L. Stalling, J.D. Petty, G.R. Dubay and R.A. Smith. *J. Chromatogr. Sci. 18*, 107–108,
112–115, (1980).

80TH4'
LC/MS Application to Tannery Effluent Analysis. A.D. Thruston and J.M. McGuire. Presented at
28th Ann. Conf. on Mass Spectrom. and Allied Topics, New York, NY, U.S.A. May 25–30,
1980. Abstr. RPMP 14, p. 614.

80TS85
Characteristics of Mass Spectra of Organic Compounds Obtained by the Liquid Ionization Method.
M. Tsuchiya, K. Seita and T. Taira. *Shitsuryo Bunseki 28*, 235–241, (1980).

80TS41
A New Method of Ionization for Organic Compounds in the Liquid Phase under Atmospheric
Pressure. M. Tsuchiya and T. Taira. *Int. J. Mass Spectrom. Ion Phys. 34*, 351–359,
(1980).

80TS01
Current Status of Directly Coupled Liquid Chromatography-Mass Spectrometry. S. Tsuge and Y.
Hirata. *Bunseki 10*, 751–754, (1980).

80TS05
A Directly Coupled Micro Liquid Chromatograph and Mass Spectrometer with Vacuum Nebulizing
Interface. S. Tsuge, Y. Yoshida and T. Takeuchi. *Chem. Biomed. Environ. Instrum. 10*, 405–
418, (1980).

80WE5'
Mass Spectroscopic Analysis of Organic Substances. R. Wechsung. Ger. Offen. 2,837,715 (Cl.
G0IN27/62) March 1980; *Chem Abstr. 93*, 36488z (1980).

80YA87
Direct Coupling Interface for High Sensitivity LC/MS. E. Yamauchi, T. Mizuno and K. Azuma.
Shitsuryo Bunseki 28, 227–234, (1980).

80YO36
Direct Measurement of Mass Fragmentograms for Eluents from a Micro-Liquid Chromatograph Using an Improved Nebulizing Interface. Y. Yoshida, H. Yoshida, S. Tsuge, T. Takeuchi and K. Mochizuki. *J. High Resolut. Chromatogr. Chromatogr. Commun.* 3, 16–20, (1980).

81AL24
Mass Spectrometric Analyzer for Individual Aerosol Particles. J. Allen and R.K. Gould. *Rev. Sci. Instrum.* 52, 804–809, (1981).

81AM53
Structural Elucidation, Using HPLC-MS and GLC-MS, of the Acidic Polysaccharide Secreted by *Rhizobium meliloti* Strain 1021. P. Aman, M. McNeil, L.-E. Franzen, A.G. Darvill and P. Albersheim. *Carbohydr. Res.* 95, 263–282, (1981).

81AR37
Optimization of the Instrumental Parameters of a Combined Liquid Chromatograph Mass Spectrometer Coupled by an Interface for Direct Liquid Introduction. II. Nebulization of Liquids by Diaphragms. P.J. Arpino, P. Krien, S. Vajta and G. Devant. *J. Chromatogr.* 203, 117–130, (1981).

81AY1
Liquid Chromatography-Mass Spectrometry Abstracts. (I.C. Aylott, ed.) PRM Science and Technology, London. Vols. 1–2, (Apr. 1981-Nov. 1983), Abstrs. 1–192.

81BA90
Direct Liquid Chromatography-Mass Spectrometry Analysis of Analgesics and Steroids. J.D. Baty and R.G. Willis. *J. Clin. Chem. Clin. Biochem.* 19, 610, (1981).

81BL5'
Simplified LC-MS Systems Using the Thermospray Technique. C.R. Blakley and M.L. Vestal. Presented at 29th Ann. Conf. on Mass Spectrom. and Allied Topics, Minneapolis, MN, U.S.A. May 24–29, 1981. Abstr. TPA 1, p. 275.

81BR8'
Device for Pretreatment of Dissolved Substances for Mass Spectroscopy Studies. C. Brunnee. Ger. Offen. 3,007,538 (Cl. GOIN1/28) Sept. 1981; *Chem Abstr.* 95 196892w (1981).

81BR0'
LC/MS Interface for EI and DCI Technique. C. Brunnee, L. Delgmann, G. Dielmann, W. Meyer and P. Thorenz. Presented at 29th Ann. Conf. on Mass Spectrom. and Allied Topics, Minneapolis, MN, U.S.A. May 24–29, 1981. Abstr. TP 4, p. 280.

81CA6'
Characterization and Optimization of Thermospray Ionization. J.J. Carmody, C.R. Blakley and M.L. Vestal. Presented at Ann. Conf. on Mass Spectrom. and Allied Topics, Minneapolis, MN, U.S.A. May 24–29, 1981. Abstr. MPC 7, p. 156.

81CA77
Atmospheric Pressure Ionization Mass Spectrometry. D.I. Carroll, I. Dzidic, E.C. Horning and R.N. Stillwell. *Appl. Spectrosc. Rev.* 17, 337–406, (1981).

81CH31
Liquid Chromatograph/Mass Spectrometer Interface with Continuous Sample Preconcentration. R.G. Christensen, H.S. Hertz, S. Meiselman and E. White V. *Anal. Chem.* 53, 171–174, (1981).

81CH3KE'
Developments in the Analysis of Priority Pollutants by Liquid Chromatography/Mass Spectrometry and Immunoassay Procedures. R.G. Christensen, H.S. Hertz, D.J. Reeder and E. White. In: *Adv. Ident. Anal. Org. Pollut. Water* (L.H. Kieth, ed.), Vol. 1, pp. 447–453, Ann Arbor Science, Ann Arbor, MI, (1981).

81DA97
Studies on Chemical Ionization Mass Spectrometry Using Solvent-Methane Mixtures as Reagents Gas. S. Daijima and Y. Iida. *Mass Spectrom.* 29, 277–286, (1981).

81DE83
Isolation and Characterization of Two New Modified Uracil Derivatives from Human Urine. N.C.

De, A. Mittelman, S.P. Dutta, C.G. Edmonds, E.E. Jenkins, J.A. McCloskey, C.R. Blakley, M.L. Vestal and G.B. Chedda. *Carbohydr. Nucleos. Nucleot.* 8(5), 363–389, (1981).
81DI5'
The Application of the Direct Liquid Introduction LC-MS Interface to Problems in Biochemistry. D.J. Dixon. In: *Proceedings of the International Symposium on High Performance Liquid Chromatography in Protein and Peptide Chemistry* (A. Henscher and K-P. Hupe, eds.), pp. 125–142, De Gruter, Berlin (1981).
81DV3'
Protein Sequencing by Liquid Flow Mass Spectrometry. E. Dvorin, P. Tao, J. Carmody, C.R. Blakley, D. Dyckes and M.L. Vestal. Presented at 29th Ann. Conf. on Mass Spectrom. and Allied Topics, Minneapolis, MN, U.S.A. May 24–29, 1981. Abstr. RAMOB 5, p. 493.
81EC79
Combined Liquid Chromatography-Mass Spectrometry in Studies of Drugs, Pesticides and Natural Products. C. Eckers, D.E. Games, M.L. Games, W. Kuhnz, E. Lewis, N.C.A. Weerasinghe and S.A. Westwood. *Anal. Chem. Symp. Ser.* 7, 169–182, (1981).
81ED9'
Characterization of Two New Modified Uracil Derivatives from Human Urine by Combined Liquid Chromatography-Mass Spectrometry. C.G. Edmonds, E.E. Jenkins, J.A. McCloskey, C.R. Blakley, M.L. Vestal, N.C. De, A. Farber, S.P. Dutta and G.B. Chedda. Presented at 29th Ann. Conf. on Mass Spectrom. and Allied Topics, Minneapolis, MN, U.S.A. May 24–29, 1981. Abstr. RAMOA 2, p. 469–470.
81ER64
Determination of Fatty Acid Composition via Chemical Ionization-Mass Spectrometry. W.L. Erdahl, W. Beck, C. Jones, D.E. Jarvis and O.S. Privett. *Lipids 16*, 614–622, (1981).
81EV86
The Construction and Use of Simple Interfaces for Combined Liquid Chromatography Mass Spectrometry. N. Evans and J.E. Williamson. *Biomed. Mass Spectrom.* 8, 316–321, (1981).
81FO20
A Mass Spectrometric Determination of Trace Elements in Aqueous Media without Preconcentration. G.O. Foss. Univ. Microfilms Int., Ann Arbor, MI 48106, U.S.A. Order no. 828816. *Diss. Abstr. Int. 42*, 2820, (1981).
81FR61
Bibliography [in Mass Spectrometry Instrumentation and Theory]. D.A. Freer. *Dyn. Mass Spectrom. 6*, 311–318, (1981).
81GA84
Combined High Performance Liquid Chromatography Mass Spectrometry. D.E. Games. *Biomed. Mass Spectrom. 8*, 454–462, (1981).
81GA6'
Micro-Column LC/MS. D.E. Games, M.S. Lant and S.A. Westwood. Presented at 29th Ann. Conf. on Mass Spectrom. and Allied Topics, Minneapolis, MN, U.S.A. May 24–29, 1981. Abstr. RAMOA 6, p. 476–477.
81GA31
Studies of Combined Liquid Chromatography-Mass Spectrometry with a Moving-Belt Interface. D.E. Games, P. Hirter, W. Kuhnz, E. Lewis, N.C.A. Weerasinghe and S.A. Westwood. *J. Chromatogr. 203*, 131–138, (1981).
81GA4'
Studies of Ergot Alkaloids Using LC/MS and MS/MS. D.E. Games, C. Eckers, B.P. Swann and D.N.B. Mallen. Presented at 29th Ann. Conf. on Mass Spectrom. and Allied Topics, Minneapolis, MN, U.S.A. May 24–29, 1981. Abstr. RAMOA 10, p. 484–485.
81GA92
Quantitative Selected Ion Monitoring (QSIM) of Drugs and/or Drug Metabolites in Biological Matrices. W.A. Garland and M.C. Powell. *J. Chromatogr. Sci. 19*, 392–434, (1981).

81GA4MO'
Combined High Performance Liquid Chromatography-Mass Spectrometry (LC-MS). D.E. Games.
In: *Soft Ionization in Biological Mass Spectrometry, Proc. Chem. Soc. Symp.* 1980 (H.R. Morris,
ed.), p. 54–68, Heyden, London (1981).
81GI87
Electrodynamic Effects in Field Desorption Mass Spectrometry. U. Giessmann and F.W. Röllgen.
Int. J. Mass Spectrom. Ion Phys. 38, 267–279, (1981).
81GR62
The Use of an Inductively Coupled Plasma as an Ion Source for Aqueous Solution Samples. A.L.
Gray and A.R. Date. *Dyn. Mass Spectrom. 6*, 252–266, (1981).
81GR1'
The Current Status of Particulate Impact Mass Spectrometry. F.T. Greene. Presented at 29th Ann.
Conf. on Mass Spectrom. and Allied Topics, Minneapolis, MN, U.S.A. May 24–29, 1981. Abstr.
RAMOA 3, p. 471.
81HA32
Laser Ionization Mass Spectrometry of Nonvolatile Samples. E.D. Hardin and M.L. Vestal. *Anal.
Chem. 53*, 1492–1497, (1981).
81HA9'
Laser Ionization Mass Spectrometry of Nonvolatile Molecules. E.D. Hardin and M.L. Vestal.
Presented at 29th Ann. Conf. on Mass Spectrom. and Allied Topics, Minneapolis, MN, U.S.A.
May 24–29, 1981. Abstr. MAMO 85, p. 29.
81HE4'
The Construction and Use of a New DLI Micro LC/MS Diaphragm Interface. J.D. Henion. Pre-
sented at 29th Ann. Conf. on Mass Spectrom. and Allied Topics, Minneapolis, MN, U.S.A. May
24–29, 1981. Abstr. RAMOA 8, p. 474–475.
81HE97
Continuous Monitoring of Total Micro LC Eluant by Direct Liquid Introduction LC/MS. J.D.
Henion. *J. Chromatogr. Sci. 19*, 57–64, (1981).
81HE33
Micro Liquid Chromatography/Mass Spectrometry Diaphragm Probe Interface. J.D. Henion and T.
Wachs. *Anal. Chem. 53*, 1963–1965, (1981).
81HE8'
Routine Direct Liquid Introduction LC/MS Applied to Drug Analysis. J.D. Henion. Presented at
29th Ann. Conf. on Mass Spectrom. and Allied Topics, Minneapolis, MN, U.S.A. May 24–29,
1981. Abstr. RAMOA 7, p. 478–479.
81HO88
The Use of Combined High Performance Liquid Chromatography Negative Ion Chemical Ionization
Mass Spectrometry to Confirm the Administration of Synthetic Corticosteroids to Horses. E.
Houghton, M.C. Dumasia and J.K. Welby. *Biomed. Mass Spectrom. 8*, 558–564, (1981).
81HO7'
Mass Spectrometric Analysis of Solutions with an Inductively Coupled Plasma Ion Source. R.S.
Houk, H.J. Svec and V.A. Fassel. Presented at 29th Ann. Conf. on Mass Spectrom. and Allied
Topics, Minneapolis, MN, U.S.A. May 24–29, 1981. Abstr. TAMOC 1, p. 197.
81II29
CI Mass Spectra of Intact Sample Solution of Labile Compounds. Y. Iida and S. Okada. *Seikei
Daigaku Kogakubu Kogaku Hokoku 32*, 2189–2190, (1981).
81IR9'
Intense Molecular Ions from Labile and Polar Compounds by Ion Evaporation. J.V. Iribarne, P.
Dziedzic and B.A. Thomson. Presented at 29th Ann. Conf. on Mass Spectrom. and Allied Topics,
Minneapolis, MN, U.S.A. May 24–29, 1981. Abstr. RPMOA 4, p. 519–520.
81IR4'
Method and Apparatus for the Analysis of Chemical Compounds in Aqueous Solution by Mass

Spectroscopy of Evaporating Ions. J.V. Iribarne. U.S. 4,300,044 (Cl 250–282; BO1D59/44), 10 Nov. 1981. Appl. 147,45, 7 May 1980, 7p.
81JE3'
Liquid Chromatograph-Mass Spectrometer. Jeol Ltd. Jpn. Kokai Tokkyo Koho JP 81, 61,643 (CI. GO1N27/62), 27 May 1981, Appl. 79/137,951, 25 Oct. 1979, 3p.
81JE5'
Liquid Chromatograph-Mass Spectrometer. Jeol Ltd. Jpn. Kokai Tokkyo Koho JP 81, 78,055, 26 June 1981, Appl. 79/155,105, 30 Nov. 1979, 3p.
81KE6'
Use of Non-Volatile Buffer Solutions with LC/MS. P. Kelley. Presented at 29th Ann. Conf. on Mass Spectrom. and Allied Topics, Minneapolis, MN, U.S.A. May 24–29, 1981. Abstr. TPA 2, p. 276–277.
81KE56
Utilization of Direct Liquid Inlet LC/MS in Studies of Pharmacological and Toxicological Importance. C.N. Kenyon, A. Melera and F. Erni. *J. Anal. Toxicol. 5*, 216–230, (1981).
81KI39
On-Line Liquid Chromatography-Mass Spectrometry of Ion Pairs. D.P. Kirby, P. Vouros, B.L. Karger, B. Hidy and B. Petersen. *J. Chromatogr. 203*, 139–152, (1981).
81LA5'
Recent Progress in Ecdysteroid Analytical Methods. R. Lafont, P. Beydon, B. Mauchamp, G. Somme-Martin, M. Andrianjafintrimo and P. Krien. *Sci. Papers Inst. Org. Phys. Chem. Wroclaw Tech. Univ. Conf. 7*, 125–144, (1981).
81MA11
The Use of On-Line High Performance Liquid Chromatography Mass Spectrometry (HPLC-MS) for the Identification of Ranitidine and Its Metabolites in Urine. L.E. Martin, J. Oxford and R.J.N. Turner. *Xenobiotica 11*, 831–840, (1981).
81MA33
Analysis of Liquid Crystal Mixtures. T.I. Martin and W.E. Haas. *Anal. Chem. 53*, 593A–602A, (1981).
81MA51
Analysis of Juvenile Hormones by High Performance Liquid Chromatography Coupled with Mass Spectrometry. B. Mauchamp, R. Lafont and P. Krien. *Dev. Endocrinol. 15*, 21–31, (1981).
81MC63
Development and Trends in Instrumentation in Mass Spectrometry. A. McCormick. *Mass Spectrom. 6*, 153–172, (1981).
81MC39
Mass Spectrometry as a Separation Technique: Analysis of Involatile Samples. F.W. McLafferty and E.R. Lory. *J. Chromatogr. 203*, 109–116, (1981).
81MC40
Tandem Mass Spectrometry. F.W. McLafferty. *Science 214*, 280–287, (1981).
81ME66
Gas Chromatography-Mass Spectrometry and High Performance Liquid Chromatography Mass Spectrometry. F.A. Mellon. *Mass Spectrom. 6*, 196–232, (1981).
81NA7'
Application of the Ion-Drift Spectrometer to Macromass Spectrometry III. Final Report and Conclusions. K. Nakamae and M. Dole. Presented at 29th Ann. Conf. on Mass Spectrom.and Allied Topics, Minneapolis, MN, U.S.A. May 24–29, 1981. Abstr. RPMOA3, p. 517–520.
81NO34
Microcolumns in Liquid Chromatography. M. Novotny. *Anal. Chem. 53*, 1294A–1308A, (1981).
81OE14
Liquid Chromatography/Mass Spectrometry in Combination. G. Oestvold. *Kjemi 1*, 14–15, 34, (1981).

81OK97
CI Mass Spectra of Intact Labile Compounds by Solution Sample Introduction Method. S. Okada and Y. Iida. *Mass Spectrosc. 29*, 287–294, (1981).
81OR0'
Secondary Ion Mass Spectrometry of Frozen Aqueous Solutions. R.G. Orth and J. Michl. Presented at 29th Ann. Conf. on Mass Spectrom. and Allied Topics, Minneapolis, MN, U.S.A. May 24–29, 1981. Abstr. RPMOC 5, p. 560–561.
81OT66
Anionization of Volatile Molecules on the Surface of Electrolyte Solutions Exposed to High Electric Fields. K.H. Ott, F.W. Röllgen, P. Dähling, J.J. Zwinselman, R.H. Fokkens and N.M.M. Nibbering. *Org. Mass Spectrom. 16*, 336–338, (1981).
81PR26
Lipid Analysis by Liquid Chromatography-Mass Spectrometry. O.S. Privett and W.L. Erdahl. *Methods Enzymol. 72*, 56–108, (1981).
81QI28
Application of Combined LC/MS. Z. Qiao. *Mass Spectrom. 2*, 28–36, (1981).
81RA5'
A Thin Layer Chromatogram Scanner Coupled to a Mass Spectrometer for TLC-MS. L. Ramaley, M.E. Nearing, W.D. Jamieson and R.G. Ackman. Presented at 29th Ann. Conf. on Mass Spectrom. and Allied Topics, Minneapolis, MN, U.S.A. May 24–29, 1981. Abstr. MPB 8, p. 135–136.
81RA23
Direct Coupling of a Dense (Supercritical) Gas Chromatograph to a Mass Spectrometer Using a Supersonic Molecular Beam Interface. L.G. Randall and A.L. Wahrhaftig. *Rev. Sci. Instrum. 52*, 1283–1295, (1981).
81SC65
Direct Coupling of a Micro High-Performance Liquid Chromatograph and a Mass Spectrometer. K.H. Schäfer and K. Levsen. *J. Chromatogr. 206*, 245–252, (1981).
81SM2'
Analysis of Biomass Samples Using an LCMS Incorporating SIMS, Ribbon Storage Techniques, and a Triple Quadrupole Mass Spectrometer. R.D. Smith and A.L. Johnson. Presented at 29th Ann. Conf. on Mass Spectrom. and Allied Topics, Minneapolis, MN, U.S.A. May 24–29, 1981. Abstr. RAMOA 9, p. 482–483.
81SM07
The Application of Collision-Induced Dissociation Techniques in a New Liquid Chromatograph-Mass Spectrometer Using Both Electron-Impact and Ion Bombardment Ionization. R.D. Smith and A.L. Johnson. *Am. Chem. Soc. Abstr. 180*, 137, (1981).
81SM39
Deposition Method for Moving Ribbon Liquid Chromatograph-Mass Spectrometer Interface for On-Line Chromatography-Mass Spectroscopy Analysis. R.D. Smith and A.L. Johnson. *Anal. Chem. 53*, 739–740, (1981).
81SM5'
Design and Development of a New LC-MS Using SIMS and Collision Induced Dissociation. R.D. Smith. Presented at 29th Ann. Conf. on Mass Spectrom. and Allied Topics, Minneapolis, MN, U.S.A. May 24–29, 1981. Abstr. MPC 12, p. 165–166.
81SM33
Liquid Chromatography-Mass Spectrometry with Electron Impact and Fast Ion Bombardment with a Ribbon Storage Interface. R.D. Smith, J.E. Burger and A.L. Johnson. *Anal. Chem. 53*, 1603–1611, (1981).
81SM30
Ribbon Storage Techniques for Liquid Chromatography Mass Spectrometry. R.D. Smith and A.L. Johnson. *Anal. Chem. 53*, 1120–1122, (1981).

81SU7'
Characterization of Natural Mixtures of Sterol Peroxides by LC/MS Using Direct Liquid Inlet (DLI) Interface. F.R. Sugnaux, A.A.L. Gunatilaka and C. Djerassi. Presented at 29th Ann. Conf. on Mass Spectrom. and Allied Topics, Minneapolis, MN, U.S.A. May 24–29, 1981. Abstr. RAMOA 1, p. 467–468.

81TH87
Characterization of Organics in Tannery Effluents by Liquid Chromatography Mass Spectrometry. A.D. Thruston and J.M. McGuire. *Biomed. Mass Spectrom. 8*, 47–50, (1981).

81TI85
Microcapillary Liquid Chromatography in Open Tubular Columns with Diameters of 10–50 Micron: Potential Application to Chemical Ionization Mass Spectrometric Detection. R. Tijssen, J.P.A. Bleumer, A.L.C. Smit and M.E. Van Krefeld. *J. Chromatogr. 218*, 137–165, (1981).

81TS15
Characteristics of a Liquid Ionization Detector for Analyzing Minute Amounts of Organic Compounds in Solution. M. Tsuchiya, K. Kawabe, Y. Toyoura, T. Taira, S. Tanaka, Y. Saito and W. Otake. *Nippon Kagaku Kaishi 1*, 145–149, (1981).

81TS2'
Liquid Ionization Mass Spectrometry-The Effect of Solvents. M. Tsuchiya, Y. Sugano, T. Taira and Y. Saito. Presented at 29th Ann. Conf. on Mass Spectrom. and Allied Topics, Minneapolis, MN, U.S.A. May 24–29, 1981. Abstr. RAMOA 4, p. 472–473.

81TS25
On-Line System of Liquid Chromatography and Mass Spectrometry: Current Status and Future Trends of Direct Coupling of a Liquid Chromatograph and a Mass Spectrometer. S. Tsuge. *Kagaku No Ryoiki Nokan 132*, 155–170, (1981).

81TS66
Progress in Direct Connection of a Liquid Chromatograph and a Mass Spectrometer. S. Tsuge. *Kagaku (Kyoto) 36*, 226–228, (1981).

81TW8'
LC/MS of Lipophilic Compounds Using Non-Aqueous Reversed-Phase Chromatography. P.A. Tway and W.B. Caldwell. Presented at 29th Ann. Conf. on Mass Spectrom. and Allied Topics, Minneapolis, MN, U.S.A. May 24–29, 1981. Abstr. TPA 3, p. 278–279.

81UN36
Identification of Quaternary Alkaloids in Mushroom by Chromatography/Secondary Ion Mass Spectrometry. S.E. Unger, A. Vincze, R.G. Cooks, R. Chrisman and L.D. Rothman. *Anal. Chem. 53*, 976–981, (1981).

81VE09
Liquid Chromatography-Mass Spectrometry. M.L. Vestal. *Am. Chem. Soc. Abstr. 180*, 79, (1981).

81VO7'
LC-MS-NCI of Explosives. R.D. Voyksner, Y. Tondeur, C.E. Parker, J.D. Henion and J. Yinon. Presented at 29th Ann. Conf. on Mass Spectrom. and Allied Topics, Minneapolis, MN, U.S.A. May 24–29, 1981. Abstr. TPA13, p. 297–298.

81WE43
Use of Circular Dichroism as a High-Performance Liquid Chromatography Detector. S.A. Westwood, D.E. Games and L. Sheen. *J. Chromatogr. 204*, 103–107, (1981).

81WH6'
Continuous-Flow Solution Concentrator and Liquid Chromatograph/Mass Spectrometer Interface and Method for Using Both. E. White V, H.S. Hertz and R.G. Christensen. US 4,281,246 (Cl. 250–282, BO1D 59/44) 28 July 1981, (EPA, U.S.A.) Appl. 84,273, 12 Oct. 1979, 7p.

81WI40
Rapid Combined Liquid Chromatography/Mass Spectrometry Method for Quantitative Analysis of

Atrazine. J.E. Williamson and N. Evans. *J. High Resolut. Chromatogr. Chromatogr. Commun. 4*, 130–131, (1981).

81WR85

The Determination of Underivatized Chlorophenols in Human Urine by High Performance Liquid Chromatography Mass Spectrometry and Selected Ion Monitoring. L.H. Wright, T.R. Edgerton, S.J. Arbes, Jr. and E.M. Lores. *Biomed. Mass Spectrom. 8*, 475–479, (1981).

81YU89

Analysis of N-Acetyl-N,O,S-Permethylated Peptides by Combined Liquid Chromatography-Mass Spectrometry. T.J. Yu, H. Schwartz, R.W. Giese, B.L. Karger and P. Vouros. *J. Chromatogr. 218*, 519–533, (1981).

82AB47

High Performance Liquid Chromatographic Resolution and Quantification of a Dilactonic Antibiotic Mixture (Antimycin A). S.L. Abidi. *J. Chromatogr. 234*, 187–200, (1982).

82AL15

High Performance Liquid Chromatography-Mass Spectrometry with Transport Interfaces. N.J. Alcock, C. Eckers, D.E. Games, M.P.L. Games, M.S. Lant, M.A. McDowall, M. Rossiter, R.W. Smith, S.A. Westwood and H.Y. Wong. *J. Chromatogr. 251*, 165–174, (1982).

(306)

82AL99

Liquid Chromatography/Mass Spectrometry Using Glass Lined Stainless Steel Microbore Columns. N.J. Alcock, L. Corbelli, D.E. Games, M.S. Lant and S.A. Westwood. *Biomed. Mass Spectrom. 9*, 499–504, (1982).

82AR3'

The Injection of Liquid Solution into a CI Source under LC/MS Conditions. Application to Quaternary Ammonium Salts in an Acetonitrile Solution. P.J. Arpino, G. Guiochon, J.P. Bounine and M. Dedieu. Presented at 30th Ann. Conf. on Mass Spectrom. and Allied Topics, Honolulu, HI, U.S.A. June 6–11, 1982. Abstr. ROC 4, p. 603.

82AR14

On-Line Liquid Chromatography/Mass Spectrometry? An Odd Couple. P.J. Arpino. *Trends Anal. Chem. 1*, 154–158, (1982).

82AR13

Optimization of the Instrumental Parameters of a Combined Liquid Chromatograph Mass Spectrometer Coupled by an Interface for Direct Liquid Introduction. III. Why the Solvent Should Not Be Removed in Liquid Chromatographic-Mass Spectrometric Interfacing Methods. P.J. Arpino and G. Guiochon. *J. Chromatogr. 251*, 153–164, (1982).

82AR96

Trying to Arrange a Difficult Marriage: A Report on the Workshop on Liquid Chromatography-Mass Spectrometry. P. Arpino. *Biomed. Mass Spectrom. 9*, 176–180, (1982).

82AR51

Chromatography Combined with Mass Spectrometry. P. Arpino, M. Dedieu, J.P. Bounine and G. Guiochon. *Int. J. Mass Spectrom. Ion Phys. 45*, 161–172, (1982).

82BA91

LC-MS Detection of Steroids and Drugs at Low and Medium Resolution. J.D. Baty and R.G. Willis. *Anal. Proc. (London) 19*, 251–253, (1982).

82BE79

A Study of the Formation of High Molecular Weight Water Cluster Ions *J. Chem. Phys. 77*, 2549–2557, (1982).

82BL7'

Design and Performance of LC-MS Systems Using the Thermospray Technique. C.R. Blakley and M.L. Vestal. Presented at 30th Ann. Conf. on Mass Spectrom. and Allied Topics, Honolulu, HI, U.S.A. June 6–11, 1982. Abstr. MPA 11, p. 117.

82BO47
LC Detectors: The Search Is On for the Ultimate Detector. S.A. Borman. *Anal. Chem. 54*, 327A–332A, (1982).

82BR44
Experiments with the Combination of Micro Liquid Chromatography and a Chemical Ionization Quadrupole Mass Spectrometer. A.P. Bruins and B.F.H. Drenth. *Pharm. Weekbl.-Sci. Ed. 4*, 204, (1982).

82BU43
Mass Spectrometry. A.L. Burlingame, A. Dell and D.H. Russell. *Anal. Chem. 54*, 363R–409R, (1982).

82BU87
Mass Spectrometry of Large, Fragile and Involatile Molecules. K.L. Busch and R.G. Cooks. *Science 218*, 247–253, (1982).

82CA46
Characterization of Glucuronides by Chemical Ionization Mass Spectrometry with Ammonia as Reagent Gas. T. Cairns and E.G Siegmund. *Anal. Chem. 54*, 2456–2461, (1982).

82CA43
Liquid Chromatography/Mass Spectrometry of Kepone Hydrate, Kelevan and Mirex. T. Cairns, E.G. Siegmund and G.M. Doose. *Anal. Chem. 54*, 953–957, (1982).

82CA4′
Characterization of Polymers and Polymeric Components Using Direct Inlet (DLI) Interface. D.C. Canada and D.S. Zing. Presented at 30th Ann. Conf. on Mass Spectrom. and Allied Topics, Honolulu, HI, U.S.A. June 6–11, 1982. Abstr. MPA 14, p. 124.

82CH5′
Probing Ion-Polymer Interactions by Electrohydrodynamic Ionization Mass Spectrometry. K.W.S. Chan and K.D. Cook. Presented at 30th Ann. Conf. on Mass Spectrom. and Allied Topics, Honolulu, HI, U.S.A. June 6–11, 1982. Abstr. ROE 7, p. 645–647.

82CH41
Observation of Some Transition Metal Complexes in Solution by Electrohydrodynamic Ionization Mass Spectrometry. K.W.S. Chan and K.D. Cook. *J. Am. Chem. Soc. 104*, 5031–5034, (1982).

82CH7′
Quantitative Trace Analysis by Reversed Phase LC/MS Employing Continuous Sample Pre-Concentration. R.G. Christensen, H.S. Hertz, S. Meiselman and E. White V. Presented at 30th Ann. Conf. on Mass Spectrom. and Allied Topics, Honolulu, HI, U.S.A. June 6–11, 1982. Abstr. ROC 1, 597–598.

82CU57
Recent Developments in LC/MS Interconnection. Z.F. Curry. *J. Liq. Chromatogr. 5 (Suppl.2)*, 257–272, (1982).

82DE13
Application of a Combined Liquid Chromatographic-Mass Spectrometric Instrument Using an Interface for Direct Liquid Introduction. M. Dedieu, C. Juin, P. Arpino, J.P. Bounine and G. Guiochon. *J. Chromatogr. 251*, 203–213, (1982).

82DE5′
Application of a Direct LC/MS System Utilizing a Special Desolvation Chamber. M. Dedieu, G. Devant, C. Juin, M. Hardy, J.P. Bounine, P.J. Arpino and G. Guiochon. Presented at 30th Ann. Conf. on Mass Spectrom. and Allied Topics, Honolulu, HI, U.S.A. June 6–11, 1982. Abstr. MPA 15, p. 125.

82DE42
Soft Negative Ionization of Non-Volatile Molecules by Introduction of Liquid Solutions into a Chemical Ionization Mass Spectrometer. M. Dedieu, C. Juin, P.J. Arpino and G. Guiochon. *Anal. Chem. 54*, 2372–2375, (1982).

82DE47'
Ion Detector for Liquid Chromatography. Denshi Kagaku Ltd. Jpn. Kokai Tokkyo Koho JP 82 13,347 (Cl. G01N27/62), 23 Jan. 1982, Appl. 80/87,582, 27 June 1980, 3p.

82DE03
Liquid Chromatography Mass Spectrometry (LC-MS)-Current Problems and New Applications of the Method. G. Devant. *Anal. Biol. Clin. 40*, 373, (1982).

82DI03
Application of Direct Coupling of Liquid Phase Chromatography and of Mass Spectroscopy to Biochemical Problems. D.J. Dixon. *Analusis 10*, 343–348, (1982).

82DU2'
A Study of Sensitivity vs. Ion Source Temperature for Direct Liquid Introduction LC-MS. G.R. Dubay and M. Cadiz. Presented at 30th Ann. Conf. on Mass Spectrom. and Allied Topics. Honolulu, HI, U.S.A. June 6–11, 1982, Abstr. MPA 8, p. 112–113.

82EC2'
The Determination of Steroids and Related Compounds by Micro LC/MS. C. Eckers, D. Skrabalak and J.D. Henion. Presented at 30th Ann. Conf. on Mass Spectrom. and Allied Topics, Honolulu, HI U.S.A. June 6–11, 1982. Abstr. ROC 9, p. 612–613.

82EC82
On-Line Direct Liquid Introduction Interface for Micro-Liquid Chromatography/Mass Spectrometry: Application to Drug Analysis. C. Eckers, D. Skrabalak and J. Henion. *Clin. Chem. 28*, 1882–1886, (1982).

82EC92
Studies of Ergot Alkaloids Using High-Performance Liquid Chromatography Mass Spectrometry and B/E Linked Scans. C. Eckers, D.E. Games, D.N.B. Mallen and B.P. Swann. *Biomed. Mass Spectrom. 9*, 162–173, (1982).

82EC93
Studies of Ergot Alkaloids Using High Performance Liquid Chromatography Mass Spectrometry and Mass Spectrometry. C. Eckers, D.E. Games, D.N.B. Mallen and B.P. Swann. *Anal. Proc. (London) 19*, 133–137, (1982).

82ED6'
Nucleoside Analysis by Combined Liquid Chromatography-Mass Spectrometry: Transfer RNA Hydrolysates. C.G. Edmonds, H. Pang, J.A. McCloskey, C.R. Blakley and M.L. Vestal. Presented at 30th Ann. Conf. on Mass Spectrom. and Allied Topics, Honolulu, HI, U.S.A. June 6–11, 1982. Abstr. ROC 6, p. 606–607.

82ER98
Analysis of Lipids by High Performance Liquid Chromatography Via Flame Ionization Detector and Mass Spectrometry. W.L. Erdahl, F.O. Phillips, W.R. Anderson and O.S. Privett. *J. Am. Oil. Chem. Soc. 59*, 308H (1982).

82ER11
Liquid Chromatography-Mass Spectrometry in the Pharmaceutical Industry: Objectives and Needs. F. Erni. *J. Chromatogr. 251*, 141–151, (1982).

82ES8'
LC-MS of Nucleosides Using a Commercially Available DLI-Probe. E.L. Esmans, Y. Luyten, F.C. Alderweireldt, P. Krien and G. Devant. Presented at 30th Ann. Conf. on Mass Spectrom. and Allied Topics, Honolulu, HI, U.S.A. June 6–11, 1982. Abstr. ROC 7, p. 608–609.

82FE33
Combined Chromatography and Tandem Mass-Spectrometry GC-MS-MS and LC-MS-MS. D.D. Fetterolf and R.A. Yost. *Am. Chem. Soc. Abstr. 183*, 93, (1982).

82FR11
First Workshop on Liquid Chromatography-Mass Spectrometry, Montreux. R.W. Frei, Ed. *J. Chromatogr. 251*, 1, (1982).

82FU27
Mass Spectrometric Analysis of Involatile Samples. J.H. Futrell, A.L. Wahrhaftig and L.G. Randall. *Govt. Rep. Announce. Index (U.S.) 82*, 4627, (1982).

82GA93
Combined Liquid Chromatography-Mass Spectrometry of Pesticides and Natural Products. D.E. Games, C. Eckers, M.S. Lant, E. Lewis, N.C.A. Weerasinghe and S.A. Westwood. *Anal. Proc. (London) 19*, 253–256, (1982).

82GA02
Extra Column Band Spreading in High-Performance Liquid Chromatography-Mass Spectrometry Using a Moving Belt Interface: Numerical Evaluation of System Variance. D.E. Games, M.J. Hewlins, S.A. Westwood and D.J. Morgan. *J. Chromatogr. 250*, 62–67, (1982).

82GA27
Liquid Chromatography Mass Spectrometry Workshop: Report. D.E. Games. *LC/MS Abstr. 2*, 17–19, (1982).

82GA95
Microbore High-Performance Liquid Chromatography Mass Spectrometry. D.E. Games, M.S. Lant, S.A. Westwood, M.J. Cocksedge, N. Evans, J. Williamson and B.J. Woodhall. *Biomed. Mass Spectrom. 9*, 215–234, (1982).

82GR4'
Mixture Analysis by New Mass Spectrometric Techniques: A Survey. H.F. Gruetzmacher. *Aether. Oele, Ergeb. Int. Arb. 1979–1980*, 1–24, (1982).

82HA0'
LC-MS System Employing Thermospray Sample with Belt Transport and Laser Desorption. E.D. Hardin and M.L. Vestal. Presented at 30th Ann. Conf. on Mass Spectrom. and Allied Topics, Honolulu, HI, U.S.A. June 6–11, 1982. Abstr. WPB 26, p. 570.

82HE0'
Applications of Micro LCMS to the Determination of Drug Residues Found in Equine Plasma and Urine. J.D. Henion, C. Eckers and G.A. Maylin. Presented at 30th Ann. Conf. on Mass Spectrom. and Allied Topics, Honolulu, HI, U.S.A. June 6–11, 1982. Abstr. ROC 8, p. 610–611.

82HE41
Determination of Sulfa Drugs in Biological Fluids by Liquid Chromatography/Mass Spectrometry. J.D. Henion, B.A. Thomson and P.A. Dawson. *Anal. Chem. 54*, 451–456, (1982).

82HE99
Direct Liquid Introduction Micro Liquid-Chromatography Mass Spectrometry. J.D. Henion. *Acta Pharm. Suec. 19*, 69–70, (1982).

82HI4'
Ion Detector for Liquid Chromatography. Hitachi Ltd. Jpn. Kokai Tokkyo Koho JP 57 53,054 [82 53,054] (Cl. HO1J49/30), 29 Mar. 1982, Appl. 80/127,255, 16 Sept. 1980, 4p.

82HI64'
Mass Spectrometer for Liquid Chromatography-Mass Spectrometry. Hitachi Ltd. Jpn. Kokai Tokkyo Koho JP 151,164 [82,151,164], 18 Sept. 1982.

82HO5'
Mass Spectra of Polar Organic Compounds in Aqueous Solutions Introduced into an Inductively Coupled Plasma (ICP). R.S. Houk and J.A. Olivares. Presented at 30th Ann. Conf. on Mass Spectrom. and Allied Topics, Honolulu, HI, U.S.A. June 6–11, 1982. Abstr. ROC 5, p. 605.

82HO70
Mass Spectra of Polar Organic Compounds in Aqueous Solutions Introduced into an Inductively Coupled Plasma. R.S. Houk, V.A. Fassel and H.J. Svec. *Org. Mass. Spectrom. 17*, 240–244, (1982).

82JA8'
Mass Spectrometer Sample Inlet System. Japan Spectroscopic Co. Ltd. Jpn. Kokai Tokkyo Koho JP
    82,101,328 (Cl. HO1J49/04), 23 June 1982, Appl. 80/176,925, 15 Dec. 1980, 7p.
82JE6'
Liquid Chromatograph Mass Spectrometer. Jeol Ltd. Jpn. Kokai Tokkyo Koho JP 82,00836 (Cl.
    H01J49/04), 5 Jan. 1982, Appl. 80/74,697, 3 June 1980, 3p.
82JE3'
Liquid Chromatograph-Mass Spectrometer Tandem System. Jeol Ltd. Jpn. Kokai Tokkyo Koho JP
    82,53,653 (Cl. G0IN27/62), 30 Mar. 1982, Appl. 80/128,760, 17 Sept. 1980, 3p.
82JE2'
Ion Source of Mass Spectrometer. Jeol Ltd. Jpn. Kokai Tokkyo Koho JP 82,88,662 (Cl.
    H01J49/16), 02 June 1982, Appl. 80/164,383, 21 Nov. 1980, 3p.
82JU79
Combined Liquid Chromatography Time-of-Flight Mass Spectrometry: An Application of Cf Frag-
    ment Induced Desorption Mass Spectrometry. H. Jungclas, H. Danigel, L. Schmidt and J. Dell-
    brugge. *Org. Mass Spectrom. 17*, 499–502, (1982).
82JU76
Quantitative Californium–252 Plasma Desorption Mass Spectrometry for Pharmaceuticals. A New
    Approach to Coupling Liquid Chromatography with Mass Spectrometry. H. Jungclas, H. Danigel
    and L. Schmidt. *Org. Mass Spectrom. 17*, 86–90, (1982).
82KA43
Sample Introduction System for Atmospheric Pressure Ionization Mass Spectrometry of Nonvolatile
    Compounds. H. Kambara. *Anal. Chem. 54*, 143–146, (1982).
82KE6'
Analysis of Mixtures by HPLC and FAB. T.R. Kemp, I.A.S. Lewis and J.C. Bill. Presented at 30th
    Ann. Conf. on Mass Spectrom. and Allied Topics, Honolulu, HI, U.S.A. June 6–11, 1982, Abstr.
    TOD 11, p. 246.
82KE5'
Novel-Ionization and Sample Introduction Techniques for the Analysis of Thermally-Labile and
    Non-Volatile Compounds by a Quadrupole Mass Spectrometer. C.N. Kenyon and P.C. Goodley.
    Presented at 30th Ann. Conf. on Mass Spectrom. and Allied Topics, Honolulu, HI, U.S.A. June
    6–11, 1982. Abstr. MPB 3, p. 145–146.
82KI25
Studies in On-Line Reversed Phase Liquid Chromatography/Mass Spectrometry. D.P. Kirby. Univ.
    Microfilms Int. Ann Arbor, MI 48106, U.S.A. Order no. DA 8205041. *Diss. Abstr. Int.B 42*,
    3675–3676, (1982).
82KN96
Advances in Detector Technology for High Performance Liquid Chromatography. Plenary Lecture.
    J.H. Knox. *Anal. Proc. (London) 19*, 166–170, (1982).
82KR19
Application of Microbore Columns to Liquid Chromatography-Mass Spectrometry. P. Krien, G.
    Devant and M. Hardy. *J. Chromatogr. 251*, 129–139, (1982).
82LE9CO'
High Performance Semi-Preparative Liquid Chromatography and Liquid Chromatography-Mass
    Spectrometry of Diesel Engine Emission Particulate Extracts. S.P. Levine, L.M. Skewes, L.D.
    Abrams and A.G. Palmer III. In: *Aromatic Hydrocarbons: Physical Biological Chemical, 6th
    International Symposium, 1981* (M. Cooke, A.J. Dennis and G.L. Fisher, eds.), p. 439–448,
    Battelle Press, Columbus, OH, (1982).
82LE39
Coupling of HPLC with Mass Spectrometry. K. Levsen. *Comm. Eur. Communities [REP] EUR
    1982 EUR 7623* (Anal. Org. Micropolit. Water), 149–158, (1982).

82LO5'
Sample Detection Using Liquid Chromatograph/Mass Spectrometer Interface with Stationary Con-
centrator. H.H. Lo and M.W. Siegel. Presented at 30th Ann. Conf. on Mass Spectrom. and Allied
Topics, Honolulu, HI, U.S.A. June 6–11, 1982. Abstr. MPA 10, p. 115–116.

82LU02
On-Line HPLC/GC Techniques for the Analyses of Biomass Derived Hydrocarbons. D.J. Luzbetak
and J.J. Hoffmann. *J. Chromatogr. Sci. 20*, 132–135, (1982).

82MA43
Column Liquid Chromatography. R.E. Majors, H.G. Barth and C.H. Lochmuller. *Anal. Chem. 54*,
323R–363R, (1982).

82MA34
Analysis of Liquid Crystal Mixtures with and without Dyes by Liquid Chromatography Mass Spectrometry (LCMS). T.I. Martin and W.E. Haas. *Am. Chem. Soc. Abstr. 183*, 54, (1982).

82MA15
Use of High-Performance Liquid Chromatography-Mass Spectrometry for the Study of the Metabolism of Ranitidine in Man. L.E. Martin, J. Oxford and R.J.N. Tanner. *J. Chromatogr. 251*, 215–
224, (1982).

82MA67
Influence of the Packing Material and the Column Filters on the Reliability of a High Performance
Liquid Chromatograph-Mass Spectrometer Interface Based on Direct Liquid Inlet Principle. B.
Mauchamp and P. Krien. *J. Chromatogr. 236*, 17–24, (1982).

82MC7AL'
Techniques for the Structure Elucidation of Complex Nucleosides by Mass Spectrometry. J.A.
McCloskey. In: *Proceedings of the 4th International Round Table on Nucleosides, Nucleotides
and Their Biological Applications*, Univ. Antwerp, Antwerp, Belgium, 4–6 Feb. 1981. (F.C.
Alderweireldt and E.L. Esmans, eds.), Univ. Antwerp, 1982. p. 47–67.

82MC98
Problems and Solutions in Combined Liquid Chromatography-Mass Spectrometry. W.H. McFadden.
Anal. Proc. (London) 19, 258–261, (1982).

82MC33
Structural Analysis of Complex Carbohydrates Using High Performance Liquid Chromatography,
Gas Chromatography and Mass Spectrometry. M. McNeil, A.G. Darvill, P. Aman, L.E. Franzen
and P. Albersheim. *Methods Enzymol. 83*, 3–45, (1982).

82NI13
Survey of Ionization Methods with Emphasis on Liquid Chromatography-Mass Spectrometry.
N.M.M. Nibbering. *J. Chromatogr. 251*, 93–104, (1982).

82OR14
Analysis of Liquids by Using Spark Source Mass Spectrometers. K.G. Ordzhonikidze, D.S
Nabichvrishvili, Yu.P. Osei and A.A. Lysyakov. *Prib. Tekh. Eksp. 1*, 164–166, (1982).

82PA75
Analysis of Explosives by Liquid Chromatography-Negative Ion Chemical Ionization Mass Spec-
trometry. C.E. Parker, R.D. Voyksner, Y. Tondeur, J.D. Henion, D.J. Harvan, J.R. Hass and J.
Yinon. *J. Forensic Sci. 27*, 495–505, (1982).

82PA73
High Performance Liquid Chromatography-Negative Chemical Ionization Mass Spectrometry of
Organophosphorous Pesticides. C.E. Parker, C.A. Haney and J.R. Hass. *J. Chromatogr. 237*,
233–248, (1982).

82PA27
High Performance Liquid Chromatography-Mass Spectrometry of Triazine Herbicides. C.E. Parker,
C.A. Haney, D.J. Harvan and J.R. Hass. *J. Chromatogr. 242*, 77–96, (1982).

82PO31
Developments in Selective Detectors for HPLC. H. Poppe. *Comm. Eur. Communities [REP] EUR 1982, EUR 7623* (Anal. Org. Micropollut. Water), 141–148, (1982).

82RA7′
Laser Desorption TLC/CI Mass Spectrometry. L. Ramaley, M.A. Vaughan, W.D. Jamieson and N.H. Burnett. Presented at 30th Ann. Conf. on Mass Spectrom. and Allied Topics, Honolulu, HI, U.S.A. June 6–11, 1982. Abstr. ROD 8, p. 627.

82RI84
Fast Atom Bombardment Mass Spectrometry. K.L. Rinehart, Jr. *Science 218*, 254–260, (1982).

82SC33
Assessment of a Moving Belt Type HPLC-MS Interface with Respect to Its Use in Organic Water Pollution Analysis. H. Schaunburg, H. Schlitt and H. Knoeppel. *Comm. Eur. Communities [REP] EUR 1982 EUR 7623* (Anal. Org. Micropollut. Water) 193–198, (1982).

82SC85
Combined Liquid Chromatography-Mass Spectrometry for Trace Analysis of Pharmaceuticals. L. Schmidt, H. Danigel and H. Jungclas. *Nucl. Instrum. Methods Phys. Res. 198*, 165–167, (1982).

82SM4′
Capillary Column Supercritical Fluid Chromatography-Mass Spectrometry. R.D. Smith, J. Fjeldsted and M.L. Lee. Presented at 30th Ann. Conf. on Mass Spectrom. and Allied Topics, Honolulu, HI, U.S.A. June 6–11, 1982. Abstr. MOF 3, p. 84–85.

82SM43
Capillary Column Supercritical Fluid Chromatography/Mass Spectrometry. R.D. Smith, W.D. Felix, J.C. Fjeldsted and M.L. Lee. *Anal. Chem. 54*, 1883–1885, (1982).

82SM71
Direct Fluid Injection Interface for Capillary Supercritical Fluid Chromatography-Mass Spectrometry. R.D. Smith, J.C. Fjeldsted and M.L. Lee. *J. Chromatogr. 247*, 231–243, (1982).

82ST5′
Secondary Ion Mass Spectrometry of Frozen Solution of Inorganic Salts. D. Stulik, R.G. Orth and J. Michl. Presented at 30th Ann. Conf. on Mass Spectrom. and Allied Topics, Honolulu, HI, U.S.A. June 6–11, 1982. Abstr. RPA 19, p. 695.

82SU19
Liquid Chromatography-Mass Spectrometry of Polar Lipids: Comparison of On-Line Method Using Direct Liquid Introduction Interface with Off-Line Method Using Desorption/Chemical Ionization. F.R. Sugnaux and C. Djerassi. *J. Chromatogr. 251*, 189–201, (1982).

82TA51
Direct Coupling of an Ultra-Micro High Performance Liquid Chromatograph and a Mass Spectrometer. T. Takeuchi, D. Ishii, A. Saito and T. Ohki. *J. High Resolut. Chromatogr. Chromatogr. Commun. 5*, 91–92, (1982).

82TA0′
A Study on Techniques for Micro-HPLC/MS. T. Takeuchi, K. Matsuoka and D. Ishii. Presented at 30th Ann. Conf. on Mass Spectrom. and Allied Topics, Honolulu, HI, U.S.A. June 6–11, 1982. Abstr. MPA 13, p. 120–121.

82TH9′
Ion Evaporation/Mass Spectrometry of Labile Compounds: Analysis by MS/MS and LC/MS. B.A. Thomson, J.V. Iribarne and P.J. Dziedzic. Presented at 30th Ann. Conf. on Mass Spectrom. and Allied Topics, Honolulu, HI, U.S.A. June 6–11, 1982. Abstr. ROC 2, p. 599–600.

82TH49
Liquid Ion Evaporation/Mass Spectrometry/Mass Spectrometry for the Detection of Polar and Labile Molecules. B.A. Thomson, J.V. Iribarne and P.J. Dziedzic. *Anal. Chem. 54*, 2219–2224, (1982).

82TS05
Desorption by Electric Field Observed in Liquid Ionization Mass Spectrometry. M. Tsuchiya, T. Nonaka, T. Taira and S. Tanaka. *Mass Spectrom. 30*, 95–97, (1982).

82TS6'
Liquid Ionization Field Desorption Mass Spectrometry-Effect of Electric Field on a Sample Probe. M. Tsuchiya, T. Nonaka and T. Taira. Presented at 30th Ann. Conf. on Mass Spectrom. and Allied Topics, Honolulu, HI, U.S.A. June 6–11, 1982. Abstr. WPB 6, p. 536–537.

82TS73
Advances of Ionization Methods in Organic Mass Spectrometry. S. Tsuge, K. Matsumoto and Y. Hirata. *Kagaku 37*, 243–246, (1982).

82TS1'
Application of Directly Coupled Liquid Chromatograph-Mass Spectrometer Using a New Vacuum Nebulizing Interface Incorporated with a Cooling Jacket to the Analysis of Non Volatile and/or Thermally Unstable Compounds. S. Tsuge K. Matsumoto, H. Yoshida, K. Ohta and Y. Hirata. Presented at 30th Ann. Conf. on Mass Spectrom. and Allied Topics, Honolulu, HI, U.S.A. June 6–11, 1982. Abstr. ROC 3, p. 601–602.

82VE9'
Speculations on the Mechanism of Thermospray and Other Soft Ionization Techniques. M.L. Vestal. Presented at 30th Ann. Conf. on Mass Spectrom. and Allied Topics, Honolulu, HI, U.S.A. June 6–11, 1982. Abstr. MOD 8, p. 59.

82VO6'
Approaches to On-Line Ion-Pair Extraction and Derivatization for High Performance Liquid Chromatography-Mass Spectrometry. P. Vouros, E.P. Lankmayr, M.J. Hayes, B.L. Karger and J.M. McGuire. Presented at 30th Ann. Conf. on Mass Spectrom. and Allied Topics, Honolulu, HI, U.S.A. June 6-ll, 1982. Abstr. ROC 11, p. 616.

82VO15
New Approaches to On-Line Ion-Pair Extraction and Derivatization for High Performance Liquid Chromatography-Mass Spectrometry. P. Vouros, E.P. Lankmayr, M.J. Hayes, B.L. Karger and J.M. McGuire. *J. Chromatogr. 251*, 175–188, (1982).

82VO45
Analysis of the Chemical Ionization Reagent Plasma Produced in Liquid Chromatography/Mass Spectrometry by Direct Liquid Introduction Mass Spectrometry/Mass Spectrometry. R.D. Voyksner, J.R. Hass and M.M. Bursey. *Anal. Chem. 54*, 2465–2470, (1982).

82VO43
Effects of Pressure, Temperature, and Solvent Composition in Analysis by Direct Liquid Introduction Liquid Chromatography/Mass Spectrometry. R.D. Voyksner, C.E. Parker, J.R. Hass and M.M. Bursey. *Anal. Chem. 54*, 2583–2586, (1982).

82VO2'
A Micro Computer Controlled Tandem Quadrupole Mass Spectrometer Using a Direct Liquid Insertion Probe for Screening Analysis. R.D. Voyksner, J.R. Hass and M.M. Bursey. Presented at 30th Ann. Conf. on Mass Spectrom. and Allied Topics, Honolulu, HI, U.S.A. June 6–11, 1982. Abstr. FOD 8, p. 802–803.

82VO51
An On-Line Liquid Chromatograph/Mass Spectrometer/Mass Spectrometry Experiment. R.D. Voyksner, J.R. Hass and M.M. Bursey. *Anal. Lett. 15*, 1–12, (1982).

82WE91
Micro Column High Performance Liquid Chromatography. S.A. Westwood, D.E. Games, M.S. Lant and B.J. Woodhall. *Anal. Proc. (London) 19*, 121–123, (1982).

82WI29
Combining Liquid Chromatography with Mass Spectrometry. R.C. Willoughby and R.F. Browner. *Trace Anal. 2*, 69–109, (1982).

82WR01
Combined Liquid Chromatographic/Mass Spectrometric Analysis of Carbamate Pesticides. L.H. Wright. *J. Chromatogr. Sci. 20*, 1–6, (1982).

82WR80
Determination of Aldicarb Residues in Water by Combined High Performance Liquid Chromatography-Mass Spectrometry. L.H. Wright, M.D. Jackson and R.G. Lewis. *Bull. Environ. Contam. Toxicol.* 28, 740–747, (1982).

82YE8'
Performance Characteristics of the Thermospray LC-MS Interface. A.L. Yergey, M.L. Vestal and C.R. Blakley. Presented at 30th Ann. Conf. on Mass Spectrom. and Allied Topics, Honolulu, HI, U.S.A. June 6–11, 1982. Abstr. MPA 12, p. 118–119.

82YI4'
High-Pressure Mass Spectrometry of Some LC-MS Liquid Reagents. J. Yinon and A. Cohen. Presented at 30th Ann. Conf. on Mass Spectrom. and Allied Topics, Honolulu, HI U.S.A. June 6–11, 1982. Abstr. ROC 10, p. 614–615.

82YO29
Liquid Chromatography-Mass Spectrometry. D.A. Yorke. *Proc. Inst. Pet. (London) 2* (Petroanal. 1981), 159–70, (1982).

82YO14
Improvement of Vacuum Nebulizing Interface for Direct Coupling Microliquid Chromatograph with Mass Spectrometer and Some Applications to Polar Natural Organic Compounds. H. Yoshida, K. Matsumoto, K. Itoh, S. Tsuge, Y. Hirata, K. Mochizuki, N. Kokubun and Y. Yoshida. *Fresenius' Z. Anal. Chem. 311*, 674–680, (1982).

82ZO23
Determination of the Molecular Weight of a Liquid-Phase Chemical Reaction Product by the Mass Spectrographic Method of Field Evaporation of Ions from Solutions. N. Zolotoi, G.V. Karpov and V.E. Skurat. *Khim. Fiz. 1982*(7), 893–6, (1982).

82ZO25
Determination of the Molecular Weight of Nucleic Acid Bases and Nucleotides by Mass Spectroscopy of the Field Evaporation of Ions from Solutions. N.B. Zolotoi, G.V. Karpov and V.E. Skurat. *Khim. Fiz. 1982*(5), 575–80, (1982).

83AL83
LC-MS and MS-MS Studies of Natural Oxygen Heterocyclic Compounds. N.J. Alcock, W. Kuhnz and D.E. Games. *Int. J. Mass Spectrom. Ion Phys. 48*, 153–156, (1983).

83AP50
Gas-Nebulized Direct Liquid Introduction Interface for Liquid Chromatography/Mass Spectrometry. J. A. Apffel, U.A.Th. Brinkman, R.W. Frei and E.A.I.M. Evers. *Anal. Chem. 55*, 2280–2284, (1983).

83AR13
Optimization of the Instrumental Parameters of a Combined Liquid Chromatograph Mass Spectrometer Coupled by an Interface for Direct Liquid Introduction. IV. A New Desolvation Chamber for Droplet Focusing or Townsend Discharge Ionization. P.J. Arpino, J.P. Bounine, M. Dedieu and G. Guiochon. *J. Chromatogr. 271*, 43–50, (1983).

83AR4'
Liquid Nebulizers for Direct LC/MS Interfaces. P.J. Arpino, G. Guiochon, G. Devant and C. Beaugrand. *Am. Chem. Soc. Abstr. 186*, ANYL0004, (1983).

83AR33
Mass-Spectrometric Detectors. P.J. Arpino. *Chromatogr. Sci. 23*, 243–322, (1983).

83BE63
Ionization of Nonvolatile Molecules and Their Analysis by MS/MS. C. Beaugrand and G. Devant. *Int. J. Mass Spectrom. Ion Phys. 46*, 23–26, (1983).

83BL50
Thermospray Interface for Liquid Chromatography/Mass Spectrometry. C.R. Blakley and M.L. Vestal. *Anal. Chem. 55*, 750–754, (1983).

83BR6'
Calculations on Capillary Interfaces for LC/MS. A.P. Bruins and B.F.H. Drenth. Presented at The 31st Ann. Conf. on Mass Spectrom. and Allied Topics, Boston, MA, U.S.A. May 8–13, 1983. Abstr. FOE9, p. 866–867.

83BR11
Experiments with the Combination of a Micro Liquid Chromatography and a Chemical Ionization Quadrupole Mass Spectrometer Using a Capillary Interface for Direct Liquid Introduction. A.P. Bruins and B.F.H. Drenth. *J. Chromatogr. 271*, 71–82, (1983).

83BR63
Experiments with the Combination of a Jasco Micro Liquid Chromatograph and a Quadrupole Mass Spectrometer. A.P. Bruins and B.F.H. Drenth. *Int. J. Mass Spectrom. Ion Phys. 46*, 213–216, (1983).

83BR98
Experiments with the Coupling of a Jasco Micro Liquid Chromatograph to a Finnigan MAT 3300 Quadrupole Mass Spectrometer. A.P. Bruins and B.F.H. Drenth. *Spectra 9*, 18–22, (1983).

83CA03
Liquid Chromatography Mass Spectrometry of Dexamethasone and Betamethasone. T. Cairns, E.G. Siegmund, J.J. Stamp and J.P. Skelly. *Biomed. Mass Spectrom. 10*, 203–208, (1983).

83CA04
Liquid Chromatography Mass Spectrometry of Thermally Labile Pesticides. T. Cairns, E.M. Siegmund and G.M. Doose. *Biomed. Mass Spectrom. 10*, 24–29, (1983).

83CA55
Permeable Membrane/Mass Spectrometric Measurement of an Enzymatic Kinetic Isotope Effect: Alpha-Chymotrypsin-Catalyzed Transesterification. K.C. Calvo, C.R. Weisenberger, L.B. Anderson and M.H. Klapper. *J. Am. Chem. Soc. 105*, 6935–6941, (1983).

83CH56
Factors Affecting Mass Spectral Sensitivity for Ions Sampled by Field Evaporation from a Liquid Matrix. K.W.S. Chan and K.D. Cook. *Anal. Chem. 55*, 1306–1309, (1983).

83CH2'
Factors Affecting Sampling of Ions from a Liquid Matrix. K.W. Chan and K.D. Cook. *Am. Chem. Soc. Abstr. 186*, ANYL0002, (1983).

83CH4'
A Direct Liquid Introduction LCMS Interface for Use on Sector Mass Spectrometers. J.R. Chapman, S. Evans and K.R. Compson. Presented at the 31st Ann. Conf. on Mass Spectrom. and Allied Topics, Boston, MA, U.S.A., May 8–13, 1983. p. 854–855.

83CH61
LCMS Interfacing of Sector Mass Spectrometers. J.R. Chapman, E. Harden, S. Evans and L.E. Moore. *Int. J. Mass Spectrom. Ion Phys. 46*, 201–204, (1983).

83CH8'
Approaches to Direct Liquid Introduction with Sector Mass Spectrometers. J.R. Chapman, E.H. Harden, S. Evans and K.R. Compson. *Am. Chem. Soc. Abstr. 186*, ANYL0018, (1983).

83CH7'
Quantitative Trace Organic Analysis by Combined LC/MS. R.G. Christensen, H.S. Hertz, S. Meiselman and E. White. *Am. Chem. Soc. Abstr. 186*, ANYL0007, (1983).

83CH11
Quantitative Trace Analysis by Reversed-Phase Liquid Chromatography-Mass Spectrometry. R.G. Christensen, E. White V, S. Meiselman and H.S. Hertz. *J. Chromatogr. 271*, 61–70, (1983).

83CO23
Mass Spectrometry: Analytical Capabilities and Potentials. R.G. Cooks, K.L. Busch and G.L. Glish. *Science (Washington, D.C.) 222*, 273–291, (1983).

83CO55
Direct Liquid Introduction/Thermospray Interface for Liquid Chromatography/Mass Spectrometry.
T. Covey and J. Henion. *Anal. Chem. 55*, 2275–2280, (1983).

83DA23
A Californium-252 Fission Fragment-Induced Desorption Mass Spectrometer: Design, Operation and
Performance. H. Danigel, H. Jungclas and L. Schmidt. *Int. J. Mass Spectrom. Ion Phys. 52*, 223–
240, (1983).

83DE33
New Ionization Methods in Mass Spectrometry. E. De Pauw. *Chim. Nouv. 3*, 43–45, (1983).

83DI2'
Analysis of Ranitidine and Metabolites Using DLI LC-MS. D.J. Dixon, R. Schuster, L.E. Martin
and J. Oxford. *Am. Chem. Soc. Abstr. 186*, ANLY0022, (1983).

83DO65
Investigation of an LC/MS Interface for EI-CI and FAB Ionization. P. Dobberstein, E. Korte, G.
Meyerhoff and R. Pesch. *Int. J. Mass Spectrom. Ion Phys. 46*, 185–188, (1983).

83EC65
Micro LC/MS Applications: Steroids, Antibiotics and Other Biologically Active Compounds. C.
Eckers, J.D. Henion, G.A. Maylin, D.S. Skrabalak, J. Vessman, A.M. Tivert and J.C. Green-
field. *Int. J. Mass Spectrom. Ion Phys. 46*, 205–208, (1983).

83ED6'
LC/MS Studies of Nucleic Acid Constituents Using the Thermospray Technique. C.G. Edmonds and
J.A. McCloskey. *Am. Chem. Soc. Abstr. 186*, ANYL0006, (1983).

83ED07
A Bibliography of Combined Liquid Chromatography Mass Spectrometry. C.G. Edmonds, J.A.
McCloskey and V.A. Edmonds. *Biomed. Mass Spectrom. 10*, 237–252, (1983).

83ED9'
Analysis of Modified Nucleosides by Combined Liquid Chromatography-Thermopsray Mass Spectrometry. C.G. Edmonds and J.A. McCloskey. Presented at 31st Ann. Conf. on Mass Spectrom.
and Allied Topics, Boston, MA, U.S.A. May 13–18, 1983. Abstr. RPB23, p. 789–790.

83ES98
Combination of Liquid Chromatography and Mass Spectrometry. Is It the One Marriage in a Thou-
sand? E.L. Esmans. *Chem. Mag. (Ghent) 9*(5), 18–20, (1983).

83ES07
Liquid Chromatography-Mass Spectrometry of Nucleosides Using a Commercially Available Direct
Liquid Introduction Probe. E.L. Esmans, Y. Luyten and F.C. Alderweireldt. *Biomed. Mass
Spectrom. 10*, 347–351, (1983).

83EV0'
Analysis of Phospholipids by Positive and Negative Ion CI-MS via a Moving Belt LC/MS Interface.
J.E. Evans, F.B. Jungalwala and R.H. McCluer. Presented at 31st Ann. Conf. on Mass Spectrom.
and Allied Topics, Boston, MA, U.S.A. May 8–13, 1983. Abstr. MPB 14, p. 160–161.

83EV3'
Analysis of Underivatized Glycosphingolipids by Ammonia CI-MS via a Moving Belt LC/MS
Interface. J.E. Evans, R.H. McCluer and H. Kadowaki. Presented at 31st Ann. Conf. on Mass
Spectrom. and Allied Topics, Boston, MA, U.S.A. May 8–13, 1983. Abstr. RPB25, p. 793–
794.

83EV94
First Steps in LC/MS with Simple Interfaces for the Finnigan MAT 44. N. Evans. *Spectra 9*, 14–17,
(1983).

83FE13
Ionization of Middle Mass Molecules: Ejection of Ions from Solution. C. Fenselau, R.J. Cotter, D.
Heller and J. Yergey. *J. Chromatogr. 271*, 3–12, (1983).

83FE1'
Desorption of Ions from Solution: Fast Atom Bombardment and the Thermospray Technique. C.

Fenselau, D. Liberato, J. Yergey, A.L. Yergey and R.J. Cotter. *Am. Chem. Soc. Abstr. 186*, ANYL0001, (1983).
83FE9'
Mechanism for Producing Pseudo-Molecular Ions from Nonvolatile Molecules. G.J. Fergusson and M.L. Vestal. Presented at 31st Ann. Conf. on Mass Spectrom. and Allied Topics, Boston, MA, U.S.A. May 8–13, 1983. Abstr. WPC 28, p. 609–610.
83FO7'
Studies of Environmental Test Well Samples Using GC/MS and LC/MS. M.G. Foster, O. Meresz and D.E. Games. Presented at 31st Ann. Conf. on Mass Spectrom. and Allied Topics, Boston MA, U.S.A. May 8–13, 1983. Abstr. WOC11, p. 477–478.
83FO08
Capillary Gas Chromatographic Mass Spectrometric and Microbore Liquid Chromatographic Mass Spectrometric Studies of a Test Well Sample from a Landfill Site. M.G. Foster, O. Meresz, D.E. Games, M.S. Lant and S.A. Westwood. *Biomed. Mass Spectrom. 10*, 338–342, (1983).
83FR11
Preface: Second Workshop on Liquid Chromatography-Mass Spectrometry and Mass Spectrometry-Mass Spectrometry, Oct. 21–22, 1982. R.W. Frei (ed.). *J. Chromatogr. 271*, 1, (1983).
83GA61
LC-MS Studies with Moving Belt Interfaces. D.E. Games, N.J. Alcock, L. Cocbelli, C. Eckers, M.P.L. Games, A. Jones, M.S. Lant, M.A. McDowall, M. Rossiter, R.A. Smith, S.A. Westwood and H.Y. Wong. *Int. J. Mass Spectrom. Ion Phys. 46*, 181–184, (1983).
83GA7'
Problem Solving with LC/MS. D.E. Games. Presented at the 31st Ann. Conf. on Mass Spectrom. and Allied Topics, Boston MA, U.S.A., May 8–13, 1983. Abstr. MOE4, p. 117–120.
83GA02
Combined High-Performance Liquid Chromatography-Mass Spectrometry. D.E. Games. *Anal. Proc. (London) 20*, 352–354, (1983).
83GA93
Combined Liquid Chromatography/Mass Spectrometry (LC/MS). D.E. Games. *Spectra 9*, 3–8, (1983).
83GA11
High-Performance Liquid Chromatography/Mass Spectrometry (HPLC/MS). D.E. Games. *Adv. Chromatogr. 21*, 1–39, (1983).
83GA8'
Recent Progress in LC/MS. D.E. Games, M.A. McDowall, M.G. Foster and O. Meresz. In: *Proceedings of the 3rd European Symposium on the Analysis of Organic Micropollutants in Water, Oslo, Norway*, Sept. 19–21, 1983, p. 68–76, Reidel Publ., Boston (1983).
83GA3'
Adventures in Liquid Chromatography-Mass Spectrometry Using Moving Belt Systems. D.E. Games, M.A. McDowall and K. Levsen. *Am. Chem. Soc. Abstr. 186*, ANYL0003, (1983).
83GA84
Microbore High-Performance Liquid Chromatography and Its Utility for Liquid Chromatography/Mass Spectrometry. D.E. Games and S.A. Westwood. *Eur. Spectrosc. News 48*, 14, 16–18, (1983).
83GO1'
Experimental Approaches to and Evaluation of LC/MS Interfaces Based upon Direct Liquid Introduction. P.C. Goodley and C.N. Kenyon. *Am. Chem. Soc. Abstr. 186*, ANYL0021, (1983).
83GO4'
Experimental Approaches to LC/MS. P.C. Goodley, C.N. Kenyon, J.-L. Truche, G. Mathers, J.F. Mahoney and J. Perel. Presented at 31st Ann. Conf. on Mass Spectrom. and Allied Topics, Boston, MA, U.S.A. May 8–13, 1983. p. 864–865.
83GU13
How to Interface a Chromatographic Column to a Mass Spectrometer. G. Guiochon and P.J. Arpino. *J. Chromatogr. 271*, 13–25, (1983).

83HA5'
LC-MS by Thermospray Sample Deposition and Laser Desorption. E.D. Hardin, T.P. Fan and M.L. Vestal. Presented at 31st Ann. Conf. on Mass Spectrom. and Allied Topics, Boston, MA, U.S.A. May 8–13, 1983. Abstr. RPB 26, p. 795.

83HA1'
HPLC-MS Using a Moving Belt Interface. M.J. Hayes, B.L. Karger and P. Vouros. Presented at 31st Ann. Conf. on Mass Spectrom. and Allied Topics, Boston, MA, U.S.A. May 8–13, 1983. Abstr. MOE 5, p. 121–122.

83HA55
Moving Belt Interface with Spray Deposition for Liquid Chromatography/Mass Spectrometry. M.J. Hayes, E.P. Lankmayer, P. Vouros, B.L. Karger and J.M. McGuire. *Anal. Chem. 55*, 1745–1752, (1983).

83HE11
Experimental Methods for On-Line Mass Spectrometry in Fermentation Technology. E. Heinzle, K. Furukawa, I.J. Dunn and J.R. Bourne. *Bio/Techniques 1*, 181–188, (1983).

83HE3'
Direct Liquid Introduction LC/MS. J.D. Henion. Presented at 31st Ann. Conf. on Mass Spectrom. and Allied Topics, Boston, MA, U.S.A. May 8–13, 1983. Abstr. MOE3, p. 113–114.

83HE2'
Thermospray LC/MS Progress with a Commercially Available GC/MS. J.D. Henion. Presented at 31st Ann. Conf. on Mass Spectrom. and Allied Topics, Boston, MA, U.S.A. May 8–13, 1983. Abstr. FOE 7, p. 862–863.

83HE8'
Practical DLI LC/MS with Improved Detection Limits. J.D. Henion. Presented at 31st Ann. Conf. on Mass Spectrom. and Allied Topics, Boston, MA, U.S.A. May 8–13, 1983. Abstr. FOE 10, p. 868–869.

83HE5'
Quantitative Micro LC/MS of Drugs in Biological Samples. J.D. Henion and D.S. Skrabalak. Presented at 31st Ann. Conf. on Mass Spectrom. and Allied Topics, Boston, MA, U.S.A. May 8–13, 1983. Abstr. FOB 8, p. 815–6.

83HU17
Mechanized Off-Line Combination of Microbore High-Performance Liquid Chromatography and Laser Mass Spectrometry. J.F.K. Huber, T. Dzido and F. Heresch. *J. Chromatogr. 271*, 27–33, (1983).

83IR01
Atmospheric Pressure Ion Evaporation-Mass Spectrometry. J.V. Iribarne, P.J. Dziedzic and B.A. Thomson. *Int. J. Mass Spectrom. Ion Phys. 50*, 331–347, (1983).

83JE0'
Argon Ion Source for Mass Spectrographic Device. Jeol Ltd. Jpn. Kokai Tokkyo Koho JP 58 54,540 [83 54,540] (Cl. HO1J49), 31 Mar. 1983, Appl. 81/154,066, 29 Sept. 1981, 3pp.

83JU49
Liquid Chromatography Combined with Mass Spectrometry Using Fission Fragment Induced Ionization. H. Jungclas, H. Danigel and L. Schmidt. *INIS Atomindex 14*, 748989, (1983).

83JU15
Fractional Sampling Interface for Combined Liquid Chromatography/Mass Spectrometry with Californium–252 Fission Fragment-Induced Ionization. H. Jungclas, H. Danigel and L. Schmidt. *J. Chromatogr. 271*, 35–41, (1983).

83JU67
Liquid Chromatography/Mass Spectrometry with Californium–252 Fission Fragment Induced Ionization. H. Jungclas, H. Danigel and L. Schmidt. *Int. J. Mass Spectrom Ion Phys. 46*, 197–200, (1983).

83KE4'
Processing of Comparative DLI-LC/MS, DCI and PID Data and of Capillary GC/MS Data Using Customized Procedures Based on the Rpn Data Processing Stack. C.N. Kenyon, R.H. Ellis and P.C. Goodley. Presented at 31st Ann. Conf. on Mass Spectrom. and Allied Topics, Boston, MA, U.S.A. May 8–13, 1983. Abstr. WPA 28, p. 564–565.

83KE58
Direct Liquid Inlet LC/MS and Negative Chemical Ionization MS. C.N. Kenyon, P.C. Goodley, D.J. Dixon, J.O. Whitney, K.F. Faull and J.D. Barchas. *Am. Lab. (Fairfield, Conn.) 15*, 38, 41–49, (1983).

83KE05
Sequencing of Underivatized Peptides by Direct Liquid Inlet Liquid Chromatography-Mass Spectrometry. C.N. Kenyon. *Biomed. Mass Spectrom. 10*, 535–543, (1983).

83KI9HR'
Amino Acid Sequence of Polypeptides by Enzymic Hydrolysis and Direct Detection Using a Thermospray LC/MS. H.Y. Kim, D. Pilosof, M.L. Vestal, D. F. Dyckes and J.P. Kitchell. In: *Peptides: Structure and Function, Proceedings of the 8th American Peptide Symposium.* (V.J. Hruby and D.H. Rich, eds.), p. 719–722, Pierce Chem. Co., Rockford, Ill. (1983).

83KR83
Dust Particle Induced Absorption: A New Method of Ion Generation from Organic Solids. F.R. Krueger and W. Knabe, *Org. Mass Spectrom. 18*, 83–84 (1983).

83KU05
Direct Liquid Inlet LC/MS of Natural Triacylglycerols and Steryl Esters. A. Kuksis, J.J. Myher and L. Marai. *J. Am. Oil Chem. Soc. 60*, 735, (1983).

83KU59
Chromatographic-Mass Spectrometric Instrumentation for Studying Aqueous Media. A.S. Kuzema and A.T. Pilipenko. *Khim. Tekhnol. Vody 5*, 79–85, (1983).

83LA67
Band Broadening with the Moving Belt LC/MS Interface. E. Lankmayer, M.J. Hayes, B.L. Karger, P. Vouros and J.M. McGuire. *Int. J. Mass Spectrom. Ion Phys. 46*, 177–180, (1983).

83LA69
Microbore High Performance Liquid Chromatography-Mass Spectrometry. M.S. Lant, D.E. Games, S.A. Westwood and B.J. Woodhall. *Int. J. Mass Spectrom. Ion Phys. 46*, 189–192, (1983).

83LE69
Direct Coupling of a Micro High Performance Liquid Chromatograph and a Mass Spectrometer. K. Levsen and K.H. Schäfer. *Int. J. Mass Spectrom. Ion Phys. 46*, 209–212, (1983).

83LE11
Determination of Phenlyureas by On-Line Liquid Chromatography-Mass Spectrometry. K. Levsen, K.H. Schäfer and J. Freudenthal. *J. Chromatogr. 271*, 51–60, (1983).

83LE12
High-Performance Liquid Chromatograph-Mass Spectrometer (HPLC-MS) Coupling. K. Levsen. *Nachr. Chem. Tech. Lab. 31*, 782–784, 786, (1983).

83LE6'
Determination of Poly-Aromatic Hydrocarbons in Natural Marine Sediments by Use of a Simple LC/MS Interface. E. Lewis, R. Guevremont and W.D. Jamieson. Presented at 31st Ann. Conf. on Mass Spectrom. and Allied Topics, Boston, MA, U.S.A. May 8–13, 1983. Abstr. TPB 22, p. 416–417.

83LE0'
Soft Ionization LC-MS. I.A.S. Lewis and P.W. Brooks. Presented at 31st Ann. Conf. on Mass Spectrom. and Allied Topics, Boston, MA, U.S.A. May 8–13, 1983. Abstr. FOE 1, p. 850–851.

83LE20'
Fast Atom Bombardment LC-MS. I.A.S. Lewis and P.W. Brooks. *Am. Chem. Soc. Abstr. 186*, ANYL0020, (1983).

83LI7'
Characterization of Glucuronides by Thermospray LC-MS. D.J. Liberato, C. Fenselau and A.L. Yergey. Presented at 31st Ann. Conf. on Mass Spectrom. and Allied Topics, Boston, MA, U.S.A. May 8–13, 1983. Abstr. RPB 22, p. 787–788.

83LI51
Characterization of Glucuronides with a Thermospray Liquid Chromatography/Mass Spectrometry Interface. D.J. Liberato, C.C.Fenselau, M.L. Vestal and A.L. Yergey. *Anal. Chem. 55*, 1741–1744, (1983).

83MA18
Special Report on LC/MS. R.E. Majors. *Liq. Chromatogr. HPLC Mag. 1*, 488, (1983).

83MA10
Analysis of Triacylglycerols by Reversed-Phase High-Pressure Liquid Chromatography with Direct Liquid Inlet Mass Spectrometry. L. Marai, J.J. Myher and A. Kuksis. *Can. J. Biochem. Cell Biol. 61*, 840–849, (1983).

83MA16
A Bibliography of Combined Liquid Chromatography Mass Spectrometry: A Most Useful Compilation of References. I. Matsumoto. *GC-MS News 11*, 96–97, (1983).

83MC1'
Analysis of Perbenzoylated Neutral Glucosphingolipids by Ammonia CI LC/MS. R.H. McCluer, S.K. Gross and J.E. Evans. Presented at 31st Ann. Conf. on Mass Spectrom. and Allied Topics, Boston, MA, U.S.A., May 8–13, 1983. Abstr. RPB 24, p. 791–792.

83MC87
Mass Spectral and LC-MS Studies of Beta-Lactam Antibiotics and Pseudomonic Acids. M.A. McDowall, D.E. Games and J.L. Gower. *Int. J. Mass Spectrom. Ion Phys. 48*, 157–160, (1983).

83MC93
Thermospray LC/MS: Supplement or Substitute for Existing Techniques. W.H. McFadden. *Spectra 9*, 23–28, (1983).

83MI13
Mass Spectrometry of Biogenic Catecholamine Ion Pairs by Direct Liquid Introduction. H. Milon and H. Bur. *J. Chromatogr. 271*, 83–92, (1983).

83MI55
Atmospheric Pressure Ionization Mass Spectrometry. R.K. Mitchum and W.A. Korfmacher. *Anal. Chem. 55*, 1485–1499, (1983).

83NE07
New Trends in Analytical Mass Spectrometry. D. Nelson. *Chem. Aust. 50*, 227–32, (1983).

83PA6'
LC/MS of Pesticides. C.E. Parker, C.A. Haney and J.R. Hass. Presented at 31st Ann. Conf. on Mass Spectrom. and Allied Topics, Boston, MA, U.S.A. May 8–13, 1983. Abstr. RPB 27, p. 796–797.

83PI8'
Peptide Sequencing by Enzymatic Hydrolysis Directly Coupled to Thermospray LCMS. D. Pilosof, H.-Y. Kim, D.F. Dyckes and M.L. Vestal. *Am. Chem. Soc. Abstr. 186*, ANYL0008, (1983).

83PO7'
EI Spectra from a DLI LC/MS Interface: Optimization of the Mass Spectrometer. F. Poeppel, J. Buchner and H.H. Lo. Presented at 31st Ann. Conf. on Mass Spectrom. and Allied Topics, Boston, MA, U.S.A. May 8–13, 1983. Abstr. RPB 28, p. 797–798.

83RA55
Thin Layer Chromatographic Plate Scanner Interfaced with a Mass Spectrometer. L. Ramaley, M.E. Nearing, M. Vaughan, R.G. Ackman and W.D. Jamieson. *Anal. Chem. 55*, 2285–2289, (1983).

83RA7PA'
Analysis of Dense (Supercritical) Gas Systems. L.G. Randall. In: *Chemical Engineering at Supercritical Conditions* (M.E. Paulaitis, ed.), p. 477–498, Ann Arbor Science, Ann Arbor, MI (1983).

83RO87

Peptide Sequencing by Combined Liquid Chromatography-Mass Spectrometry. P. Roepstorff, D.E. Games, M.P.L. Games and M.A. McDowall. *Int. J. Mass Spectrom. Ion Phys. 48*, 197–200, (1983).

83SA17

An Interface for Liquid Chromatograph/Mass Spectrometer Using Atmospheric Pressure Ionization Coupled with a Nebulizing Sample Introduction System. M. Sakairi and H. Kambara. *Shitsuryo Bunseki 31*, 87–95, (1983).

83SC68

Analysis of Surfactants by Newer Mass Spectrometric Techniques. Part II: Anionic Surfactants. E. Schneider, K. Levsen, P. Dähling and F.W. Röllgen. *Fresenius' Z. Anal. Chem. 316*, 488–492, (1983).

83SM1'

Application of Capillary Supercritical Fluid Chromatography-Mass Spectrometry. R.D. Smith, H.R. Udseth and B.W. Wright. Presented at 31st Ann. Conf. on Mass Spectrom. and Allied Topics, Boston, MA, U.S.A. May 8–13, 1983. Abstr. MOE 2, p. 111–112.

83SM26

New Method for the Direct Analysis of Supercritical Fluid Coal Extraction and Liquefaction. R.D. Smith and H.R. Udseth. *Fuel 62*, 466–468, (1983).

83SM85

Direct Mass Spectrometric Analysis of Supercritical Fluid Extraction Products. R.D. Smith and H.R. Udseth. *Sep. Sci. Technol. 18*, 245–252, (1983).

83SM07

Direct Supercritical Fluid Injection Mass Spectrometry of Trichothecenes. R.D. Smith and H.R. Udseth. *Biomed. Mass Spectrom. 10*, 577–580, (1983).

83SM56

Mass Spectrometry with Direct Supercritical Fluid Injection. R.D. Smith and H.R. Udseth. *Anal. Chem. 55*, 2266–2272, (1983).

83SM22

New Approaches Combining Chromatography and Mass Spectrometry for Synfuel Analysis. R.D. Smith. In: Advances in Techniques for Synthetic Fuels Analysis, [Technical Report] PNL-SA– 11562, p. 332–352, Pacific Northwest Labs, DE83–015528 (1983).

83SM67

Supercritical Fluid Chromatography-Mass Spectrometry. R.D. Smith, J. Fjeldsted and M.L. Lee. *Int. J. Mass Spectrom. Ion Phys. 46*, 217–220, (1983).

83SM87

LC-MS of Polynuclear Aromatic and Heterocyclic Compounds. R.W. Smith, D.E. Games and S.F. Noel. *Int. J. Mass Spectrom. Ion Phys. 48*, 327–330, (1983).

83ST15

Liquid Chromatography/Mass Spectrometry Using a Direct Liquid Introduction Interface on a Sector Mass Spectrometer. R.S. Stradling and A.E. Ashcroft. *Kratos Bull. 1*, 5–8, (1983).

83SU47

Direct Liquid Introduction Micro-Liquid Chromatography-Mass Spectrometry Coupling. Optimization of Droplet Desolvation and Instrumental Parameters for High Sensitivity. F.R. Sugnaux, D.S. Skrabalak and J.D. Henion. *J. Chromatogr. 264*, 357–376, (1983).

83TH9'

LC/MS/MS with an Atmospheric Pressure Chemical Ionization Source. B.A. Thomson, L. Danylewych–May and J.D. Henion. *Am. Chem. Soc. Abstr. 186*, ANYL0019, (1983).

83TH5'

Analysis of Complex Mixtures for Trace Involatile Species by LC/MS/MS. B.A. Thomson, L. Danylewych–May, A. Ngo and J.D. Henion. Presented at 31st Ann. Conf. on Mass Spectrom. and Allied Topics, Boston, MA, U.S.A. May 8–13, 1983. p. 65–66.

83TH2'
Design and Performance of a New Total Liquid Introduction LC/MS Interface. B.A. Thomson and L. Danylewych–May. Presented at 31st Ann. Conf. on Mass Spectrom. and Allied Topics, Boston, MA, U.S.A. May 8–13, 1983. Abstr. FOE 2, p. 852–3.

83TS3'
Determination of Active Hydrogen in Amino Acids and Nucleosides by Liquid Ionization Mass Spectrometry. M. Tsuchiya, H. Kuwabara, Y. Sugano and T. Taira. Presented at 31st Ann. Conf. on Mass Spectrom. and Allied Topics, Boston, MA, U.S.A. May 8–13, 1983. p. 373–374.

83TS6'
Liquid Ionization-Mass Spectrometry of Phospholipids. M. Tsuchiya, H. Kuwabara and T. Taira. Presented at 31st Ann. Conf. on Mass Spectrometry and Allied Topics, Boston, MA, U.S.A. May 8–13, 1983. p. 306–307.

83TS65
Application of Liquid Ionization Mass Spectrometry to Involatile Compounds. M. Tsuchiya, T. Taira, H. Kuwabara and T. Nonaka. *Int. J. Mass Spectrom. Ion Phys. 46*, 355–358, (1983).

83UD8'
Process Development and Analysis Using Supercritical Fluid Extraction-Mass Spectrometry. H.R. Udseth and R.D. Smith. Presented at 31st Ann. Conf. on Mass Spectrom. and Allied Topics, Boston, MA, U.S.A. May 8–13, 1983. Abstr. FOE 5, p. 858–859.

83VE27
Ionization Techniques for Nonvolatile Molecules. M.L. Vestal. *Mass Spectrom. Rev. 2*, 447–480, (1983).

83VE56
Ion Emissions from Liquids. M.L. Vestal. *Springer Ser. Chem. Phys. 25*, 246–263, (1983).

83VE7'
Thermospray LC-MS of Large Molecules. M.L. Vestal. *Am. Chem. Soc. Abstr. 185*, ANYL0017, (1983).

83VE5'
Thermospray LC-MS. M.L. Vestal. *Am. Chem. Soc. Abstr. 186*, ANYL0005, (1983).

83VE8'
Recent Advances in Thermospray LC-MS. M.L. Vestal. Presented at 31st Ann. Conf. on Mass Spectrom. and Allied Topics, Boston, MA, U.S.A. May 8–13, 1983. Abstr. MOE 1, p. 108–110.

83VE63
Studies of Ionization Mechanisms Involved in Thermospray LC-MS. M.L. Vestal. *Int. J. Mass Spectrom. Ion Phys. 46*, 193–196, (1983).

83VO3'
HPLC-MS Using a Moving Belt Interface. P. Vouros and B.L. Karger. *Am. Chem. Soc. Abstr. 186*, ANYL0023, (1983).

83VO99
The Moving Belt as an Interface for HPLC/MS. P. Vouros and B.L. Karger. *Spectra 9*, 9–13, (1983).

83VO49
Development and Application of Liquid Chromatography/Mass Spectrometry and Kinetic Energy Release Measurements for Trace Organic Analysis. R. D. Voyksner. *Diss. Abstr. Int. B 44*, 1109, (1983).

83VO67'
Liquid Chromatography/Mass Spectrometry Analysis of Candidate Anti-Malarials. R.D. Voyksner, J.T. Bursey, J.W. Hines and E.D. Pellizzari. Presented at 31st Ann. Conf. on Mass Spectrom. and Allied Topics, Boston, MA, U.S.A. May 8–13, 1983. Abstr. FOE 4, p. 856–857.

83VO65
A Microcomputer Controlled Tandem Quadrupole Mass Spectrometer Using a Direct Liquid Introduction Probe for Screening Analysis. R.D. Voyksner, J.R. Hass and M.M. Bursey. *Anal. Lett. 16*, 1165–1175, (1983).

83VO6'
LC/MS/MS and LC/MS Development and Applications in Environmental Chemistry. R.D. Voyskner, C.E. Parker, G.D. Marbury, J.R. Hass and M.M. Bursey. *Am. Chem. Soc. Abstr. 186*, ANYL0016, (1983).

83WH68
Comparison of Gas Chromatography/Mass Spectrometry and Liquid Chromatography/Mass Spectrometry for Confirmation of Coumaphos and Its Oxygen Analog in Eggs and Milk. K.D. White, Z. Min, W.C. Brumley, R.T. Krause and J.A. Sphon. *J. Assoc. Off. Anal. Chem. 66*, 1358–1364, (1983).

83WI21
Hyphenated Techniques for Analysis of Complex Organic Mixtures. C.L. Wilkins. *Science 222*, 291–296, (1983).

83WO19
Progress in Coupling Systems Boosts LC/MS Prospects. W. Worthy. *Chem. Eng. News 61*(39), 19–20, (1983).

83YE7'
Direct Quantitative Analysis of Intractable Molecules by Thermospray LC/MS. A.L. Yergey, D.J. Liberato and M.L. Vestal. *Am. Chem. Soc. Abstr. 186*, ANYL0017, (1983).

83YE0'
Direct Analysis of Intractable Endogenous Molecules by Thermospray LC/MS. A.L. Yergey, D.J. Liberato and M.L. Vestal. Presented at 31st Ann. Conf. on Mass Spectrom. and Allied Topics, Boston, MA, U.S.A. May 8–13, 1983. Abstr. FOE 6, p. 860–861.

83YI7'
Analysis of Explosives by LC/MS. J. Yinon. In: *Proceedings of the International Symposium on the Analysis and Detection of Explosives*, p. 227–234, Federal Bureau of Investigation, Washington D.C. (1983).

83YI83
Forensic Applications of LC-MS. J. Yinon. *Int. J. Mass Spectrom. Ion Phys. 48*, 253–256, (1983).

83YI85
High Performance Liquid Chromatography-Mass Spectrometry of Explosives. J. Yinon and D.-G. Hwang. *J. Chromatogr. 268*, 45–53, (1983).

83YI87
High Pressure Mass Spectrometry of Some Liquid Chromatographic-Mass Spectrometric Reagents. J. Yinon and A. Cohen. *Org. Mass Spectrom. 18*, 47–51, (1983).

83YU03
High Performance Liquid Chromatography/Mass Spectrometry Examination of C-Methylation Artifacts from the Permethylation Reaction of Peptides. T.J. Yu, B.L. Karger and P. Vouros. *Biomed. Mass Spectrom. 10*, 633–640, (1983).

83ZA87
Chromatography and Mass Spectrometry of Organic Compounds. V.G. Zaikin. *Zh. Vses. Khim. O-va. im D.I. Mendeleeva 28*, 67–72, (1983).

84AB6'
Application of the Monodisperse Aerosol Generation/Liquid Chromatography Interface to Samples of Environmental and Biochemical Importance. L.E. Abbey, D.E. Bostwick, P.C. Winkler, R.C. Willoughby and R.F. Browner. Presented at 32nd Ann. Conf. on Mass Spectrom. and Allied Topics, San Antonio, TX, U.S.A. May 27–June 1, 1984. p. 96–97.

84AL96
Mechanism of Ion Formation during Electrohydrodynamic Sputtering of a Liquid in a Vacuum. M.L. Aleksandrov, L.N. Gal, N.V. Krasnov, V.I. Nikolaev, V.A. Pavlenko and V.A. Shkurov. *Zh. Anal. Khim. 39*, 1596–1602, (1984).

84AL00
Direct Coupling of a Microcolumn Liquid Chromatograph and a Mass Spectrometer. M.L. Alek-

sandrov, L.N. Gal, N.V. Krasnov, V.I. Nikolaev, V.A. Pavlenko, V.A. Shkurov, G.I. Baram, M.A. Grachev, V.D. Knorre and Y.S. Kusner. *Bioorg. Khim. 10*, 710–712, (1984).

84AN20
Advances in Liquid Chromatography-Mass Spectrometry Interfacing on Kratos Instrumentation. Anonymous. *Kratos Bull. 2*, 10–13, (1984).

84AN27
Instrumentation '84. Anonymous. *Chem. Eng. News 62*, 87–88, (1984).

84AO43
Development and Evaluation of a New Universal Detector for Liquid Chromatography Based on Thermospray Vaporization/Ionization and an Rf Only Quadrupole Mass Filter. J.N. Ao. *Diss. Abstr. Int. B 44*, 2413, (1984).

84AP23
Micro Post-Column Extraction System for Interfacing Reversed Phase Micro Liquid Chromatography and Mass Spectrometry. J.A. Apffel, U.A.Th. Brinkman and R.W. Frei. *J. Chromatogr. 312*, 153–164, (1984).

84AR9'
Fast Injection of Liquid Solutions into a Quadrupole Mass Spectrometer Using Short Fused Silica Capillaries. P.J. Arpino and C. Beaugrand. Presented at 32nd Ann. Conf. on Mass Spectrom. and Allied Topics, San Antonio, TX, U.S.A. May 27–June 1, 1984. p. 199–200.

84AS89
Detection Systems. S.P. Assenza. *Chromatogr. Sci. 28*, 139–160, (1984).

84AZ32
Use of Liquid Chromatography Mass Spectrometry for the Quantification of Dethiobiotin and Biotin in Biological Samples. M. Azoulay, P.L. Desbene, F. Frappier and Y. Georges. *J. Chromatogr. 303*, 272–276, (1984).

84BA7'
FAB-MS Directly from Electrophoretic Plates. Applications to Quaternary Organoarsenic Compounds. D.F. Barafsky and M.L. Zimmerman. Presented at 32nd Ann. Conf. on Mass Spectrom. and Allied Topics, San Antonio, TX, U.S.A. May 27–June 1, 1984. p. 747–748.

84BA45
Modern (Soft) Ionization Sources. M. Barber, R.S. Bordoli and G.J. Elliot. *Gazz. Chim. Ital. 144*, 305–311, (1984).

84BE64
Identification of Dyes by Thermospray Ionization and Mass Spectrometry/Mass Spectrometry. L.D. Betowski and J.M. Ballard. *Anal. Chem. 56*, 2604–2607, (1984).

84BL1'
Ion Vapor Source for MS of Liquids. C.R. Blakley and M.L. Vestal. CA 1,162,331 (Cl. B01D59/44),14 Feb. 1984. (Res. Corp. Can.), US 152,767, 23 May 1980.

84BR1'
Isolation of a Plant Cell Culture Product and Its Identification by Combined LC/MS. A.P. Bruins and N. Pras. Presented at 32nd Ann. Conf. on Mass Spectrom. and Allied Topics, San Antonio, TX, U.S.A. May 27–June 1, 1984. p. 111–112.

84BR31
Isolation of 3,4-Dihydroxyphenylacetic Acid Produced from p-Hydroxyphenylacetic Acid by Immobilized Plant Cells of *Mucuna pruriens* and Its Identification by LC/MS. A.P. Bruins and N. Pras. *Anal. Chim. Acta 163*, 91–100, (1984).

84BR62
Isolation of a Plant Cell Culture Product and Its Identification by Combined Liquid Chromatography-Mass Spectrometry. A.P. Bruins and N. Pras. *Pharm. Weekbl.-Sci. Ed. 6*, 252, (1984).

84BR98
Liquid Chromatography-Mass Spectrometry. A.P. Bruins. *Tijdschr. Ned. Ver. Klin. Chem. 9*, 8–10, (1984).

84BU28
New Biological Dimension in Mass Spectromtery. K.L. Busch and G.L. Glish. *Bio/Techniques 2*, 128–139, (1984).

84CH6'
LCMS: Direct Liquid Introduction and Thermospray Operation on a Magnetic Sector Instrument. J.R. Chapman, S. Sayyid and S. Evans. Presented at 32nd Ann. Conf. on Mass Spectrom. and Allied Topics, San Antonio, TX, U.S.A. May 27–June 1, 1984. p. 106–107.

84CH8'
Quantitative Trace Analysis by LC/MS. R.G. Christensen and E. White V. Presented at 32nd Ann. Conf. on Mass Spectrom. and Allied Topics, San Antonio, TX, U.S.A. May 27–June 1, 1984. p. 308–309.

84CO6'
A Direct Liquid Introduction Micro Liquid Chromatography/Mass Spectrometry and Gas Chromatography/Mass Spectrometry Procedure for Determination of Diethylstilbestrol and Its Metabolites. T. Corey, J. Crowther, G. Maylin and J. Henion. Presented at 32nd Ann. Conf. on Mass Spectrom. and Allied Topics, San Antonio, TX, U.S.A. May 27–June 1, 1984. p. 446.

84CO10
Membrane Inlet Mass Spectrometry: A Universal Monitor for Dissolved Gases in Microbial Physiology. R.P. Cos, B.B. Jensen, D. Joergensen and H. Degn. *Microbiol. Sci. 1*, 200–203, (1984).

84CR61
Liquid Chromatographic/Mass Spectrometric Determination of Optically Active Drugs. J.B. Crowther, T.R. Corey, E.A. Dewey and J.D. Henion. *Anal. Chem. 56*, 2921–2926, (1984).

84DE61
Analysis of Neuropeptides by LC and MS. D.M. Desiderio. *Techn. Instrum. Anal. Chem. 6*, 211–231, (1984).

84DE77
A Review of Combined LC and MS. D.M. Desiderio and G.H. Fridland. *J. Liq. Chromatogr. 7*, 317–352, (1984).

84DE83
Microbore Columns. D. Dezaro and R.A. Hartwick. In: *HPLC in Nucleic Acid Research* (P.R. Brown, ed.), *Chromatogr. Sci. Series 28*, 113–138, Dekker, New York, (1984).

84DO67
Solution Introduction Mass Spectrometry: A Simple Alternative to Desorption Techniques. R.C. Dougherty and J. De Kanel. *Anal. Chem. 56*, 2977–2979, (1984).

84ED1'
Analysis of Potent Mutagens in Broiled Fish by Combined Liquid Chromatography/Thermospray Mass Spectrometry. C.G. Edmonds, S.K. Sethi, J.A. McCloskey, Z. Yamaizumi, H. Kasai and S. Nishimura. Presented at 32nd Ann. Conf. on Mass Spectrom. and Allied Topics, San Antonio, TX, U.S.A. May 27–June 1, 1984. p. 201–202.

84EL97
Analysis of the Ozonation Products of Phthalic Acid in Water Using Combined High Performance-Liquid Chromatography/Mass Spectrometry. A.S. El Dine, A. Bermond, D.N. Rutledge and C. Ducauze. *Analyst (London) 109*, 817–821, (1984).

84ER16
HPLC-MS-Computer Analysis of Lipids. W.L. Erdahl, F.C. Phillips, D.E. Jarvis and O.S. Privett. *J. Am. Oil Chem. Soc. 61*, 256, (1984).

84ES5'
Direct Liquid Introduction/Liquid Chromatography-Mass Spectrometry of Nucleosides Using a Self-Made Desolvation Chamber. E.L. Esmans, P. Geboes, Y. Luyten and F.C. Alderweireldt. Presented at 32nd Ann. Conf. on Mass Spectrom. and Allied Topics, San Antonio, TX, U.S.A. May 27–June 1, 1984. p. 115–116.

84EV3'
LC/MS Analysis of Underivatized Glycosphingolipids from Tissue Extracts. J.E. Evans and R.H. McCluer. Presented at 32nd Ann. Conf. on Mass Spectrom. and Allied Topics, San Antonio, TX, U.S.A. May 27–June 1, 1984. p. 683–684.

84FA0'
Direct Comparison of Secondary Ion and Laser Desorption Mass Spectrometry in a Moving Belt LC-MS System. T.P. Fan, E.D. Hardin and M.L. Vestal. Presented at the 32nd Ann. Conf. on Mass Spectrom. and Allied Topics, San Antonio TX, U.S.A. May 27–June 1, 1984. p. 840–841.

84FA60
Direct Comparison of Secondary Ion and Laser Desorption Mass Spectrometry on Bioorganic Molecules in a Moving Belt Liquid Chromatography/Mass Spectrometry System. T.P. Fan, E.D. Hardin and M.L. Vestal. *Anal. Chem. 56*, 1870–1876, (1984).

84FE69
A Comparison of Thermospray and Fast Atom Bombardment Mass Spectrometry as Solution Dependent Ionization Techniques. C. Fenselau, D.J. Liberato, J.A. Yergey, R.J. Cotter and A.L. Yergey. *Anal. Chem. 56*, 2759–2762, (1984).

84FR8'
Moving Belt Interface Applicability with Microbore Liquid Chromatograph/Mass Spectrometry. D. Fraisse, M. Emmelin, J.L. Rocca and S. Alamercery. Presented at 32nd Ann. Conf. on Mass Spectrom. and Allied Topics, San Antonio, TX, U.S.A. May 27–June 1, 1984. p. 88–89.

84GA49
Electrohydrodynamic Introduction of Liquid Substances into a Mass Spectrometer. L.N. Gal, N.V. Krasnov, Y.S. Kusner, V.I. Nikolaev, V.G. Prokod'ko and G.V. Simonova. *Zh. Tekh. Fiz. 54*, 1559–1571, (1984).

84GA18
The State of the Art of LC/MS and Compatible Ionization Methods. D.E. Games. *Kem. Kemi 11*, 8–10, (1984).

84GA499
Analysis of Pepper and Capsicum Oleoresins by High Performance Liquid Chromatography/Mass Spectrometry and Field Desorption Mass Spectrometry. D.E. Games, N.J. Alcock, J. Van Der Greef, L.M. Nyssen, H. Maarse and M.C. Ten Noever De Brauw. *J. Chromatogr. 294*, 269–279, (1984).

84GA13
Combined LC/MS and Its Application in Food Science. D.E. Games, N.J. Alcock, I. Horman, E. Lewis, M.A. McDowall and A. Van Moncur. In: *Chromatography and Mass Spectrometry in Nutrition Science and Food Safety* (A. Frigerio and H. Milon, eds.), *Anal. Chem. Symp. Ser. 21*, 263–280, (1984).

84GA90
Combined LC/MS and Its Utility in Studies of Natural Products. D.E. Games. *Pharm. Weekbl. 119*, 30–33, (1984).

84GA5'
Studies with Moving Belt and Thermospray LC/MS. D.E. Games, N.J. Alcock, I.G. Beattie, M.A. McDowall, A. Van Moncur, E.D. Ramsey and P.W. Skett. Presented at 32nd Ann. Conf. on Mass Spectrom. and Allied Topics, San Antonio, TX, U.S.A. May 27–June 1, 1984. p. 195–196.

84GA4'
Combined LC/MS of Amino Acids and Peptides. D.E. Games, E.D. Ramsey, D.A. Catlow, B.J. Woodhall and D. Roepstorff. Presented at 32nd Ann. Conf. on Mass Spectrom. and Allied Topics, San Antonio, TX, U.S.A. May 27–June 1, 1984. p. 34–35.

84GA14
Application of LC-MS to the Analysis of Water. D.E. Games, M.G. Foster and O. Meresz. *Anal. Proc. (London) 21*, 174–177, (1984).

84GA17
A Comparison of Moving Belt Interfaces for Liquid Chromatography/Mass Spectrometry. D.E. Games, M.A. McDowall, K. Levsen, K.H. Schäfer, P. Dobberstein and J.L. Gower. *Biomed. Mass Spectrom. 11*, 87–95, (1984).
84GA48
Recent Progress in LC/MS. D.E. Games, M.A. McDowall, M.G. Foster and O. Meresz. *Comm. Eur. Communities [Rep.] EUR 1984 EUR 8518* (Anal. Org. Micropollut. Water), 68–76, (1984).
84GI19
Electrospray Mass Spectroscopy of Macromolecules: Application of an Ion-Drift Spectrometer. J. Gieniec, L. Mack, K. Nakamae, C. Gupta, V. Kumar and M. Dole. *Biomed. Mass Spectrom. 11*, 259–268, (1984).
84GO8'
Mass Analysis of Molecules of Molecular Weight Greater than 1500 Daltons Using a Thermospray Ion Source. P.C. Goodley and C.N. Kenyon. Presented at 32nd Ann. Conf. on Mass Spectrom. and Allied Topics, San Antonio, TX, U.S.A. May 27–June 1, 1984. p. 108–109.
84HA62
Laser Desorption Mass Spectrometry with Thermospray Sample Deposition for Determination of Non-Volatile Biomolecules. E.D. Hardin, T.P. Fan, C.R. Blakley and M.L. Vestal. *Anal. Chem. 56*, 2–7, (1984).
84HA20
Metabolism of Magnolol from Magnoliac Cortex. 1-Application of Liquid Chromatography/Mass Spectrometry to the Analysis of Metabolites of Magnolol in Rats. M. Hattori, T. Sakamoto, Y. Endo, N. Kakiuchi, K. Kobashi, T. Mizuno and T. Namba. *Chem. Pharm. Bull. 32*, 5010–5017, (1984).
84HA5'
High Performance Gradient Liquid Chromatography/Mass Spectrometry Using Microbore Columns with a Moving Belt Interface. M.J. Hayes, H.E. Schwartz, P. Vouros, B.L. Karger, A.D. Thruston, Jr. and J.M. McGuire. Presented at 32nd Ann. Conf. on Mass Spectrom. and Allied Topics, San Antonio, TX, U.S.A. May 27–June 1, 1984. p. 205–206.
84HA69
Gradient Liquid Chromatography/Mass Spectrometry Using Microbore Columns and a Moving Belt Interface. M.J. Hayes, H.E. Schwartz, P. Vouros, B.L. Karger, A.D. Thruston, Jr. and J.M. McGuire. *Anal. Chem. 56*, 1229–1236, (1984).
84HE80
Micro LC/MS Coupling. J. Henion. In: *Theory, Practice and Application of Microcolumn High Performance Liquid Chromatography*. (P. Kucera, ed.), *J. Chromatogr. Libr. 28*, 260–300, (1984).
84HE0'
On Line Chiral LC/MS for the Identification of d,l-Enantiomer Pairs. J. Henion, J. Crowther and T. Covey. Presented at 32nd Ann. Conf. on Mass Spectrom. and Allied Topics, San Antonio, TX, U.S.A. May 27–June 1, 1984. p. 170–171.
84HE3'
An Improved Thermospray LC/MS System for Determining Drugs in Urine. J. Henion, J. Crowther and T. Covey. Presented at 32nd Ann. Conf. on Mass Spectrom. and Allied Topics, San Antonio, TX, U.S.A. May 27–June 1, 1984. p. 203–204.
84HE1'
Progress in Rapid Urine Analysis for Drugs by On-Line LC/MS/MS. J. Henion and T. Covey. Presented at 32nd Ann. Conf. on Mass Spectrom. and Allied Topics, San Antonio, TX, U.S.A. May 27–June 1, 1984. p. 791–792.
84HE8'
Advances in Microbore LC/MS. J. Henion. *Am. Chem. Soc. Abstr. 188*, ANYL0078, (1984).

84HI91
Mass Spectrometry and Clinical Chemistry. R.E. Hill and D.T. Whelan. *Clin. Chim. Acta 139*, 231–294, (1984).

84HI7'
Liquid Chromatograph-Mass Spectrometer. Hitachi Ltd. Jpn. Kokai Tokkyo Koho JP 59,210,357 [84,210,357] (Cl. G01N27/62), 29 Nov. 1984. Appl. 84/83,721, 27 Apr. 1984.

84HI8'
Liquid Chromatograph-Atmospheric Pressure Ionization Mass Spectrometer. Hitachi Ltd. Jpn. Kokai Tokkyo Koho JP 59,210,358 [84,210,358] (Cl. G01N27/62), 29 Nov. 1984, Appl. 84/83,722, 27 Apr. 1984.

84IN13
Protonation in Organic Secondary Ion Mass Spectrometry: Correlation with Solution Chemistry Involved in Sample Preparation. J. Inchaouh, J.C. Blais, G. Bolbach and A. Brunot. *Int. J. Mass Spectrom. Ion Processes 61*, 153–156, (1984).

84JE4'
Liquid Chromatograph-Mass Spectrograph. Jeol Ltd. Jpn. Kokai Tokkyo Koho JP 59,153,164 [84,153,164] (Cl. G01N27/62), 1 Sept. 1984, Appl. 83/27,335, 21 Feb. 1983.

84JE8'
Ion Source. Jeol Ltd. Jpn. Kokai Tokkyo Koho JP 59,112,558 [84,112,558] (Cl. H01J49), 29 Jun. 1984, Appl. 82/221,599, 17 Dec. 1982.

84JG1'
Liquid Chromatographic Mass Analyzer. JGC Corp. Jpn. Kokai Tokkyo Koho JP 59,32,861 [84 32,861], 22 Feb. 1984.

84JO45
Evaluation of High Performance Liquid Chromatography-Mass Spectrometry for Forensic Analyses. J.R. Joyce. *J. Forensic Sci. Soc. 24*, 155, (1984).

84JU58
Compositional and Molecular Species Analysis of Phospholipids by High Performance Liquid Chromatography Coupled with Chemical Ionization Mass Spectrometry. F.B. Jungalwala, J.E. Evans and R.H. McCluer. *J. Lipid Res. 25*, 738–749, (1984).

84JU59
High Performance Liquid Chromatography-Chemical Ionization Mass Spectrometry of Sphingoid Bases Using Moving Belt Interface. F.B. Jungalwala, J.E. Evans, J. Kadowaki and R.H. McCluer. *J. Lipid Res. 25*, 209–216, (1984).

84KA59
New Ionization Methods for Nonvolatile Compounds. H. Kambara. *Bunseki 5*, 349–351, (1984).

84KA2'
High Performance Liquid Chromatography/Mass Spectrometry Using Microbore Columns with a Moving Belt Interface. B.L. Karger, P. Vouros, M. J. Hayes and H. Schwartz. *Am. Chem. Soc. Abstr. 187*, ANYL0002, (1984).

84KI64
On-Line Peptide Sequencing by Enzymatic Hydrolysis, High Performance-Liquid Chromatography and Thermospray Mass Spectrometry. H.Y. Kim, D. Pilosof, D.F. Dyckes and M.L. Vestal. *J. Am. Chem. Soc. 106*, 7304–7309, (1984).

84KI3'
Peptide Sequencing by On-Line Enzymatic Hydrolysis and Thermospray LC/MS. H.Y. Kim, D. Pilosof, D.F. Dyckes and M.L. Vestal. Presented at 32nd Ann. Conf. on Mass Spectrom. and Allied Topics, San Antonio, TX, U.S.A. May 27–June 1, 1984. p. 353–354.

84KU23
Liquid Ionization Mass Spectrometry. H. Kuwabara and M. Tsuchiya. *Mass Spectrosc. 32*, 263–266, (1984).

84LA2'
Method and Apparatus for the Mass Spectrometric Analysis of Solutions. M.J. Labowsky, J.B. Fenn and M. Yamashita. Eur. Pat. Appl. EP 123,552 (Cl. H01J49/00), 31 Oct. 1984, US Appl. 486,645, 20 Apr. 1983.

84LA4'
A Thermospray Ionization on a Sector Instrument. B.S. Larsen, R.J. Cotter and C.N. Fenselau. Presented at 32nd Ann. Conf. on Mass Spectrom. and Allied Topics, San Antonio, TX, U.S.A. May 27–June 1, 1984. p. 84–85.

84LE23
Coupling of High-Performance Chromatography-Mass Spectrometry (HPLC-MS). K. Levsen. Gewässerschutz, Wasser, Abswasser 72, 253–285, (1984).

84LE18
On-Line Liquid Chromatographic/Mass Spectrometric Studies Using a Moving Belt Interface and Negative Chemical Ionization. K. Levsen, K.H. Schäfer and P. Dobberstein. Biomed. Mass Spectrom. 11, 308–309, (1984).

84LE3'
The Analysis of Trace Quantities of Thermally Labile Compounds by Thermospray LC/MS. I.A.S. Lewis, D.C. Smith and M. Veares. Presented at 32nd Ann. Conf. on Mass Spectrom. and Allied Topics, San Antonio, TX, U.S.A. May 27–June 1, 1984. p. 803–804.

84LI6'
Identification and Quantitative Analysis of Acylcarnitines in Human Urine by Thermospray HPLC/MS. D.J. Liberato, A.L. Yergey and D.S. Millington. Presented at 32nd Ann. Conf. on Mass Spectrom. and Allied Topics. San Antonio, TX, U.S.A. May 27–June 1, 1984. p. 76–77.

84LI13
The Use of Surfactants to Modify Molar Response Factors in the Secondary Ion Mass Spectrometry of Liquid Surfaces. W.V. Ligon, Jr. and S.B. Dorn. Int. J. Mass Spectrom. Ion Processes 61, 113–122, (1984).

84MA4'
Crossed Beam Technique for Coupling LC/MS. J.F. Mahoney, J. Perel and M. Siegel. Presented at 32nd Ann. Conf. on Mass Spectrom. and Allied Topics, San Antonio, TX, U.S.A. May 27–June 1, 1984. p. 94–95.

84MA60
Column Liquid Chromatography. R.E. Majors, H.G. Barth and C.H. Lochmuller. Anal. Chem. 56, 300R–349R, (1984).

84MA41
A Comparison of the Moving Belt and Direct Liquid Introduction Interfaces for High Performance Liquid Chromatography-Mass Spectrometry of Ranitidine and Its Metabolites. L.E. Martin, J. Oxford, D. Dixon and B. Schuster. Methodol. Surv. Biochem. Anal. 14, 191–194, (1984).

84MA76
HPLC Mass Spectrometry and LC/MS of Gossypol and Its Derivatives. J.A. Martin, R.H. Zhou, D.E. Games, A. Jones and E.D. Ramsey. J. High Resolut. Chromatogr. Chromatogr. Commun. 7, 196–202, (1984).

84MC2'
Analysis of Phospholipids by HPLC-Mass Spectrometry. R.H. McCluer, F.B. Jungalwala and J.E. Evans. Presented at 32nd Ann. Conf. on Mass Spectrom. and Allied Topics, San Antonio, TX, U.S.A. May 27–June 1, 1984. p. 102–103.

84MC04
New LC/MS Hardware for Finnigan MAT Users. B. McFadden. Anal. News 10, 4–5, (1984).

84MC0'
Thermospray: New Developments and Applications. W.H. McFadden and M.G. Tucker. Presented at 32nd Ann. Conf. on Mass Spectrom. and Allied Topics, San Antonio, TX, U.S.A. May 27–June 1, 1984. p. 100–101.

84MC61
Trends in Analytical Instruments. F.W. McLafferty. *Science 226*, 251–253, (1984).
84MI25
LC/MS Application to the Analysis of Liposoluble Vitamins. H. Milon and H. Bur. *LC Mag. 2*, 455–457, (1984).
84MI256
Development of a New Type LC/MS Interface and Applicability for Nonvolatile Compounds. T. Mizuno, A. Koushi and K. Yoshihiro. *Shitsuryo Bunseki 32*, 285–296, (1984).
84MU65
Sampling of Ions from Volatile Solutions by Electrohydrodynamic Mass Spectrometry. S.L. Murawski and K.D. Cook. *Anal. Chem. 56*, 1015–1020, (1984).
84MU8′
Liquid Ionization Mass Spectrometry of Organic Compounds. K. Musha, H. Kuwabara and M. Tsuchiya. Presented at 32nd Ann. Conf. on Mass Spectrom. and Allied Topics, San Antonio, TX, U.S.A. May 27–June 1, 1984. p. 558–559.
84MY39
Quantitation of Natural Triacylglycerols by Reversed Phase Liquid Chromatography with Direct Liquid Inlet Mass Spectrometry. J.J. Myher, A. Kuksis, L. Marai and F.J. Mangamaro. *J. Chromatogr. 283*, 289–301, (1984).
84NO0′
Open Tubular Supercritical Fluid Chromatography. M. Novotny, M.L. Lee, P.A. Peaden, J.C. Fjeldsted and S.R. Springston. US 4,479,380 (Cl. 73–61. 1C; GO1N31/08), 30 Oct. 1984, Appl. 352,890, 26 Feb. 1982.
84OK67
A New Type Surface Ionization Source with an Additional Mode of Electrohydrodynamic Ionization for SIMS. T. Okutani, M. Fukuda, T. Noda, H. Tamura and H. Watanabe. *Springer Ser. Chem. Phys. 36*, 127–129, (1984).
84PA42
The Characterisation of LSD in Illicit Preparations by HPLC and Microbore HPLC/MS. A.C. Patel. *J. Forensic Sci. Soc. 24*, 162, (1984).
84PI13
Direct Monitoring of Sequential Enzymatic Hydrolysis of Peptides by Thermospray Mass Spectrometry. D. Pilosof, H.Y. Kim, M.L. Vestal and D.F. Dyckes. *Biomed. Mass. Spectrom. 11*, 403–407, (1984).
84PI66
Determination of Nonderivatized Peptides by Thermospray Liquid Chromatograph Mass Spectrometry. D. Pilosof, H.Y. Kim, D.F. Dyckes and M.L. Vestal. *Anal. Chem. 56*, 1236–1239, (1984).
84PI3′
Mechanistic Studies on Thermospray Ionization. D. Pilosof, H.Y. Kim and M.L. Vestal. Presented at 32nd Ann. Conf. on Mass Spectrom. and Allied Topics, San Antonio, TX, U.S.A. May 27–June 1, 1984. p. 193–194.
84PI21
Resolution and Quantitation of Diacylglycerol Moieties of Natural Glycerophospholipids by Reversed-Phase Liquid Chromatography with Direct Liquid Introduction/Mass Spectrometry. S. Pind, A. Kuksis, J.J. Myher and L. Marai. *Can. J. Biochem. Cell Biol. 62*, 301–311, (1984).
84PO8′
Optimization of Ion Source in Thermospray LCMS. F. Poeppel, J. Buchner, H.H. Lo and J. Dulak. Presented at 32nd Ann. Conf. on Mass Spectrom. and Allied Topics, San Antonio, TX, U.S.A. May 27–June 1, 1984. p. 98–99.
84QU7′
HPLC-MS Determination of Polycyclic Aromatic Hydrocarbons in Marine Sediments. M.A. Quilliam, R.J. Gergely, M.S. Lant, W.D. Jamieson, E. Lewis and R.M. Gershey. Presented at the

32nd Ann. Conf. on Mass Spectrom. and Allied Topics. San Antonio TX, U.S.A. May 27–June 1, 1984. p. 277–278.

84RI48
Admission of Volatile Organic Impurities in Water into a Mass Spectrometer. A.A. Rivlin, V.G. Karachevtsev, N. Ledneva, K.K. Man'kovskii and K.I. Pruger. *Prib. Tekh. Eksp. 3*, 218–219, (1984).

84RO76
Gas Chromatography-Mass Spectrometry and High-Performance Liquid Chromatography/Mass Spectrometry. M.E. Rose. *Mass Spectrom. 7*, 196–292, (1984).

84SC47
Field Effects in Molecular Ion Formation by the Thermospray Technique. G. Schmelzeisen-Redeker, U. Giessmann and F.W. Röllgen. *J. Phys. Colloq. C9*, 297–302, (1984).

84SC83
Thermospray LC/MS. R. Schubert. *GIT Fachz. Lab. 28*, 323–325, (1984).

84SC59
The Use of Microbore Columns in Liquid Chromatography and On-Line Liquid Chromatography/Mass Spectrometry. H.E. Schwartz. *Diss. Abstr. Int. B 45*, 1769, (1984).

84SK56
Equine Betamethasone Metabolism and Determination of Corticosteroids in Equine Biological Fluids by Micro-LC/MS. D.S. Skrabalak. *Diss. Abstr. Int. B 45*, 1446, (1984).

84SK59
Qualitative Detection of Corticosteroids in Equine Biological Fluids and the Comparison of Relative Dexamethasone Metabolite/Dexamethasone Concentration in Equine Urine by Micro-Liquid Chromatography-Mass Spectrometry. D.S. Skrabalak, T.R. Covey and J.D. Henion. *J. Chromatogr. 315*, 359–372, (1984).

84SM7'
New Developments in High Performance Capillary Supercritical Fluid Chromatography-Mass Spectrometry. R.D. Smith, B.W. Wright and H.R. Udseth. Presented at 32nd Ann. Conf. on Mass Spectrom. and Allied Topics, San Antonio, TX, U.S.A. May 27–June 1, 1984. p. 207–208.

84SM66
Rapid and Efficient Capillary Column Supercritical Fluid Chromatography with Mass Spectrometric Detection. R.D. Smith, H.T. Kalinoski, H.R. Udseth and B.W. Wright. *Anal. Chem. 56*, 2476–2480, (1984).

84SM0'
Characterization of Supercritical Fluids Using Direct Fluid Injection-Mass Spectrometry. R.D. Smith, H.T. Kalinoski and H.R. Udseth. Presented at 32nd Ann. Conf. on Mass Spectrom. and Allied Topics, San Antonio, TX, U.S.A. May 27–June 1, 1984. p. 560–561.

84SM95
Supercritical Fluid Methods in Analytical Chemistry. R.D. Smith, B.W. Wright and H.R. Udseth. *Anal. Chem. Symp. Ser. 19*, 375–380, (1984).

84SM61
Capillary Supercritical Fluid Chromatography/Mass Spectrometry with Electron Impact Ionization. R.D. Smith, H.R. Udseth and H.T. Kalinoski. *Anal. Chem. 56*, 2971–2973, (1984).

84SM10'
Selected Application of LC/MS to Problems in Biomedical Analysis. R.W. Smith, K. Yamaguchi, C.E. Parker, H.B. Matthews, J.R. Hass, T. Nagano, I. Fridovich, I.P. Lee, I.A.S. Lewis and I. Parr. Presented at 32nd Ann. Conf. on Mass Spectrom. and Allied Topics, San Antonio, TX, U.S.A. May 27–June 1, 1984. p. 110.

84ST97
The Operation and Application of Thermospray LC/MS [Liquid Chromatography/Mass Spectrometry] in Biochemistry. M.S. Story, W.H. McFadden and M. Tucker. *Iyo Masu Kenkyukai Koenshu 9*, 207–212, (1984).

84ST3'
A Simplified Moving Belt Interface for Liquid Chromatography-Mass Spectrometry. S.J. Stout and A.R. Da Cunha. Presented at 32nd Ann. Conf. on Mass Spectrom. and Allied Topics, San Antonio, TX, U.S.A. May 27–June 1, 1984. p. 113–114.

84ST9'
Liquid Chromatography/Fast Atom Bombardment Mass Spectrometry (LC/FABMS). J.G. Stroh, J.C. Cook, K.L. Rinehart, Jr. and I.A.S. Lewis. Presented at 32nd Ann. Conf. on Mass Spectrom. and Allied Topics, San Antonio, TX, U.S.A. May 27–June 1, 1984. p. 119–120.

84SU91
Mass Spectrometric Investigation on Hydration of Nucleic Acid Components in Vacuum. II. N-Methylated Adenines. L.F. Sukhodub, V.S. Shelkovskii and K.L. Wierzchowski. *Biophys. Chem. 19*, 191–200, (1984).

84TA59
Thermospray LC/MS of Nonvolatile Biomolecules. P.I. Tao. *Diss. Abstr. Int. B 45*, 2899–2900, (1984).

84TH0'
Ion Emission Processes at Atmospheric Pressure: A Comparison of Thermospray and Ion Evaporation. B.A. Thomson. Presented at 32nd Ann. Conf. on Mass Spectrom. and Allied Topics, San Antonio, TX, U.S.A. May 27–June 1, 1984. p. 190–191.

84TH44
Application of a Spray Deposition Method for Reversed Phase Liquid Chromatography-Mass Spectrometry. A.D. Thruston. *Govt. Rep. Announce. Index (U.S.) 84*, 74, (1984).

84TS23
Ionization at a Liquid Surface and Ionization under Normal Pressure. M. Tsuchiya and H. Kawahara. *Shitsuryo Bunseki 32*, 443R–477R, (1984).

84TS00
Studies on the Structure and Physical and Chemical Properties of Organic Compounds by Liquid Ionization Mass Spectrometry. M. Tsuchiya, H. Kuwabara and A. Hasegawa. *Nippon Kagaku Kaishi 10*, 1550–1557, (1984).

84UD2'
Studies of Supercritical Fluid Extraction and Fractionation Processes for Polar and Mixed Fluids by DFI-MS. H.R. Udseth and R.D. Smith. Presented at 32nd Ann. Conf. on Mass Spectrom. and Allied Topics, San Antonio, TX, U.S.A. May 27–June 1, 1984. p. 542–543.

84VA4AD'
The Applicability of Field Desorption Mass Spectrometry and Liquid Chromatography/Mass Spectrometry for the Analysis of Pungent Principles of Capsicum and Black Pepper. J. Van Der Greef, L.M. Nijssen, H. Maarse, M. Ten Noever De Brauw, D.E. Games and N.J. Alcock. In: *Progress in Flavour Research 1984, Proc. 4th Weurman Flavour Res. Symp.* , *Dourdan, France, May 9– 11, 1984* (J. Adds, ed.), pp. 603–612, Elsevier, Amsterdam, 1984.

84VE60
Thermospray Liquid Chromatographic Interface for Magnetic Mass Spectrometers. M.L. Vestal. *Anal. Chem. 56*, 2590–2592, (1984).

84VE65
High Performance Liquid Chromatography-Mass Spectrometry. M.L. Vestal. *Science 226*, 275–281, (1984).

84VO62
Modification and Optimization of Source and Interface Parameters for Direct Liquid Introduction Liquid Chromatography/Mass Spectrometry of Pesticides. R.D. Voyksner and J.T. Bursey. *Anal. Chem. 56*, 1582–1587, (1984).

84VO7'
Enhancement of the Amount of Information Obtained by Thermospray HPLC/MS. R.D. Voyksner, J.T. Bursey and C.A. Haney. Presented at 32nd Ann. Conf. on Mass Spectrom. and Allied Topics, San Antonio, TX, U.S.A. May 27–June 1, 1984. p. 197–198.

84VO78'
Analysis of Candidate Anti-Cancer Drugs by Thermospray HPLC/MS. R.D. Voyksner, J.T. Bursey
and J. Hines. Presented at 32nd Ann. Conf. on Mass Spectrom. and Allied Topics, San Antonio,
TX, U.S.A. May 27–June 1, 1984. p. 117–118.

84VO16
A Comparison of Thermospray and Direct Liquid Introduction High Performance Liquid
Chromatography/Mass Spectrometry for the Analysis of Candidate Antimalarials. R.D. Voyksner,
J.T. Bursey, J.W. Hines and E.D. Pellizzari. *Biomed. Mass Spectrom.* 11, 616–621, (1984).

84VO5'
Post-Column Addition of Buffer for Thermospray Liquid Chromatography Mass Spectrometry. R.D.
Voyksner, J.T. Bursey and E.D. Pellizzari. Presented at 32nd Ann. Conf. on Mass Spectrom. and
Allied Topics, San Antonio, TX, U.S.A. May 27–June 1, 1984. p. 695–696.

84VO21
Analysis of Selected Pesticides by High-Performance LC/MS. R.D. Voyksner, J.T. Bursey and E.D.
Pellizzari. *J. Chromatogr.* 312, 221–235, (1984).

84VO67
Postcolumn Addition of Buffer for Thermospray Liquid Chromatography/Mass Spectrometric Identification of Pesticides. R.D. Voyksner, J.T. Bursey and E.D. Pellizzari. *Anal. Chem.* 56, 1507–
1514, (1984).

84VO3'
A Comparison of Direct Liquid Introduction and Thermospray HPLC/MS for the Analysis of
Pesticides. R.D. Voyksner, J.T. Bursey and E.D. Pellizzari. Presented at the 32nd Ann. Conf. on
Mass Spectrom. and Allied Topics. San Antonio TX, U.S.A. May 27–June 1, 1984. p. 3–4.

84WE9'
Direct Determination of Acetylcholine in Mouse Brain by Thermospray-Mass Spectrometry. S.T.
Weintraub, D.J. Liberato and A.L. Yergey. Presented at 32nd Ann. Conf. on Mass Spectrom. and
Allied Topics, San Antonio, TX, U.S.A. May 27–June 1, 1984. p. 129–130.

84WE18
Microbore HPLC and Its Uses with Mass Spectrometry. S.A. Westwood. *Anal. Proc. (London) 21*,
418–420, (1984).

84WH93
Recent Developments in Detection Techniques for HPLC. Part II. Other Detectors. P.C. White.
Analyst (London) 109, 973–974, (1984).

84WH2'
The Electrospray Ion (ESPI) Source as an LC/MS Interface. C.M. Whitehouse, J.B. Fenn and R.N.
Dreyer. Presented at 32nd Ann. Conf. on Mass Spectrom. and Allied Topics, San Antonio, TX,
U.S.A. May 27–June 1, 1984. p. 92–93.

84WH8'
An Electrospray Ion Source for Mass Sepctrometry of Fragile Organic Species. C.M. Whitehouse,
J.B. Fenn and M. Yamashita. Presented at 32nd Ann. Conf. on Mass Spectrom. and Allied
Topics, San Antonio, TX, U.S.A. May 27–June 1, 1984. p. 188–189.

84WH2OG'
Ion Pumping and Free Jet Expansion. C.M. Whitehouse, M. Yamashita, C.K. Meng and J.B. Fenn.
In: *Rarified Gas Dynamics, Proceedings of the 14th International Symposium* (H. Oguchi, ed.),
Vol. 2, p. 857–864, Tokyo Univ. Press, Tokyo (1984).

84WI65
Monodisperse Aerosol Generation Interface for Combining Liquid Chromatography with Mass Spectrometry. R.C. Willoughby and R.F. Browner. *Anal. Chem.* 56, 2625–2631, (1984).

84WI2'
Monodisperse Aerosol Generation Interface for Combining Liquid Chromatography with Mass Spectrometry (MAGIC). Some Theoretical Considerations. R.C. Willoughby and R.F. Browner. Presented at 32nd Ann. Conf. on Mass Spectrom. and Allied Topics, San Antonio, TX, U.S.A. May
27–June 1, 1984. p. 192–193.

84WI47
Studies with an Aerosol Generation Interface for LC-MS. R.C. Willoughby. *Diss. Abstr. Int. B. 44*, 2417, (1984).

84WO81
Peak Identification. M.J. Wojtusik. in: *HPLC in Nucleic Acid Research: Methods and Applications* (P.R. Brown, ed.) *Chromatogr. Sci. Series*, Vol. 28, pp. 81–98, Marcel Dekker, New York (1984).

84WR43
Supercritical Fluid Chromatography and Supercritical Fluid Chromatography/Mass Spectrometry of Marine Diesel Fuel. B.W. Wright, H.R. Udseth, R.D. Smith and R. N. Hazlett. *J. Chromatogr. 314*, 253–262, (1984).

84YA81
Negative Ion Production with the Electrospray Ion Source. M. Yamashita and J.B. Fenn. *J. Phys. Chem. 88*, 4671–4675, (1984).

84YA819
Electrospray Ion Source, Another Variation on the Free-Jet Theme. M. Yamashita and J.B. Fenn. *J. Phys. Chem. 88*, 4451–4459, (1984).

84YA93
Application of Electrospray Mass Spectrometry in Medicine and Biochemistry. M. Yamashita and J.B. Fenn. *Iyo Masu Kenkyukai Koenshu 9*, 203–206, (1984).

84YA62
A New Transport Detector for High-Performance Liquid Chromatography Based on Thermospray Ionization. L. Yang, G.J. Fergusson and M.L. Vestal. *Anal. Chem. 56*, 2632–2636, (1984).

84YE0'
Analysis of Intractable Endogenous Molecules by Thermospray LC/MS. A.L. Yergey and D.L. Liberato. Presented at 32nd Ann. Conf. on Mass Spectrom. and Allied Topics, San Antonio, TX, U.S.A. May 27–June 1, 1984. p.90–91.

84YE98
Thermospray LC/MS for the Analyses of L-Carnitine and Its Short-Chain Acyl Derivatives. A.L. Yergey, D.J. Liberato and D.S. Millington. *Anal. Biochem. 139*, 278–283, (1984).

84YE4'
Thermospray HPLC-MS of Prostaglandins and Oxygenated Metabolites of Arachidonic Acid (C20:4) and Docosahexaenoic Acid (C22:6). J.A. Yergey and N. Salem, Jr. Presented at 32nd Ann. Conf. on Mass Spectrom. and Allied Topics, San Antonio, TX, U.S.A. May 27–June 1, 1984. p.104–105.

84YI6'
Metabolic Studies of Explosives. LC/MS of Metabolites of 2,4,6-Trinitrotoluene. J. Yinon and D.G. Hwang. Presented at 32nd Ann. Conf. on Mass Spectrom. and Allied Topics, San Antonio, TX, U.S.A. May 27–June 1, 1984. p.86–87.

84YO88
New Methods for Characterization of Supercritical Fluid Solutions. C.R. Yonker, B.W. Wright, H.R. Udseth and R.D. Smith. *Ber. Bunsen-Ges. Phys. Chem. 88*, 908–911, (1984).

84YU15
Sequence Analysis of Derivatized Peptides by High Performance Liquid Chromatography/Mass Spectrometry. T.J. Yu, H.A. Schwartz, S.A. Cohen, P. Vouros and B.L. Karger. *J. Chromatogr. 301*, 425–440, (1984).

85AB8'
Application of an Improved Monodisperse Aerosol Generation LC/MS Interface to Selected Non-Volatile Samples. L.E. Abbey, D.E. Bostwick, P.C. Winkler and R.F. Browner. Presented at 33rd Ann. Conf. on Mass Spectrom. and Allied Topics, San Diego, CA, U.S.A. May 26–31, 1985, p.778–779.

85AL37
Direct Coupling of Packed Fused-Silica Liquid Chromatographic Columns to a Magnetic Sector
Mass Spectrometer and Application to Polar Thermolabile Compounds. H. Alborn and G. Sten-
hagen. *J. Chromatogr. 323*, 47–66, (1985).
85AL8'
Direct Coupling of Packed Fused-Silica Liquid Chromatographic Columns to a Magnetic Sector
Mass Spectrometer and Application to Polar Thermolabile Compounds. H. Alborn and G. Sten-
hagen. *Am. Chem. Soc. Abstr. 190*, AGFD0118, (1985).
85AL45
HPLC and LC-MS of Nucleosides. F.C. Alderweireldt, E.L. Esmans and P. Geboes. *Nucleos.
Nucleot. 4*, 135–137, (1985).
85AL9'
Thermospray Ionization Contributions of Gas Phase Chemical Ionization Reactions to Observed
Mass Spectrum. A.J. Alexander and P. Kebarle. Presented at 33rd Ann. Conf. on Mass Spectrom.
and Allied Topics, San Diego, CA, U.S.A. May 26–31, 1985, p.849–850.
85AL15
Use of a Novel MS Method to Sequencing of Peptides. M.L. Alexandrov, G.I. Baram, L.N. Gal,
M.A. Grachev, V.D. Knorre, N.V. Krasnov, Y.S. Kusner, O.A. Mirgrodskaya, V.I. Nikolaev
and V.A. Shkurov. *Bioorg. Khim. 11*, 705–708, (1985).
85AL10
Formation of Beams of Quasi-Molecular Ions of Peptides from Solution. M.L. Alexandrov, G.I.
Baram, L.N. Gal, N.V. Krasnov, Y.S. Kusner, O.A. Mirgrodskaya, V.I. Nikolaev and V.A.
Shkurov. *Bioorg. Khim. 11*, 700–704, (1985).
85AN13'
Successful LC/MS Seminar in Bremen. Anonymous. *Anal. News 11*, 3, (1985).
85AR33
Ten Years of Liquid Chromatography/Mass Spectrometry. P.J. Arpino. *J. Chromatogr. 323*, 3–11,
(1985).
85AR45
Optimization of the Instrumental Parameters of a Combined Liquid Chromatography/Mass Spec-
trometer Coupled by an Interface for Direct Liquid Introduction. V. Design and Construction of
LC/MS Interfaces Utilizing Fused Silica Capillary Tubes as Vacuum Nebulizers. P.J. Arpino and
C. Beaugrand. *Int. J. Mass Spectrom. Ion Processes 64*, 275–298, (1985).
85AS79
A New Instrument for GC/LC-MS. A.E. Ashcroft, J.R. Chapman, M. Kearns, M.F. Almond and
J.E. Pratt. *Am. Lab. 17*, 59–64, (1985).
85BA25
Liquid Chromatography/Mass Spectrometry of Fatty Acids as Their Anthrylmethyl Esters. J.D.
Baty, R.G. Willis and R. Tavendale. *Biomed. Mass Spectrom. 12*, 565–569, (1985).
85BE26
Analysis of Non-Volatile Nitrosamines by Moving Belt LC/MS. I.G. Beattie, D.E. Games, J.R.
Startin and J. Gilbert. *Biomed. Mass Spectrom. 12*, 616–622, (1985).
85BE2'
Coupling Microbore High Performance Liquid Chromatography to a Time-of-Flight Mass Spec-
trometer. R.C. Beavis, G. Bolbach, W. Ens, D.E Main, B. Schueler, K.G. Standing and J.B.
Westmore. Presented at 33rd Ann. Conf. on Mass Spectrom. and Allied Topics, San Diego, CA,
U.S.A. May 26–31, 1985, p.202–203.
85BE9'
LC/MS of an Aquatic Fulvic Acid Sample Hydrolyzed by Enzymes. D.J. Bertino, P.W. Albro, R.F.
Christman and J.R. Hass. Presented at 33rd Ann. Conf. on Mass Spectrom. and Allied Topics,
San Diego, CA, U.S.A. May 26–31, 1985, p.289–290.

85BE8'
Thermospray Ionization and Mass Spectrometry/Mass Spectrometry of Dyes. L.F. Betowski and J.M. Ballard. Presented at 33rd Ann. Conf. on Mass Spectrom. and Allied Topics, San Diego, CA, U.S.A. May 26–31, 1985, p.958–959.

85BI23
The Mass Spectrometer as a Detector in Chromatography. K. Biemann. In: *The Science of Chromatography* (F. Bruner, ed.), *J. Chromatogr. Libr. 32*, 43–54, (1985).

85BL3'
"Filament On" Thermospray LC/MS Analysis of Urea Herbicides. C.R. Blakley and D.A. Garteiz. Presented at 33rd Ann. Conf. on Mass Spectrom. and Allied Topics, San Diego, CA, U.S.A. May 26–31, 1985, p.563–564.

85BL9'
Identification and Quantification of Nicotine Biosynthetic Intermediates in Tobacco Roots by LC/MS and HPLC. D.L. Blume and H.L Chung. Presented at 33rd Ann. Conf. on Mass Spectrom. and Allied Topics, San Diego, CA, U.S.A. May 26–31, 1985, p.749–750.

85BR4'
On-Line Precolumn Trace Enrichment and Post-Column Extraction in HPLC with Mass Spectrometric Detection. U.A. Th. Brinkman, R.W. Frei and F.A. Maris. *Am. Chem. Soc. Abstr. 189*, ANYL0064, (1985).

85BR73
An Exceedingly Simple Mass Spectrometer Interface with Application to Reaction Monitoring and Environmental Analysis. J.S. Brodbelt and R.G. Cooks. *Anal. Chem. 57*, 1153–1155, (1985).

85BR39
Developments in Interfacing a Microbore High Performance Liquid Chromatograph with Mass Spectrometry (A Review). A.P. Bruins. *J. Chromatogr. 323*, 99–111, (1985).

85BU41
Evaluation of Reverse Phase Liquid Chromatography-Mass Spectrometry for Estimations of n-Octanol/Water Partition Coefficients for Organic Chemicals. P. Burkhard, D.W. Kuehl and G.D. Veith. *Chemosphere 14*, 1551–1560, (1985).

85BU3'
Thermospray HPLC/MS Analysis of Pesticides. J.T. Bursey and R.D. Voyksner. *Am. Chem. Soc. Abstr. 189*, PEST0013, (1985).

85BU77
Gas Phase Ionization of Selected Neutral Analytes during Thermospray Liquid Chromatography/Mass Spectrometry. M.M. Bursey, C.E. Parker, R.W. Smith and S.J. Gaskell. *Anal. Chem. 57*, 2597–2599, (1985).

85BU8'
Design and Performance of a Secondary Ion Mass Spectrometer for Chromatographic and Electrochemical Analyses. K.L. Busch, G.C. Di Donato, K.J. Kroha, R.A. Flurer and J.W. Fiola. Presented at 33rd Ann. Conf. on Mass Spectrom. and Allied Topics, San Diego, CA, U.S.A. May 26–31, 1985, p.388–389.

85CA33
Applications of Thermospray Ionization Liquid Chromatography-Mass Spectrometry to Compounds of Pharmaceutical Interest. D.A. Catlow. *J. Chromatogr. 323*, 163–170, (1985).

85CE20
Identification of Fatty Acid Esters of Chloropropanediol in Milk Fats by LC/MS. J. Cerbulis, O.W. Parks, H.M. Farrell, A. Kuksis, L. Marai and J.J. Myher. *J. Am. Oil Chem. Soc. 62*, 610, (1985).

85CH79
Parallel Mass Spectrometry for High Performance GC and LC Detection. C. Chang. *Am. Lab. 17*, 59–64, 66, (1985).

85CH7KA'
Direct Analysis of TLC Spots by FAB Mass Spectrometry. T.T. Chang and F. Andrawes. In:

Proceedings of the 3rd International Symposium on Instrumentation for High Performance Thin-Layer Chromatography. R.E. Kaiser (ed.), Institute for Chromatography, Bad Dürkheim, FRG, 1985, p.427–433.
85CH33
Liquid Chromatograph-Mass Spectrometry Interfacing on a Magnetic Sector Mass Spectrometer. Techniques and Application. J.R. Chapman. *J. Chromatogr. 323,* 153–161, (1985).
85CH7'
MS25RFA: A New Instrument for Combined Chromatography/Mass Spectrometry. J. Chapman, M. Kearns, J. Pratt and M. Almond. Presented at 33rd Ann. Conf. on Mass Spectrom. and Allied Topics, San Diego, CA, U.S.A. MAy 26–31, 1985, p.567–568.
85CH3'
Thermospray LCMS on a Magnetic Sector Instrument, Theory and Application. J.R. Chapman and J.A.E. Pratt. *Am. Chem. Soc. Abstr. 189,* ANYL0003, (1985).
85CH47
LC/MS: Developments and Applications. J.R. Chapman and J.A.E. Pratt. *Lab. Pract. 34,* 7–11, (1985).
85CH336
Determination of Dibenzothiophene in Oils by Liquid Chromatography-Tandem/Mass Spectrometry. R.G. Christensen and E. White V. *J. Chromatogr. 323,* 33–36, (1985).
85CO2'
The Role of Transport and Intrinsic Efficiency in Sampling Ions from Solutions. K.D. Cook. *Am. Chem. Soc. Abstr. 189,* ANYL0002, (1985).
85CO12'
Practical Aspects of Electrohydrodynamic Ionization. K.D. Cook. Presented at 33rd Ann. Conf. on Mass Spectrom. and Allied Topics, San Diego, CA, U.S.A. May 26–31, 1985, p.91–92.
85CO55
Observation of Transition Metal Complexes in Aqueous Solution by Fast Atom Bombardment Mass Spectrometry. F.H. Cottee, J.A. Page, D.J. Cole-Hamilton and D.W. Bruce. *J. Chem. Soc. Chem. Commun. 1985*(21), 1525–1526, (1985).
85CO1'
Mechanisms of Ion Formation in Thermospray and Other Matrix Assisted Desorption Techniques. R.J. Cotter and C. Fenselau. *Am. Chem. Soc. Abstr. 189,* ANYL0001, (1985).
85CO0'
Thermospray LC/MS as a Quantitative Tool for the Measurement of Stable Isotope Labeled Metabolites in Urine. J.E. Coutant and D.K. Satonin. Presented at 33rd Ann. Conf. on Mass Spectrom. and Allied Topics, San Diego, CA, U.S.A. May 26–31, 1985, p.780–781.
85CO74
Thermospray Liquid Chromatography/Mass Spectrometry Determination of Drugs and Their Metabolites in Biological Fluids. T.R. Covey, J.B. Crowther, E.D. Dewey and J.D. Henion. *Anal. Chem. 57,* 474–481, (1985).
85CO3'
Determination of Estrogenic Mycotoxins, Plant Estrogens and Steroidal Estrogen in Animal Tissues by Direct Insertion Probe MS/MS and HPLC/MS/MS. T.R. Covey, G.A. Maylin and J.D. Henion. Presented at 33rd Ann. Conf. on Mass Spectrom and Allied Topics, San Diego, CA, U.S.A. May 26–31, 1985, p.813.
85CR3'
SFC/MS Using a Total Effluent Direct Liquid Introduction Interface. J.B. Crowther and J.D. Henion. Presented at 33rd Ann. Conf. on Mass Spectrom. and Allied Topics, San Diego, CA, U.S.A. May 26–31, 1985. p.723–724.
85CR30
Direct Liquid Introduction LC/MS: Four Different Approaches. J.B. Crowther, T.R. Covey, D. Silvestre and J.D. Henion. *LC Mag. 3,* 240–254, (1985).

85CR71
Supercritical Fluid Chromatography of Polar Drugs Using Small-Particle Packed Columns with Mass
Spectrometric Detection. J.B. Crowther and J.D. Henion. *Anal. Chem. 57*, 2711–2716, (1985).

85DA49
Determination of Bile Acids by Combined Liquid Chromatography-Mass Spectrometry. S. Da and
D. Fraisse. *Fenxi Ceshi Tongbao 4*(3), 9–13, (1985).

85DA7′
A Simplified Moving-Belt Interface for Liquid Chromatography/Mass Spectrometry: Recent Ad-
vances. A.R. Da Cunha and S.J. Stout. Presented at 33rd Ann. Conf. on Mass Spectrom. and
Allied Topics, San Diego, CA, U.S.A. May 26–31, 1985, p.777–778.

85DA22
Combined Thin Layer Chromatography/Mass Spectrometry: An Application of 252-Cf Plasma De-
sorption Mass Spectrometry for Drug Monitoring. H. Danigel, L. Schmidt, H. Jungclas and K.H.
Pfluger. *Biomed. Mass Spectrom. 12*, 542–544, (1985).

85DA51
Drug Monitoring of Etoposide (VP16–213) 1. A Combined Method of Liquid Chromatography and
Mass Spectrometry. H. Danigel, K.H. Pfluger, H. Jungclas, L. Schmidt and J. Dellbrugge.
Cancer Chemother. Pharmacol. 15, 121–125, (1985).

85DE0′
Supercritical Fluid Chromatography (SFC) with Simultaneous Mass Spectrometric and Flame Ioniza-
tion Detection. S. Deluca, G.U. Holzer and K.J. Voorhees. Presented at 33rd Ann. Conf. on Mass
Spectrom. and Allied Topics, San Diego, CA, U.S.A. May 26–31, 1985, p.390–391.

85DE8′
Some Experiences with Thermospray Ionization. A.J. De Stefano, J.D. Henion and J. Crowther.
Presented at 33rd Ann. Conf. on Mass Spectrom. and Allied Topics, San Diego, CA, U.S.A. May
26–31, 1985, p.788–789.

85DE6′
A Comparison of Fast Atom Bombardment, Thermospray and Ammonia Chemical Ionization for the
Analysis of Glucuronides. A.J. De Stefano and T. Keough. Presented at 33rd Ann. Conf. on Mass
Spectrom. and Allied Topics, San Diego, CA, U.S.A. May 26–31, 1985, p.386–387.

85DI4′
Ammonium Hydroxide Mobile Phase for Thermospray Ionization. D.K. Dietrich, J.M. Warner, D.J.
Augustin and M.J. Palmer. Presented at 33rd Ann. Conf. on Mass Spectrom. and Allied Topics,
San Diego, CA, U.S.A. May 26–31, 1985, p.384–385.

85EC25
Combined Liquid Chromatography/Mass Spectrometry of Drugs. C. Eckers and J.D. Henion.
Chromatogr. Sci. 32, 115–149, (1985).

85ED6′
Thermospray LC-MS of Mononucleotides. C.G. Edmonds, F.F. Hsu and J.A. McCloskey. Presented at 33rd Ann. Conf. on Mass Spectrom. and Allied Topics, San Diego, CA, U.S.A. May
26–31, 1985. p.516–517.

85ED4′
Analysis of Heterocycles by Combined LC/MS. C.G. Edmonds, J.A. McCloskey, H. Kasai and S.
Nishimura. *Am. Chem. Soc. Abstr. 189*, ANYL0084, (1985).

85ED37
Thermospray Liquid Chromatography-Mass Spectrometry of Nucleosides and of Enzymic Hydrolysates of Nucleic Acids. C.G. Edmonds, M.L. Vestal and J.A. McCloskey. *Nucleic Acids Res. 13*,
8197–8206, (1985).

85ED45′
Detection of Naturally Modified Nucleosides in Enzymatic Hydrolysates of Transfer RNA by Combination HPLC/MS. C.G. Edmonds, T.C. McKee and J.A. McCloskey. Presented at 33rd Ann.

Conf. on Mass Spectrom. and Allied Topics, San Diego, CA, U.S.A. May 26–31, 1985. p.514–515.

85EG47

Mass Spectrometry in the Study of Fossil Porphyrins. G. Eglinton. In: *Mass Spectrometry in the Health and Life Sciences* (A.L. Burlingame and N Castagnoli, Jr., eds.), *Anal. Chem. Symp. Ser.* 24, 47–64, (1985).

85ER26

Analysis of Lipids by High-Performance Liquid Chromatography Chemical Ionization Mass Spectrometry. W.L. Erdahl and O.S. Privett. *J. Am. Oil Chem. Soc.* 62, 786–792, (1985).

85ES8′

DLI/LC-MS Experiments with Nucleosides. E.L. Esmans, Y. Luyten and F.C. Alderweireldt. *Am. Chem. Soc. Abstr.* 189, ANYL0068, (1985).

85ES21

Direct Liquid Introduction LC/MS Microbore Experiments for the Analysis of Nucleoside Material Present in Human Urine. E.L. Esmans, P. Geboes, Y. Luyten and F.C. Alderweireldt. *Biomed. Mass Spectrom.* 12, 241–245, (1985).

85ES3′

Measurement of Metabolic Kinetics by Thermospray LC/MS and Stable Isotopic Tracers. N.V. Esteban, D.J. Liberato and A.L. Yergey. *Am. Chem. Soc. Abstr.* 189, ANYL0083, (1985).

85FA66

Development and Evaluation of a Moving Belt Liquid Chromatograph-Mass Spectrometer with Laser and Ion Desorption. T.-P. Fan. *Diss. Abstr. Int. B.* 46, 1536–1537, (1985).

85FA1′

Studies of Matrix Effects in Laser Desorption and Secondary Ion Mass Spectrometry with a Moving Belt LC/MS. T.P. Fan and M.L. Vestal. Presented at 33rd Ann. Conf. on Mass Spectrom. and Allied Topics, San Diego, CA, U.S.A. May 26–31, 1985, p.381–382.

85FE41

Analysis of Glucuronides Using Condensed Phase Ionization Techniques. C. Fenselau. In: *Mass Spectrometry in the Health and Life Sciences* (A.L. Burlingame and N. Castagnoli, Jr., eds.), *Anal. Chem. Symp. Ser.* 24, 321–331, (1985).

85FE1′

New Instruments, New Methods and the Search for Selectivity. C. Fenselau. *Am. Chem. Soc. Abstr.* 189, PEST0011, (1985).

85FE78

Comparison of Thermospray and Fast Atom Bombardment Mass Spectrometry as Solution-Dependant Ionization Techniques-Correction. C. Fenselau, D.J. Liberato, J.A. Yergey, R.J. Cotter and A.L. Yergey. *Anal. Chem.* 57, 1168, (1985).

85FE7GE′

Fast Atom Bombardment and Thermospray Mass Spectrometry. C. Fenselau and B.S. Larsen. In: *Drug Metabolism Proc. 9th European Workshop 1984* (S. Gerard, ed.), pp.167–173, Pergamon, Oxford (1985).

85FI26

Mass Spectrometry in Chemistry. F.H. Field. *Biomed. Mass Spectrom.* 12, 626–630, (1985).

85FI4′

Thermal Injection Ion Source and Improved Method for Its Efficiency. Finnigan Corp. Jpn. Kokai Tokkyo Koho JP 60,237,354 [85,237,354] (Cl. G01N27/62), 26 Nov. 1985, US Appl. 579,211, 10 Feb. 1984; 5 pp.

85FO9′

Routine Analysis of Carbamate and Carbamoyl Pesticides via Direct Aqueous Injection HPLC/MS. D. Foerst. Presented at 33rd Ann. Conf. on Mass Spectrom. and Allied Topics, San Diego, CA, U.S.A. May 26–31, 1985. p.669–670.

85FR31
Preface: Proceedings of 3rd Symposium/Workshop on LC/MS and MS-MS, Montreux, Oct. 24–26, 1984. R.W. Frei. *J. Chromatogr. 323*, 1–2, (1985).

85FR7BR'
Combination Techniques (Gas Chromatography-Mass Spectrometry, Liquid Chromaotgraphy-Mass Spectrometry) in Biological Research. A. Frigerio, C. Marchioro and A.M. Pastorino. In: *Topics in Pharmaceutical Sciences, Proceedings of the 45th International Congress in the Pharmaceutical Sciences of the Fédération Internationale Pharmaceutical*, (D.D. Breimer and P. Speiser, eds.), p.297–310, Elsevier, Amsterdam (1985).

85GA37
High Performance Liquid Chromatography-Mass Spectrometry of Derivatized and Underivatized Amino Acids. D.E. Games and E.D. Ramsey. *J. Chromatogr. 323*, 67–69, (1985).

85GA114
A Sensible Approach for Adding Thermospray LC/MS to Your Laboratory. D.A. Garteiz. *Vestec Thermospray Newsletter 1*, 1, 4, (1985).

85GA34
Thermospray LC/MS Interface: Principles and Applications. D.A. Garteiz and M.L. Vestal. *LC Magazine 3*, 334–346, (1985).

85GA11
Thermospray LC/MS with Negative Ion Detection. D.A. Garteiz. *Vestec Thermospray Newsletter 1*, 1–2 (1985).

85GO5'
Liquid Chromatography-Mass Spectrometry Coupling for the Analysis of Both Polar and Non-Polar Compounds by Thermospray. P.C. Goodley. *Am. Chem. Soc. Abstr. 189*, ANYL0005, (1985).

85GO57
Liquid Chromatography-Mass Spectrometry: Techniques and Applications. A. Gouyette. *Drug Fate Metab. 5*, 247–272, (1985).

85HA84
Drug Metabolism, Pharmacokinetics and Toxicity. D.J. Harvey. *Mass Spectrom. 8*, 284–332, (1985).

85HA53
On-Line Liquid Chromatography/Mass Spectrometry Using Spray Deposition with a Moving Belt Interface. M.J. Hayes. *Diss. Abstr. Int. B 45*, 2133, (1985).

85HE78
State Analysis of Fermentation Using a Mass Spectrometer with Membrane Probe. E. Heinzle, H. Kramer and I.J. Dunn. *Biotechnol. Bioeng. 27*, 238–246, (1985).

85HE4'
Comparison of the Cornell and Vestec Thermospray LC/MS Interfaces. J. Henion and J. Crowther. Presented at 33rd Ann. Conf. on Mass Spectrom. and Allied Topics, San Diego, CA, U.S.A. May 26–31, 1985. p. 794–795.

85HE5'
LC/MS/MS for Rapid Screening and Identification of Drugs in Biological Samples. J. Henion and T. Covey. Presented at 33rd Ann. Conf. on Mass Spectrom. and Allied Topics, San Diego, CA, U.S.A. May 26–31, 1985. p. 575–576.

85HE7'
Recent Applications of Tandem Mass Spectrometry to Liquid Chromatography/Mass Spectrometry. J.D. Henion, T.R. Covey and J. Crowther. *Am. Chem. Soc. Abstr. 189*, ANYL0037, (1985).

85HE03
Instrumentation and Applications of Micro-Liquid Chromatography/Mass Spectrometry. J.D. Henion. In: *Microcolumn Separations* (M.V. Novotny and D. Ishii, eds.), *J. Chromatogr. Libr. 30*, 243–274, (1985).

85HE01
Forensic Mass Spectrometry in Equine Drug Testing. J. Henion, T. Covey and G. Maylin. *Spectra 10*, 21–28, (1985).

85HI41
The Effects of Oxygen on Hydrogen Production by Rumen Holotrich Protozoa as Determined by Membrane Inlet Mass Spectrometry. K. Hillman, D. Lloyd, R.I. Scott and A.G. Williams. *Spec. Publ. Soc. Gen. Microbiol. 14*, 271–277, (1985).

85HI39
Micro Column Liquid Chromatography/Mass Spectrometry Using a Capillary Interface. P. Hirter, H.J. Walther and P. Watwyler. *J. Chromatogr. 323*, 89–98, (1985).

85HI3′
Atmospheric Pressure Ionization Unit for Tandem Liquid Chromatograph-Mass Spectrometer. Hitachi Ltd. Jpn. Kokai Tokkyo Koho JP 60,127,453 (85,127,453) (Cl. G01N27/62), 8 Jul. 1985, Appl 83/233,668, 13 Dec. 1983, 4 pp.

85HO9′
The Determination of ^{15}N Enrichment of NO_2^-, and NO_3^- Using Thermospray LC/MS. L.R. Hogge, M.L. Vestal and R.K. Hynes. Presented at 33rd Ann. Conf. on Mass Spectrom. and Allied Topics, San Diego, CA, U.S.A. May 26–31, 1985. p. 629–630.

85HO2′
The Moving Belt LC/MS Interface Applied to Fat Research. M. Höhn, U. Rapp, A. Dielmann and E. Schulte. Presented at 33rd Ann. Conf. on Mass Spectrom. and Allied Topics, San Diego, CA, U.S.A. May 26–31, 1985. p. 702–703.

85HO88
Capillary Supercritical Chromatography with Simultaneous Flame Ionization and Mass Spectrometric Detection. G. Holzer, S. Deluca and K.J. Voorhees. *J. High Resolut. Chromatogr. Chromatogr. Commun. 8*, 528–531, (1985).

85HO9SA′
Capillary Supercritical Fluid Chromatography with Simultaneous Flame Ionization and Mass Spectrometric Detection. G. Holzer, S. Deluca and K.J. Voorhees. In: *Proceedings of the 6th International Symposium on Capillary Chromatography* (P. Sandra, ed.), p. 919–924, Huethig, Heidelberg (1985).

85HS14
Remedy for Blocked Thermospray Vaporizer. F.F. Hsu and C.G. Edmonds. *Vestec Thermospray Newsletter 1*, 4, (1985).

85IT61
Direct Coupling of Micro High-Performance Liquid Chromatography with Fast Atom Bombardment Mass Spectrometry. Y. Ito, T. Takeuchi, D. Ishii and M. Goto. *J. Chromatogr. 346*, 161–166, (1985).

85JE49
Cyanobacterial Dinitrogen Uptake Measurements Using Membrane-Inlet Mass Spectrometry. B.B. Jensen and R.P. Cox. *Spec. Publ. Soc. Gen. Microbiol. 14*, 279–285, (1985).

85JE8′
Tandem Liquid Chromatograph-Mass Spectrometer. Jeol Ltd. Jpn. Kokai Tokkyo Koho JP 60,129,668 [85,129,668] (Cl. G01N30/72), 10 Jul. 1985, Appl. 83/239,459, 19 Dec. 1983, 3 pp.

85JE0′
Liquid Chromatograph-Mass Spectrometer. Jeol Ltd. Jpn Kokai Tokkyo Koho JP 60, 0500 [85,01,550] (Cl. GO127/62), 07 Jan. 1985. Appl. 83/108,640 17 Jan. 1983, 4 pp.

85JO90
Clusters of Organic Molecules in a Supersonic Jet Expansion. H.T. Jonkman, U. Evens and J. Kommandeur. *J. Phys. Chem. 85*, 4240–4243, (1985).

85JO28
Thermospray LC/MS Applied to the Analysis of Complex Mixtures of Non-Polar Materials. J.R. Joyce, R.E. Ardrey and I.A.S. Lewis. *Biomed. Mass Spectrom. 12*, 588–592, (1985).

85JU1EI'
Recent Development in Techniques for Phospholipid Analysis. F.B. Jungalwala. In: *Phospholipids in Nerve Tissues* (J. Eichberg, ed.), pp. 1–44, Wiley, New York (1985).

85KA5'
Applications of Rapid Capillary SFC and SFC-MS Analysis. H.T. Kalinoski, H.R. Udseth, B.W. Wright and R.D. Smith. Presented at 33rd Ann. Conf. on Mass Spectrom. and Allied Topics, San Diego, CA, U.S.A. May 26–31, 1985. p. 725–726.

85KA36
Recent Developments in Mass Spectrometry. H. Kambara. *Yuki Gosei Kagaku Kyokaishi 43*, 1106–1117, (1985).

85KA33
A Chromatographic Perspective of High Performance-Liquid Chromatography/Mass Spectrometry. B.L. Karger and P. Vouros. *J. Chromatogr. 323*, 13–32, (1985).

85KA9'
Kinetics of Ion Evaporation from Liquid Droplets Produced by Thermospray. J. Katta, G.J. Fergusson and M.L. Vestal. Presented at 33rd Ann. Conf. on Mass Spectrom. and Allied Topics, San Diego, CA, U.S.A. May 36–31, 1985. p. 769–770.

85KE82
Developments and Trends in Instrumentation. T.R. Kemp. *Mass Spectrom. 8*, 102–122, (1985).

85KI7'
The Moving Belt Interface. A Reliable Combined LC/MS Interface, Which Is Capable of Performing Analyses in Three Ionization Modes. G. Kilpatrick I.A.S. Lewis and J.F. Smith. Presented at 33rd Ann. Conf. on Mass Spectrom. and Allied Topics, San Diego, CA, U.S.A. May 26–31, 1985. p. 377–378.

85KI60
On-Line Peptide Sequencing Using Enzymatic Hydrolysis, High Performance Liquid Chromatography and Thermospray Mass Spectrometry. H.Y. Kim. *Diss Abstr. Int. B 46*(2), 500, (1985).

85KI8'
Phospholipid Molecular Species Analysis by Thermospray LC/MS. H.Y. Kim and N. Salem, Jr. Presented at 33rd Ann. Conf. on Mass Spectrom. and Allied Topics, San Diego, CA, U.S.A. May 26–31, 1985. p. 478–479.

85KR73
Analysis of Polynuclear Aromatic Mixtures by LC/MS. K.J. Krost. *Anal. Chem. 57*, 763–765, (1985).

85KR72
Addendum/Analysis of Polynuclear Aromatic Mixtures by LC/MS. K.J. Krost. *Anal. Chem. 57*, 1792, (1985).

85KU27
Lipid Methodology-Chromatography and Beyond. Part III. Analyses of Natural Deuterium-Labeled Glycerolipids by GC/MS and LC/MS with Specific Enzymic Hdyrolyses. A. Kuksis, J.J. Myher and L. Marai. *J. Am. Oil Chem. Soc. 62*, 767–773, (1985).

85KU22
Lipid Methodology-Chromatography and Beyond. Part II. GC/MS, LC/MS and Specific Enzymic Hydrolysis of Glycerolipids. A. Kuksis, J.J. Myher and L. Marai. *J. Am. Oil Chem. Soc. 62*, 762–767, (1985).

85KU85
Analysis of Lipids on Thin Layer Plates by Matrix-Assisted Secondary Ion Mass Spectrometry. Y. Kushi and S. Handa. *J. Biochem. 98*, 265–268, (1985).

85LA2'
High Field Ionization LC/MS. W.M. Lagna and P.S. Callery. Presented at 33rd Ann. Conf. on Mass Spectrom. and Allied Topics, San Diego, CA, U.S.A. May 26–31, 1985. p. 792–793.
85LA29
Mass Spectrometry of Thermally Unstable Molecules. Evaluation of Ionization Techniques Using Glutamine as a Reference Compound. W.M. Lagna and P.S. Callery. *Biomed. Mass Spectrom.* *12*, 699–703, (1985).
85LA33
Qualitative and Quantitative Analysis of Ranitidine and Its Metabolites by High Performance Liquid Chromatography-Mass Spectrometry. M.S. Lant, L.E. Martin and J. Oxford. *J. Chromatogr.* *323*, 143–152, (1985).
85LE33
Micro-Liquid Chromatography/Mass Spectrometry with Direct Liquid Introduction. E.D. Lee and J.D. Henion. *J. Chromatogr. Sci.* *23*, 253–265, (1985).
85LE59
Separation of Large Coal Molecules Using High Resolution Supercritical Fluid Chromatography. M.L. Lee. Report GRI–85/0179; Order No. PB86–132842/GAR, (1985), 46 pp. , from *Govt. Rep. Announce. Index (U.S.)* 86(5), Abstr. No. 610,664, (1986).
85LE35
On-Line Liquid Chromatography/Mass Spectrometry Analysis of Non-Ionic Surfactants. K. Levsen, W. Wagner-Redeker, K.H. Schäfer and P. Dobberstein. *J. Chromatogr. 323*, 135–141, (1985).
85LI433
Analysis of Acylcarnitines in Human Metabolic Disease by Thermospray LC/MS. D.J. Liberato, D.S. Millington and A.L. Yergey. In: *Mass Spectrometry in the Health Sciences* (A.L. Burlingame and N. Castagnoli, Jr., eds.), *Anal. Chem. Symp. Ser. 24*, 333–348, (1985).
85LI73
Determination of Sulfate and Nitrate Anions in Rain Water by Mass Spectrometry. W.V. Ligon, Jr. and S.B. Dorn. *Anal. Chem. 57*, 1993–1995, (1985).
85LI61
Improvement of Vacuum Nebulizing Interface for Direct Coupling of High Performance Liquid Chromatography with a Mass Spectrometer and Some Applications. H. Liu, K. Matsumoto and S. Tsumie. *Zhejiang Gongxueyuan Xuebao 26*, 1–11, (1985).
85LL19
Quadrupole Mass Spectrometry in the Monitoring and Control of Fermentations. D. Lloyd, S. Bohatka and J. Szilagyi. *Biosensors 1*, 179–212, (1985).
85LU12
Identification of the N-Oxide Metabolite of a Chemotherapeutic Agent by Thermospray LC/MS. K. Lu, R.A. Newman and M.A. McLean. *Vestec Thermospray Newsletter 1*, 2–3, (1985).
85MA75
Electrohydrodynamic Mass Spectrometric Studies of Some Polyether-Cation Complexes. V.F. Man, J.D. Dale and K.D. Cook. *J. Am. Chem. Soc. 107*, 4635–4640, (1985).
85MA33
On-Line Trace Enrichment for Improved Sensitivity in Liquid Chromatography with Direct Liquid Introduction-Mass Spectrometric Detection. F.A. Maris, R.B. Geerdink, R.W. Frei and U.A.Th. Brinkman. *J. Chromatogr. 323*, 113–120, (1985).
85MA07
Analysis of Free Fatty Acids by a Micro-Liquid Chromatograph Directly Coupled with a Mass Spectrometer through a Vacuum Nebulizing Interface. K. Matsumoto, H. Yoshida, K. Ohta and S. Tsuge. *Org. Mass Spectrom. 20*, 777–780, (1985).
85MA03
Chemical Ionization Mass Spectrometry Using a Glow Discharge Ion Source Combined with a

Nebulizer Sampling System. K. Matsumoto, H. Kojima, K. Yasuda and S. Tsuge. *Org. Mass Spectrom. 20*, 243–246, (1985).
85MC41
Mass Spectrometry of Nucleic Acid Constituents and Related Compounds. J.A. McCloskey. In: *Mass Spectrometry in the Health and Life Sciences* (A.L. Burlingame and N. Castagnoli, Jr., eds.), *Anal. Chem. Symp. Series 24*, 521–546, (1985).
85MC6'
LCMS Analysis of an Antioxidant Mixture. Comparison of Results Obtained Using a Thermospray Interface and a Moving Belt Interface. W.H. McFadden, S.A. Lammert and J. Wellby. *Am. Chem. Soc. Abstr. 189*, ANYL0036, (1985).
85MC16'
Analysis of Pesticides Using Liquid Chromatography Combined with Thermospray Ionization and Triple Quadrupole MS/MS. W.H. McFadden and S.A. Lammert. *Am. Chem. Soc. Abstr. 189*, PEST0016, (1985).
85MC3'
Application of Thermospray LC/MS and Triple Stage Quadrupole MS/MS in Pesticide Analysis. W.H. McFadden, S.A. Lammert, R.D. Voyksner and J.T. Bursey. Presented at 33rd Ann. Conf. on Mass Spectrom. and Allied Topics, San Diego, CA, U.S.A. May 26–31, 1985. p. 373–374.
85MC5'
Application of Automatic Column Switching and Solvent Switching to DLI-LC/MS. M. McKellop, G. Hansen and F. Hatch. Presented at 33rd Ann. Conf. on Mass Spectrom. and Allied Topics, San Diego, CA, U.S.A. May 26–31, 1985. p. 765–766.
85MI0'
Thermospray LC/MS for the Analysis of Intermediary Metabolites. D.S. Millington, D.A. Maltby and P. Farrow. Presented at 33rd Ann. Conf. on Mass Spectrom. and Allied Topics, San Diego, CA, U.S.A. May 26–31, 1985. p. 480–481.
85MI59
Valproylcarnitine: A Novel Drug Metabolite Identified by Fast Atom Bombardment and Thermospray Liquid Chromatography/Mass Spectrometry. D.S. Millington, T.P. Bohan, C.R. Roe, A.L. Yergey and D.J. Liberato. *Clin. Chim. Acta 145*, 69–76, (1985).
85MI09
Direct Liquid Introduction Mass Spectrometry of Some Underivatized Dipeptides and Polypeptides. H. Milon and H. Bur. *J. Chromatogr. 350*, 399–406, (1985).
85MU46
Liquid Ionization Mass Spectrometry of Cationic and Amphoteric Surfactants. K. Musha and M. Tsuchiya. *Bunseki Kagaku 34*, 786–790, (1985).
85NI37
Open-Tubular Liquid Chromatography/Mass Spectrometry Using Direct Liquid Introduction. W.M.A. Niessen and H. Poppe. *J. Chromatogr. 323*, 37–46, (1985).
85NO83
Thin Layer Chromatography-Laser Mass Spectrometry (TLC/MS) of Triphenylmethane Dyes: Initial Results. F.P. Novak and D.M. Hercules. *Anal. Lett. 18*, 503–518, (1985).
85OE9FA'
Application of Negative Ion Mass Spectrometry to Medium-Size and Large Molecules. M. Oehme. In: *Mass Spectrometry of Large Molecules* (S. Facchetti, ed.), p. 249–263, Elsevier, Amsterdam (1985).
85OL37
Determination of Aldehydes and Ketones by Derivatization and Liquid Chromatography-Mass Spectrometry. K.L. Olson and S.J. Swarin. *J. Chromatogr. 333*, 337–347, (1985).
85OL5'
Determination of Aldehydes and Ketones by Derivatization and Liquid Chromatography-Mass Spectrometry. K.L. Olson and S.J. Swarin. *Am. Chem. Soc. Abstr. 189*, ANYL0085, (1985).

85PA71
Gaseous Acidity and Basicity Scales as Guides to Chemical Ionization in Reversed-Phase Liquid Chromatography-Mass Spectrometry with Direct Liquid Introduction. C.E. Parker, M.M. Bursey, R.W. Smith and S.J. Gaskell. *J. Chromatogr. 347*, 61–74, (1985).

85PA6'
Selectivity of Detection during Liquid Chromatography-Mass Spectrometry Using Thermospray and Direct Liquid Introduction Techniques. C.E. Parker, R.W. Smith, K. Yamaguchi, H.B. Matthews and L.A. Levy. Presented at 33rd Ann. Conf. on Mass Spectrom. and Allied Topics, San Diego, CA, U.S.A. May 26–31, 1985. p. 786–787.

85PA93
Liquid Chromatography-Chloride Attachment Negative Chemical Ionization Mass Spectrometry. C.E. Parker, K. Yamaguchi, D.J. Harvan, R.W. Smith and J.R. Hass. *J. Chromatogr. 319*, 273–284, (1985).

85PA48
Separation and Detection of Enantiomers of Stilbestrol Analogs by Combined High Performance Liquid Chromatography-Thermospray Mass Spectrometry. C.E. Parker, L.A. Levy, R.W. Smith, K. Yamaguchi, S.J. Gaskell and K.S. Korach. *J. Chromatogr. 344*, 378–384, (1985).

85PR3'
Product Identification of Biotransformation Reactions by Entrapped Plant Cells Employing LC/MS. N. Pras and A.P. Bruins. Presented at 33rd Ann. Conf. on Mass Spectrom. and Allied Topics, San Diego, CA, U.S.A. May 26–31, 1985. p. 953–954.

85QU89
Tert-Butyldiphenylsilyl Derivatization for Liquid Chromatography and Mass Spectrometry. M.A. Quilliam and J.A. Yaraskavitch. *J. Liq. Chromatogr. 8*, 449–461, (1985).

85RA35
Furosemide 1-O-Acyl Glucuronide: *In Vitro* Biosynthesis and pH-Dependent Isomerization to Glucuronidase-Resistant Forms. A. Rachmel, G.A. Hazelton, A.L. Yergey and D.J. Liberato. *Drug Metab.Disp. 13*(6), 705–710, (1985).

85RO80
Gas Chromatography/Mass Spectrometry and High-Performance Liquid Chromatography/Mass Spectrometry. M.E. Rose. *Mass Spectrom. 8*, 210–283, (1985).

85RO4'
Direct HPLC/MS Using a Fused Silica Capillary Interface. R.T. Rosen and J.E. Dziedzic. *Am. Chem. Soc. Abstr. 189*, ANYL0014, (1985).

85RO6'
Optimization of Direct Liquid Introduction (DLI) HPLC-MS Using a Fused Silica Interface. R.T. Rosen. Presented at 33rd Ann. Conf. on Mass Spectrom. and Allied Topics, San Diego, CA, U.S.A. May 26–31, 1985. p. 976–977.

85SA8'
Supercritical Ammonia Fluid Injection (SFI) Chemical Ionization Mass Spectrometry (CIMS) of Polar Molecules. S. Santikarn and V.N. Reinhold. Presented at 33rd Ann. Conf. on Mass Spectrom. and Allied Topics, San Diego, CA, U.S.A. May 26–31, 1985. p. 218–219.

85SC02
Thermospray Mass Spectrometry of Diazonium and Di-, Tri-, and Tetra-Quaternary Anion Salts. G. Schmelzeisen-Redeker, F.W. Röllgen, H. Wirtz and F. Vogt. *Org. Mass Spectrom. 20*, 752–756, (1985).

85SC3'
The Use of a Thermosprayed Electrolyte Solution for Chemical Ionization of Separately Introduced Samples. G. Schmelzeisen-Redeker, L. Butfering, U. Giessmann, A. Heindrichs and F. W. Röllgen. Presented at 33rd Ann. Conf. on Mass Spectrom. and Allied Topics, San Diego, CA, U.S.A. May 26–31, 1985. p. 763–764.

85SC00
Thermospray Mass Spectrometry. A New Technique for Studying the Relative Stability of Cluster Ions of Salts. G. Schmelzeisen-Redeker, S.S. Wong, U. Giessmann and F.W. Röllgen. *Z. Naturforsch. A. 40*, 430–431, (1985).

85SC37
Studies with a Laboratory-Constructed Thermospray LC-MS Interface. G. Schmelzeisen-Redeker, M.A. McDowall, U. Giessmann, K. Levsen and F.W. Röllgen. *J. Chromatogr. 323*, 127–133, (1985).

85SH21
Liquid Chromatography/Mass Spectrometry of the Thermally Labile Herbicides, Chlorsulfuron and Sulfometuron Methyl. L.M. Shalaby. *Biomed. Mass Spectrom. 12*, 261–268, (1985).

85SH3′
Application of Gradient Elution LC/MS for the Analysis of Agricultural Compounds. L.M. Shalaby. Presented at 33rd Ann. Conf. on Mass Spectrom. and Allied Topics, San Diego, CA, U.S.A. May 26–31, 1985. p. 773–774.

85SH5′
On-Line LC/MS as a Problem-Solving Tool for the Analysis of Thermally Labile Pesticides. L.M. Shalaby. *Am. Chem. Soc. Abstr. 189*, PEST0015, (1985).

85SH2′
Studies of Ethmozin Metabolism by LC/MS/MS. E. Sheehan, L. Wong, H. Hashem, M. Zemaitis and J. Buchner. Presented at 33rd Ann. Conf. on Mass Spectrom. and Allied Topics, San Diego, CA, U.S.A. May 26–31, 1985. p. 782–783.

85SH71
Structural Analysis of Underivatized Sialic Acids by Combined High Performance Liquid Chromatography-Mass Spectrometry. A.K. Shukla, R. Schaur, U. Schade, H. Moll and E.Th. Rietschel. *J. Chromatogr. 337*, 231–238, (1985).

85SI2′
Utilization of New Mass Spectrometry Techniques for the Identification of Pesticide Metabolites. B.J. Simoneaux and G.J. Marco. *Am. Chem. Soc. Abstr. 189*, PEST0012, (1985).

85SK11
Quantitative Determination of Betamethasone and Its Major Metabolite in Equine Urine by Micro-Liquid Chromatography-Mass Spectrometry. D.S. Skrabalak, K.K. Cuddy and J.D. Henion. *J. Chromatogr. 341*, 261–269, (1985).

85SM5′
Rapid and High Resolution Methods for Capillary SFC-MS; an Alternative to LC/MS. R.D. Smith, H.T. Kalinoski, E.K. Chess, B.W. Wright and H.R. Udseth. *Am. Chem. Soc. Abstr. 189*, ANYL0065, (1985).

85SM40
Diesel Fuel Marine and Sediment Analysis. Supercritical Ammonia Extraction and Direct Fluid Injection-Mass Spectrometry. R.D. Smith, H.R. Udseth and R.N. Hazlett. *Fuel 64*, 810–815, (1985).

85SM32
Rapid and High Resolution Capillary Supercritical Fluid Chromatography (SFC) and SFC/MS of Trichothecene Mycotoxins. R.D. Smith, H.R. Udseth and B.W. Wright. *J. Chromatogr. Sci. 23*, 192–199, (1985).

85SM31
Micro-Scale Methods for Characterization of Supercritical Fluid Extraction and Fractionation Processes. R.D. Smith, H.R. Udseth and B.W. Wright. *Process Technol. Proc. 3 (Supercrit. Fluid Technol.)*, 191–223, (1985).

85ST73
Simplified Moving-Belt Interface for Liquid Chromatography/Mass Spectrometry. S.J. Stout and A.R. Da Cunha. *Anal. Chem. 57*, 1783–1786, (1985).

85ST4'
Application of On-Line Liquid Chromatography/Fast Atom Bombardment. J.G. Stroh, J.C. Cook, K.L. Rinehart, Jr., T. Kihara, Z. Huang, I.A.S. Lewis and P.A. Keifer. Presented at 33rd Ann. Conf. on Mass Spectrom. and Allied Topics, San Diego, CA, U.S.A. May 26–31, 1985. p. 784– 785.

85ST75
On-Line Liquid Chromatography-Fast Atom Bombardment Mass Spectrometry. J.G. Stroh, J.C. Cook, R.M. Milberg, L. Brayton, T. Kihara, Z. Huang, K.L. Rinehart, Jr. and I.A.S. Lewis. *Anal. Chem. 57*, 985–991, (1985).

85SU8'
Quantitative Thermospray LC/MS Analysis of a Psychotropically Active 9-Ketocannabinoid. H.R. Sullivan and D.A. Garteiz. Presented at 33rd Ann. Conf. on Mass Spectrom. and Allied Topics, San Diego, CA, U.S.A. May 26–31, 1985. p. 198–199.

85SU41
252-Cf-Plasma Desorption Mass Spectrometry. B. Sundquist and R.D. Macfarlane. *Mass Spectrom. Rev. 4*, 421–460, (1985).

85TA96
Current Situation of Pesticide Instrumental Analysis. II. High-Performance Liquid Chromatography. M. Takeda. *Shokubutsu Boeki 39*, 26–33, (1985).

85TA52
Identification of Substances of Thin-layer and Paper Chromatograms by Mass Spectra of Sputtered Ions. G.D. Tantsyrev and M.I. Povolotskaya. *Mass-Spectrom. Khim. Kinet. 1985*, 252–259, (1985).

85TA59
Thermospray Liquid Chromatography-Mass Spectrometry of Nonvolatile Biomolecules. P.I. Tao. *Diss. Abstr. Int. B. 45*, 2899–2900, (1985).

85TI31
Design and Construction of an Interface for Direct Liquid Introduction Coupling of a Microbore High Performance Liquid Chromatograph to a Quadrupole Mass Spectrometer. R. Tiebach, W. Blaas and M. Kellert. *J. Chromatogr. 323*, 121–126, (1985).

85TI83
Confirmation of Nivalenol and Deoxynivalenol by On-Line Liquid Chromatography/Mass Spectrometry and Gas Chromatography/Mass Spectrometry-Comparison of Methods. R. Tiebach, W. Blaas, M. Kellert, S. Steinmeyer and R. Weber. *J. Chromatogr. 318*, 103–111, (1985).

85TS8'
Liquid Ionization Mass Spectrometry for Liquid Chromatography. M. Tsuchiya, J. Sasagawa and K. Otsuka. *Am. Chem. Soc. Abstr. 189*, ANYL0038, (1985).

85TS69
Cluster Ions of Water Observed by Liquid Ionization Mass Spectrometry. M. Tsuchiya, K. Otsuka and H. Kuwabara. *Chem. Lett. 6*, 709–712, (1985).

85TS70
Direct Liquid Introduction Interface for Capillary Column Liquid Chromatography/Mass Spectrometry. T. Tsuda, G. Keller and H.J. Stan. *Anal. Chem. 57*, 2280–2282, (1985).

85TS07
New Approaches to Interfacing Liquid Chromatography and Mass Spectrometry. S. Tsuge. In: *Microcolumn Separations* (M.V. Novotny and D. Ishii, eds.) *J. Chromatogr. Libr. 30*, 217–241, (1985).

85TS08
Present Status of Directly Coupled HPLC-MS [High-Performance Liquid Chromatography-Mass Spectroscopy]. S. Tsuge and K. Matsumoto. *Tanpakushitsu Kakusan Koso 30*, 1138–1153, (1985).

85UD3'
Supercritical Fluid Extraction and Chromatography-Mass Spectrometry of Mycotoxins of the Trichothecene Group. H.R. Udseth, B.W. Wright and R.D. Smith. Presented at the 33rd Ann. Conf.
on Mass Spectrom. and Allied Topics, San Diego, CA, U.S.A. May 26–31, 1985. p. 663–664.
85UN12
Thermospray LC/MS and LC/MS/MS of Phenylthiohydantion-Amino Acids. S.E. Unger. *Vestec
Thermospray Newsletter 1*, 2–3, (1985).
(904)
85VA03
The Applicability of Field Desorption Mass Spectrometry and Liquid Chromatography/Mass Spec-
trometry for the Analysis of the Pungent Principles of Capsicum and Black Pepper. J. Van Der
Greef, L.M. Nijssen, H. Maarse, M. Ten Noever De Brauw, D.E. Games and N.J. Alcock. *Dev.
Food Sci. 10*, 603–612, (1985).
85VA37
Determination of Progesterone by Liquid-Chromatography-Mass Spectrometry Using a Moving Belt
Interface and Isotope Dilution. J. Van Der Greef, A.C. Tas, M.A.H. Rijk, M.C. Ten Noever De
Brauw, M. Höhn, G. Meijerhoff and U. Rapp. *J. Chromatogr. 343*, 397–401, (1985).
85VA31
Performance of a Moving Belt Liquid Chromatography-Mass Spectrometry Interface. J. Van Der
Greef, A.C. Tas, M.C. Ten Noever De Brauw, M. Höhn, G. Meijerhoff and U. Rapp. *J.
Chromatogr. 323*, 81–87, (1985).
85VA72
Identification of Antioxidant and Ultraviolet Light Stabilizing Additives in Plastics by LC/MS. J.D.
Vargo and K.L. Olson. *Anal. Chem. 57*, 672–675, (1985).
85VE1'
Discharge Ionization as a Secondary Ionization Method for Use with Thermospray. C.H. Vestal,
D.A. Garteiz, R. Smit and C.R. Blakley. Presented at 33rd Ann. Conf. on Mass Spectrom. and
Allied Topics, San Diego, CA, U.S.A. May 26–31, 1985. p. 771–772.
85VE7'
Automatic Thermospray Controller for Gradient Elutions. C.H. Vestal and G.J. Fergusson. Pre-
sented at 33rd Ann. Conf. on Mass Spectrom. and Allied Topics, San Diego, CA, U.S.A. May
26–31, 1985. p. 767–768.
85VE73
Thermospray Liquid Chromatography/Mass Spectrometry Interface with Direct Electrical Heating of
the Capillary. M.L. Vestal and G.J. Fergusson. *Anal. Chem. 57*, 2373–2378, (1985).
85VE4'
Thermospray LC/MS: A Practical Solution to a Difficult Problem. M.L. Vestal. *Am. Chem. Soc.
Abstr. 189*, ANYL0004, (1985).
85VE0'
Ion Evaporation from Liquid Droplets and Solid Particles. M.L. Vestal. Presented at 33rd Ann.
Conf. on Mass Spectrom. and Allied Topics, San Diego, CA, U.S.A. May 26–31, 1985. p. 120–
121.
85VE49
Recent Applications of Thermospray LC/MS. M.L. Vestal. In: *Mass Spectrometry in the Health and
Life Sciences* (A.L. Burlingame and N. Castagnoli, Jr., eds.), *Anal. Chem. Symp. Series 24*, 99–
118, (1985).
85VO15
Combined HPLC-Mass Spectrometry: Its Challenges, Present and Future. P. Vouros and B.L.
Karger. *New Methods Drug Res. 1*, 45–69, (1985).
85VO6'
Post-Column Chemical Derivatization in HPLC/MS Using a Moving Belt Interface. P. Vouros, A.
Sahil, C.P Tsai and B. L. Karger. *Am. Chem. Soc. Abstr. 189*, ANYL0066, (1985).

85VO9'
Post Column Chemical Derivatization in HPLC/MS Using a Moving Belt Interface. P. Vouros, A. Sahil, C.P. Tsai and B.L. Karger. Presented at 33rd Ann. Conf. on Mass Spectrom. and Allied Topics, San Diego, CA, U.S.A. May 26–31, 1985. p. 379–380.

85VO77
Analysis of Candidate Anticancer Drugs by Thermospray HPLC-MS and by High Resolution Mass Spectrometry. R.D. Voyksner, F.P. Williams and J.W. Hines. *J. Chromatogr.* 347, 137–146, (1985).

85VO70
Characterization of Dyes in Environmental Samples by Thermospray High Performance Liquid Chromatography/Mass Spectrometry. R.D. Voyksner. *Anal. Chem.* 57, 2600–2605, (1985).

85VO0'
Analysis of Candidate Anticancer Drugs by Thermospray HPLC/MS and High Resolution MS. R.D. Voyksner, F. Williams and J. Hines. Presented at 33rd Ann. Conf. on Mass Spectrom. and Allied Topics, San Diego, CA, U.S.A. May 26–31, 1985. p. 460–461.

85VO3'
Thermospray HPLC/MS Analysis of Dyes. R.D. Voyksner and T. Pack. Presented at 33rd Ann. Conf. on Mass Spectrom. and Allied Topics, San Diego, CA, U.S.A. May 26–31, 1985. p. 573–574.

85VO86'
Applications of Thermospray HPLC/MS to the Analysis of Dyes. R.D. Voyksner and J.T. Bursey. *Am. Chem. Soc. Abstr.* 189, ANYL0086, (1985).

85VO5'
Applications of Thermospray HPLC/MS. R.D. Voyksner and J.T. Bursey. Presented at the 33rd Ann. Conf. on Mass Spectrom. and Allied Topics, San Diego, CA, U.S.A. May 26–31, 1985. p. 95–96.

85VO33
Analysis of Candidate Anticancer Drugs by Thermospray High-Performance Liquid Chromatography-Mass Spectrometry. R.D. Voyksner, J.T. Bursey and J.W. Hines. *J. Chromatogr.* 373, 383–394, (1985).

85VO71
Optimization and Application of Thermospray High Performance Liquid Chromatography/Mass Spectrometry. R.D. Voyksner and C.A Haney. *Anal. Chem.* 57, 991–996, (1985).

85VO89
Thermospray High-Performance Liquid Chromatographic/Mass Spectrometric Analysis of Some Fusarium Mycotoxins. R.D. Voyksner, W.M. Hagler, Jr., K. Tyczkowska and C.A. Haney. *J. High Resolut. Chromatogr. Chromatogr. Commun.* 8, 119–125, (1985).

85WA34
Steroid Conjugates: Analysis by Thermospray Liquid Chromatography/Mass Spectrometry. D. Watson, S. Murray and G.W. Taylor. *Biochem. Soc. Trans.* 13, 1224, (1985).

85WA20
Thermospray Liquid Chromatography Negative Ion Mass Spectrometry of Steroid Sulphate Conjugates. D. Watson, G.W. Taylor and S. Murray. *Biomed. Mass Spectrom.* 12, 610–615, (1985).

85WH2'
Performance of the Electrospray Ion Source. C.M. Whitehouse, J.B. Fenn, C.-K. Meng and R.N. Dreyer. Presented at 33rd Ann. Conf. on Mass Spectrom. and Allied Topics, San Diego, CA, U.S.A. May 26–31, 1985. p. 392–394.

85WH75
An Electrospray Interface for Liquid Chromatographs and Mass Spectrometers. C.M. Whitehouse, R.N. Dreyer, M. Yamashita and J.B. Fenn. *Anal. Chem.* 57, 675–679, (1985).

85WH5FA'
Mass Spectrometric Analysis of Bile Acids in Neonatal Liver Diseases. J.D. Whitney and A.L.

Burlingame. In: *Mass Spectrometry of Large Molecules* (S. Facchetti, ed.), p. 185–207. Elsevier, Amsterdam, 1985.

85WI7'
Monodisperse Aerosol Generation Interface for Combining LC with MS (MAGIC): General Applications. R.C. Willoughby and R.F. Browner. *Am. Chem. Soc. Abstr. 189*, ANYL0067, (1985).

85WI4'
MAGIC-LC/MS: Instrumental Developments and Applications to Environmental Samples. P. Winkler, L. Abbey and R.F. Browner. *Am. Chem. Soc. Abstr. 189*, ANYL0034, (1985).

85WI3'
Development and Optimization of an Improved MAGIC-LC/MS Interface. P.C. Winkler, L.E. Abbey, N.W. Lorber and R.F. Browner. Presented at 33rd Ann. Conf. on Mass Spectrom. and Allied Topics, San Diego, CA, U.S.A. May 26–31, 1985. p. 383–384.

85YA5'
Optimization and Application of Thermospray HPLC/MS with Direct Electrical Heating of the Capillary. L. Yang and G.J. Fergusson. Presented at 33rd Ann. Conf. on Mass Spectrom. and Allied Topics, San Diego, CA, U.S.A. May 26–31, 1985. p. 775–776.

85YE0'
Solvent Selection for Thermospray LC/MS Direct Ionization. A.L. Yergey and D.J. Liberato. Presented at 33rd Ann. Conf. on Mass Spectrom. and Allied Topics, San Diego, CA, U.S.A. May 26–31, 1985. p. 790–791.

85YI05
Forensic Analysis of Explosives by Mass Spectrometry. J. Yinon. *Spectra 10*, 5–8, (1985).

85YI65
Identification of Urinary Metabolites of 2,4,6-Trinitrotoluene in Rats by LC/MS. J. Yinon and D.-G. Hwang. *Toxicol. Lett. 26*, 205–209, (1985).

85YI97
Metabolic Studies of Explosives. II. High Performance-Liquid Chromatography/Mass Spectrometry of Metabolites of 2,4,6-Trinitrotoluene. J. Yinon and D.-G. Hwang. *J. Chromatogr. 339*, 127–137, (1985).

85YI0'
Identification of TNT Metabolites in Urine and Blood by LC/MS. J. Yinon and D.-G. Hwang. Presented at 33rd Ann. Conf. on Mass Spectrom. and Allied Topics, San Diego, CA, U.S.A. May 26–31, 1985. p. 200–201.

85YI7'
Forensic Analysis of Explosives by LC/MS. J. Yinon and D.-G. Hwang. *Am. Chem. Soc. Abstr. 189*, ANYL0077, (1985).

85YI87'
Application of LC/MS in the Analysis of Explosives and Their Metabolites. J. Yinon and D.-G. Hwang. *Am. Chem. Soc. Abstr. 189*, ANYL0087 (1985).

85YU83
New Application of Mass Spectrometry in Environmental Analysis. A. Yusuhara. *Suishitsu Odaku Kenkyu 8*, 773–780, (1985).

86AB2'
Gradient Elution Liquid Chromatography/Mass Spectrometry Using a Monodispersed Aerosol Generation (MAGIC) Interface. L.E. Abbey, D.E. Bostwick, P.C. Winkler, D.D. Perkins and R.F. Browner. Presented at 34th Ann. Conf. on Mass Spectrom. and Allied Topics, Cincinnati, OH, U.S.A. June 8–13, 1986. Paper WPB9. p. 582–583.

86AB0'
Quantitative Aspects of LC/MS Analysis of Phospholipids. H.M. Abdel-Maksoud, J.E. Evans, F.B. Jungalwala and R.H. McCluer. Presented at 34th Ann. Conf. on Mass Spectrom. and Allied Topics, Cincinnati, OH, U.S.A. June 8–13, 1986. Paper MOE11. p. 120–121.

86AL7'
Effect of Gas Phase Ion Molecule Reactions on Thermospray Spectra. A.J. Alexander and P.

Kebarle. Presented at 34th Ann. Conf. on Mass Spectrom. and Allied Topics, Cincinnati, OH, U.S.A. June 8–13, 1986. Paper WOF7. p. 537–538.

86AL81

Thermospray Mass Spectrometry. Use of Gas-Phase Ion/Molecule Reactions to Explain Features of Thermospray Mass Spectra. A.J. Alexander and P. Kebarle. *Anal. Chem.* 58, 471–478, (1986).

86AL05

Coupling of a Liquid Chromatograph with a Mass Spectrometer. M.L. Alexandrov, L.N. Gal, N.V. Krasnov, V.I. Nikolaev, V.A. Shkurov, G.I. Baram, M.A. Grachev and Y.S. Kusner. *Adv. Mass Spectrom. 1985 (Pub. 1986), 10th* (Part B), 605–606, (1986).

86AL07

Method of Mass Spectrometry with Ion Extraction from Solutions at Atmospheric Pressure. M.L. Alexandrov, L.N. Gal, N.V. Krasnov, V.I. Nikolaev and V.A. Shkurov. *Adv. Mass Spectrom. 1985 (Pub. 1986), 10th* (Part B), 607–608, (1986).

86AN21

Variations on a Theme by Vestec. Anonymous. *Vestec Thermospray Newsletter* 2(5), 1–3, (1986).

86AP7′

The Application of Thermospray to the Study of the Mechanism of the Conversion of Modified Acv-Tripeptides to Penicillins and Cephams by Cell-Free Enzymes. R.T. Aplin, M. McDowell and K. Rollins. Presented at 34th Ann. Conf. on Mass Spectrom. and Allied Topics, Cincinnati, OH, U.S.A. June 8–13, 1986. Paper TPE13. p. 467.

86AR86

Recent LC/MS [Liquid Chromatography/Mass Spectrometry] Methods for Water Anaylsis. P.J. Arpino. *Comm. Eur. Communities [Rep.] EUR EUR 10388* (Org. Micropollut. Aquat. Environ.) 1, 26–39, (1986).

86AR01

Evaluation and Comparison of DLI and Thermospray Interfaces. P.J. Arpino, M. Lesieur, C. Beaugrand and F. De Maack. *Adv. Mass Spectrom. 1985 (Pub. 1986), 10th* (Part B), 621–622, (1986).

86BA15

Thermospray Ionization and Tandem Mass Spectrometry of Dyes. J.M. Ballard and L.D. Betowski. *Org. Mass Spectrom.* 21(9), 575–588, (1986).

86BA81

Column Liquid Chromatography. H.G. Barth, W.E. Barber, C.H. Lochmuller, R.E. Majors and F.E. Regnier. *Anal. Chem.* 58, 211R–250R, (1986).

86BE03

Analysis of Non-Volatile Nitrosamines by Thermospray and Moving Belt LC/MS. J.G. Beattie, D.E. Games, J. Gilbert and J.R. Startin. *Adv. Mass Spectrom. 1985 (Pub. 1986), 10th* (Part B), 573–574, (1986).

86BE99

Automated Dry Fraction Collection for Microbore High-Performance Liquid Chromatography-Mass Spectrometry. R.C. Beavis, G. Bolbach, W. Ens, D.E. Main, B. Schueler and K.G. Standing. *J. Chromatogr. 359*, 489–498, (1986).

86BE5′

Collision Induced Decompositions in the Thermospray Ion Source. F.A. Bencsath and F.H. Field. Presented at 34th Ann. Conf. on Mass Spectrom. and Allied Topics, Cincinnati, OH, U.S.A. June 8–13, 1986. Paper MOB4. p. 5–6.

86BE86

Liquid Chromatography at the 1986 Pittsburgh Conference: Detectors. V. Berry. *Am. Lab. (Fairfield, Conn.)* 18(5), 36–48, (1986).

86BE37

Supercritical Fluid Chromatographic and Supercritical Fluid Chromatographic-Mass Spectrometric

Studies of Some Polar Compounds. A.J. Berry, D.E. Games and J.R. Perkins. *J. Chromatogr.* *363*(2), 147–158, (1986).

86BE1'
Supercritical Fluid Chromatography/Mass Spectrometry Using a Moving Belt Interface. A.J. Berry, D.E. Games and J.R. Perkins. Presented at 34th Ann. Conf. on Mass Spectrom. and Allied Topics, Cincinnati, OH, U.S.A. June 8–13, 1986. Paper MOB10. p. 11–12.

86BE31
Supercritical Fluid Chromatography and Its Combination with Mass Spectrometry. A.J. Berry, D.E. Games and J.R. Perkins. *Anal. Proc. (London)* *23*(12), 451–453, (1986).

86BE5'
Thermospray LC/MS/MS Analysis of Azo Dyes Spiked in Wastewater. L.D. Betowski, S.M. Pyle, J.M. Ballard and G. Shaul. Presented at 34th Ann. Conf. on Mass Spectrom. and Allied Topics, Cincinnati, OH, U.S.A. June 8–13, 1986. Paper MPF3. p. 185.

86BI34
Short-Chain Acylcarnitines: Identification and Quantification. L.L. Bieber and J. Kerner. *Methods Enzymol.* *123*, 264–276, (1986).

86BO89
Ionization in the Gas Phase with Fast Atom Bombardment. Formation of Molecular Ions. G. Bojesen and J. Moller. *Int. J. Mass Spectrom. Ion Processes* *68*, 239–248, (1986).

86BO4'
The Use of FAB-LCMS for the Analysis of High Molecular Weight Peptides. R.S. Bordoli, A.E. Fairbrother, S.T. Krolik and H.J. Wright. Presented at 34th Ann. Conf. on Mass Spectrom. and Allied Topics, Cincinnati, OH, U.S.A. June 8–13, 1986. Paper RPE15. p. 864–865.

86BR45
Aerosols as Microsample Introduction Media for Mass Spectrometry. R.F. Browner, P.C. Winkler, D.D. Perkins and L.E. Abbey. *Microchem. J.* *34*(1), 15–24, (1986).

86BR5'
Electrospray LC/MS with an API Tandem Quadrupole MS. A. Bruins, T. Covey and J. Henion. Presented at 34th Ann. Conf. on Mass Spectrom. and Allied Topics, Cincinnati, OH, U.S.A. June 8–13, 1986. Paper WPB11. p. 585–586.

86BR9'
Comparison between Gas Phase and Liquid Phase Ionization for LC/MS/MS Using an Atmospheric Pressure Ion Source. A. Bruins, T. Covey and J. Henion. Presented at 34th Ann. Conf. on Mass Spectrom. and Allied Topics, Cincinnati, OH, U.S.A. June 8–13, 1986. Paper TOE2. p. 379–380.

86BU85
Mass Spectrometry. A.L. Burlingame, T.A. Baillie and P.J. Derrick. *Anal. Chem.* *58*, 165R–211R, (1986).

86BU05
Design and Performance of a Chromatography/SIMS Instrument. K.L. Busch, G.C. Di Donato, K.J. Kroha, R.A. Flurer and J.W. Fiola. *Adv. Mass Spectrom. 1985 (Pub. 1986), 10th* (Part B), 855–856, (1986).

86BU69
A Dual Beam Thermospray Vaporizer for On Line HPLC-Mass Spectrometry Coupling. L. Butfering, G. Schmelzeisen-Redeker and F.W. Röllgen. *J. Chem. Soc. Chem. Commun. 1986*(7), 579–582, (1986).

86BU1'
Thermospray Mass Spectrometry Applying a Dual Beam Vaporizer. L. Butfering, G. Schmelzeisen-Redeker and F.W. Röllgen. Presented at 34th Ann. Conf. on Mass Spectrom. and Allied Topics, Cincinnati, OH, U.S.A. June 8–13, 1986. Paper MOB13. p. 15–16.

86CA43
Regulatory Pesticide Analysis by Mass Spectrometry. T. Cairns and E.G. Siegmund. *Anal. Methods Pestic. Plant Growth Regul.* *14*, 193–253, (1986).

86CA1'
A Continuous Flow FAB Probe for the Introduction of Aqueous Solutions. R.M. Caprioli, T. Fan and J. Cottrell. Presented at 34th Ann. Conf. on Mass Spectrom. and Allied Topics, Cincinnati, OH, U.S.A. June 8–13, 1986. Paper TOE3. p. 381–382.

86CA18
High Sensitivity Mass Spectrometric Determination of Peptides: Direct Analysis of Aqueous Solutions. R.M. Caprioli and T. Fan. *Biochem. Biophys. Res. Commun. 141*(3), 1058–1065, (1986).

86CA89
Continuous-Flow Sample Probe for Fast Atom Bombardment Mass Spectrometry. R.M. Caprioli, T. Fan and J.S. Cottrell. *Anal. Chem. 58*(14), 2949–2953, (1986).

86CH01
MS25RFA: A New Instrument for Combined Chromatography-Mass Spectrometry. J. Chapman, M. Kearns and J. Pratt. *Adv. Mass Spectrom. 1985 (Pub. 1986), 10th* (Part B), 901–902, (1986).

86CH5'
Extensions to Combined Chromatography-Mass Spectrometry on a Double Focussing Magnetic Sector Instrument. J.R. Chapman, E.D. Ramsey and J.A.E. Pratt. Presented at 34th Ann. Conf. on Mass Spectrom. and Allied Topics, Cincinnati, OH, U.S.A. June 8–13, 1986. Paper FOB19. p. 965–966.

86CO57
Electrohydrodynamic Mass Spectrometry. K.D. Cook. *Mass Spectrom. Rev. 5*(4), 467–520, (1986).

86CO45
Supercritical Fluid Chromatography (SFC)-Mass Spectrometry (MS) Coupling. J. Cousin and P.J. Arpino. *Analusis 14*(5), 215–221, (1986).

86CO9'
Quantitation of Novel Amino Acids and Sugars by Thermospray LC/MS/MS in Biological Samples. J.E. Coutant and K.Y. Chan. Presented at 34th Ann. Conf. on Mass Spectrom. and Allied Topics, Cincinnati, OH, U.S.A. June 8–13, 1986. Paper FPA11. p. 1019–1020.

86CO81
Liquid Chromatography/Mass Spectrometry. T.R. Covey, E.D. Lee, A.P. Bruins and J.D. Henion. *Anal. Chem. 58*(14), 1451A–1461A, (1986).

86CO83
High-Speed Liquid Chromatograph/Tandem Mass Spectrometry for the Determination of Drugs in Biological Samples. T.R. Covey, E.D. Lee and J.D. Henion. *Anal. Chem. 58*(12), 2453–2460, (1986).

86CO7'
Fast LC/MS and LC/MS/MS for the Determination of Phytoestrogens and Drugs in Food Material. T. Covey and J. Henion. Presented at 34th Ann. Conf. on Mass Spectrom. and Allied Topics, Cincinnati, OH, U.S.A. June 8–13, 1986. Paper MOE1. p. 107–108.

86CO03
Heated Nebulizer APCI LC/MS and LC/MS/MS Analysis of Environmental Estrogens. T.R. Covey, G.A. Maylin and J.D. Henion. *Adv. Mass Spectrom. 1985 (Pub. 1986), 10th* (Part B), 603–604, (1986).

86CR92
Mass Spectrometry as an On-Line Detector for HPLC. T.B. Crowther, T.R. Covey and J.D. Henion. *Chem. Anal. (N.Y.) 89*, 292–300, (1986).

86CU85
Determination of Acrylamide in Sugar by Thermospray Liquid Chromatography/Mass Spectrometry. S. Cutié and G.J. Kallos. *Anal. Chem. 58*(12), 2425–2428, (1986).

86DE2'
Chromatographic Methods Combined with Mass Spectrometry. Technology and Applications in the

Fields of Environmental Studies, Pharmacology and Biochemistry. J. De Graeve, F. Berthou, M. Frost, P. Arpino and J.C. Prome. Lavoisier, Paris, (1986), p. 382.
86DE23
LC/MS Tuning and Calibration. F. De Maack, D. Wang-Iverson and J. Ashraf. *Vestec Thermospray Newsletter 2,* 3–4, (1986).
86DE25
Stopped Flow Mass Spectrometry: Application to the Carbonic Anhydrase Reaction. H. Degn and B. Kristensen. *J. Biochem. Biophys. Methods 70*(2), 185–194, (1986).
86DE8′
Development and Performance of a Disposable Moving Belt LC-MS Interface. W. De Maio and M.L. Vestal. Presented at 34th Ann. Conf. on Mass Spectrom. and Allied Topics, Cincinnati, OH, U.S.A. June 8–13, 1986. Paper WPB7. p. 578–579.
86DE46
Mass Spectroscopy at High Mass. P.J. Derrick. *Fresenius' Z. Anal. Chem. 324*(5), 486–491, (1986).
86DE97
Profiling of Amino Acids in Body Fluids and Tissues by Means of Liquid Chromatgraphy. Z. Deyl, J. Hyanek and M. Horakova. In: *Profiling of Body Fluids and Tissues* (Z. Deyl and C.C. Sweeley, eds.), *J. Chromatogr. 379,* 177–249, (1986).
86DI7′
A New Instrument for Chromatography/SIMS. G.C. Di Donato and K.L. Busch. Presented at 34th Ann. Conf. on Mass Spectrom. and Allied Topics, Cincinnati, OH, U.S.A. June 8–13, 1986. Paper TOE11. p. 397–398.
86DI81
Phase-Transition Matrix for Chromatography/Secondary Ion Mass Spectrometry. G.C. Di Donato and K.L. Busch. *Anal. Chem. 58,* 3231–3232, (1986).
86DI09
Improving the Flexibility and Productivity of the Thermospray LC-MS Interface. D. Dixon, C. Eckers, P. Sakkers and M. Lant. *Adv. Mass Spectrom, 1985 (Pub. 1986). 10th* (Part B), 609–610, (1986).
86ED6′
The Role of LC/MS in the Characterization of New Nucleosides for Transfer RNA. C.G. Edmonds, T. Hashizume, J.A. McCloskey, R. Gupta and K.O. Stetter. Presented at 34th Ann. Conf. on Mass Spectrom. and Allied Topics, Cincinnati, OH, U.S.A. June 8–13, 1986. Paper FOA1. p. 896–897.
86ED07
Thermospray Liquid Chromatography-Mass Spectrometry of Nucleosides and Nucleotides. C.G. Edmonds, F.F. Hsu and J.A. McCloskey. *Adv. Mass Spectrom. 1985 (Pub. 1986), 10th* (Part B), 627–628, (1986).
86ED75
Analysis of Mutagens from Cooked Foods by Directly Combined Liquid Chromatography-Mass Spectrometry. C.G. Edmonds, S.K. Sethi, Z. Yamaizumi, H. Kasai, S. Nishimura and J.A. McCloskey. *Environ. Health Perspect. 67,* 35–40, (1986).
86ES5′
Analysis of Some Substituted Pyridine C-Nucleosides by Microbore DLI/LC-MS. E.L. Esmans, M. Belmans, Y. Vrijen, Y. Luyten, F.C. Alderweireldt, L.L. Wotring and L.B. Townsend. Presented at 34th Ann. Conf. on Mass Spectrom. and Allied Topics, Cincinnati, OH, U.S.A. June 8–13, 1986. Paper TPE1. p. 445–446.
86FA66
Development and Evaluation of a Moving Belt Liquid Chromatograph Mass Spectrometer with Laser and Ion Desorption. T.P. Fan. *Diss. Abstr. Int. B 46*(5), 1536–1537, (1986).

86FA01
The Analysis of Penicillin by Thermospray LC/MS. P. Farrow, N. Lynaugh, M.A. McDowall and J.L. Gower. *Adv. Mass Spectrom. 1985 (Pub. 1986), 10th* (Part B), 611, (1986).
86FA92
Advances in Mass Spectrometry: New Horizons for Neuroscience. K.F. Faull. G. Feistner, K.A. Greene and O. Beck. *Trends Neurosci. 9*, 102–109, (1986).
86FE99
Glucuronic Acid, Sulfate Ester and Glutathione Xenobiotic Conjugates: Analysis by Mass Spectrometry. C. Fenselau and L. Yellet. *Am. Chem. Soc. Symp. Ser. 299*, 159–176, (1986).
86FI01
Ionization Methods in Mass Spectrometry. F.H. Field. *Adv. Mass Spectrom. 1985 (Pub. 1986), 10th* (Part A), 271–286, (1986).
86FI74
Modular Instrument for Organic Secondary Ion Mass Spectrometry and Direct Chromatographic Analysis. J.W. Fiola, G.C. Di Donato and K.L. Busch. *Rev. Sci. Instrum. 57*(9), 2294–2302, (1986).
86FL8'
Environmental Applications of the Thermospray LC/MS Interface. Part I: Dyes and Selected Pollutants. D.A. Flory and M.A. McLean. Presented at 34th Ann. Conf. on Mass Spectrom. and Allied Topics, Cincinnati, OH, U.S.A. June 8–13, 1986. Paper MPF5. p. 188.
86FO09
Applications Using the Thermospray LC/MS Interface in Biomedical and Environmental Analysis. D.L. Foerst and P.C. Goodley. *Adv. Mass Spectrom. 1985 (Pub. 1986), 10th* (Part B), 599, (1986).
86FR1'
A Comparison of Thermospray Conditions for Organophosphate Pesticides. T.J. Francl, P.J.K. French and M.L. Thomson. Presented at 34th Ann. Conf. on Mass Spectrom. and Allied Topics, Cincinnati, OH, U.S.A. June 8–13, 1986. Paper MPF7. p. 191–192.
86FR4'
The Use of Combined Supercritical Fluid Chromatography-Mass Spectrometry to Characterize Carotenoid Pigments. N.M. Frew, C.G. Johnson and R.H. Bromund. Presented at 34th Ann. Conf. on Mass Spectrom. and Allied Topics, Cincinnati, OH, U.S.A. June 8–13, 1986. Paper FPB15. p. 1054–1055.
86FU85
Supercritical Fluid for Sample Introduction in Supersonic Jet Spectrometry. H. Fukuoka, T. Imasaka and N. Ishibashi. *Anal. Chem. 58*, 375–379, (1986).
86GA03
The Combination of Liquid Chromatography and Mass Spectrometry. D.E. Games. *Adv. Mass Spectrom. 1985 (Pub. 1986), 10th* (Pt. A), 323–342, (1986).
86GA53
Mixtures of Cyclic Oligomers of Poly(lactic Acid) Analysed by Negative Chemical Ionization and Thermospray Mass Spectrometry. D. Garozzo, M. Giuffrida and G. Montaudo. *Polym. Bull. (Berlin) 15*(4), 353–358, (1986).
86GA21
The Thermospray Vaporizer Probe. D.A. Garteiz. *Vestec Thermospray Newsletter 2*, 1–3, (1986).
86GA1'
Thermospray LC/MS Analysis Using Ammonia Gas Chemical Ionization. D.A. Garteiz, C.R. Blakley and R.M. Osterman. Presented at 34th Ann. Conf. on Mass Spectrom. and Allied Topics, Cincinnati, OH, U.S.A. June 8–13, 1986. Paper WPB14. p. 591–592.
86GE6'
The Development of Thermospray Mass Spectrometry Combined with Electrochemical Redox Pro-

cesses. T.A. Getek. Presented at 34th Ann. Conf. on Mass Spectrom. and Allied Topics, Cincinnati, OH, U.S.A. June 8–13, 1986. Paper RPC12. p. 756–757.
86GE7'
Thermospray Mass Spectrometry of Aminoglycoside Antibiotics. T.A. Getek and T.G. Alexander. Presented at 34th Ann. Conf. on Mass Spectrom. and Allied Topics, Cincinnati, OH, U.S.A., June 8–13, 1986. Paper FOA9. p. 907–908.
86GO9'
Antibiotics: Low Level Detection Using Thermospray Liquid Chromaotography-Mass Spectrometry. P.C. Goodley. Presented at 34th Ann. Conf. on Mass Spectrom. and Allied Topics, Cincinnati, OH, U.S.A. June 8–13, 1986. Paper WPB13. p. 589–590.
86GO09
Thermospray Ionization LC/MS Analysis of the Thermal Degradation Products of Metoclopramide. J.L. Gower, M.J. Redrup, I. O'Brien and J.A.E. Pratt. *Adv. Mass Spectrom. 1985 (Pub. 1986), 10th* (Part B), 589–599, (1986).
86GR7'
Thermospray Liquid Chromatography-Mass Spectrometry Using a Double Focussing Magnetic Sector Instrument. R.J. Greathead, R.J. Parillo, J.R. Chapman, J.A.E. Pratt, M.R. Clench and F.A. Mellon. Presented at 34th Ann. Conf. on Mass Spectrom. and Allied Topics, Cincinnati, OH, U.S.A. June 8–13, 1986. Paper WPB12. p. 587–588.
86HA87
Electrochemical Thermospray Mass Spectrometry. G. Hambitzer and J. Heitbaum. *Anal. Chem. 58*(6), 1067–1070, (1986).
86HA21
The Use of Thermospray LC/MS for Drug Identification in Human Gastric Lavage. M. Hassan, A.J. Pesce, M. McLean and D.A. Garteiz. *Vestec Thermospray Newsletter 2*(3), 1–2, (1986).
86HA3'
Analysis of Natural Waxes Using Supercritical Fluid Chromatography/Mass Spectrometry (SFC/MS). S.B Hawthorne and D.J. Miller. *Am. Chem. Soc. Abstr. 192*, AGFD0063, (1986).
86HA17
A Rapid Method for Analysis of Phospholipids by Combined Thin Layer Chromatography/Fast Atom Bombardment Mass Spectrometry/Densitometry. A. Hayashi, T. Matsubara, Y. Nishizawa, T. Hattori and M. Morita. *Proc. Jpn. Soc. Med. Mass Spectrom. 11*, 147–150, (1986).
86HA9'
Quantitative Analysis of Terbutaline by Thermospray LC/MS. M. Hayes, M. Bax and D. Alkalay. Presented at 34th Ann. Conf. on Mass Spectrom. and Allied Topics, Cincinnati, OH, U.S.A. June 8–13, 1986. Paper FPA6. p. 1009–1010.
86HE1'
Applications of LC/MS/MS to the Determination of Azo Dyes in Environmental Samples. J. Henion and A. Bruins. Presented at 34th Ann. Conf. on Mass Spectrom. and Allied Topics, Cincinnati, OH, U.S.A. June 8–13, 1986. Paper TOD3. p. 371–372.
86HE09
LC/MS/MS Screening and Identification of Drugs and Their Metabolites in Biological Samples. J. Henion and T. Covey. *Adv. Mass Spectrom. 1985 (Pub. 1986), 10th* (Part B), 629–630, (1986).
86HE4'
Supercritical Fluid Chromatography-Fourier Transform Mass Spectroscopy. J. Henion, E. Lee, R. Cody and J. Kinsinger. Presented at 34th Ann. Conf. on Mass Spectrom. and Allied Topics, Cincinnati, OH, U.S.A. June 8–13, 1986. Paper TOA7. p. 224–225.
86HE34'
Analysis of Complex Carbohydrate Mixtures by HPLC-FABMS. G. Her, S. Santikarn and V.N. Reinhold. Presented at 34th Ann. Conf. on Mass Spectrom. and Allied Topics, Cincinnati, OH, U.S.A. June 8–13, 1986. Paper TPC7. p. 353–354.

86HE11
Coal Tar Analysis by LC/MS. A.A. Herod, W.R. Ladner, B.J. Stokes, A.J. Berry and D.E. Games. *Prepr. Am. Chem. Soc., Div. Fuel Chem. 31*, 181–190, (1986).

86HI1'
Ion Mobility Spectrometry of Underivatized Drugs after Supercritical Fluid Chromatography. H.H. Hill, Jr., R.L. Eatherton, M.A. Morrissey and W.F. Siems. *Am. Chem. Soc. Abstr. 192*, ANYL0031, (1986).

86HI01
The Application of Moving Belt LC/MS and LC/FAB/MS to Material Analysis. H. Hill and W.A. Westall. *Adv. Mass Spectrom. 1985 (Pub. 1986), 10th* (Part B), 581–582, (1986).

86HI10
Recent Advances in Supercritical Fluid Chromatography. Y. Hirata and S. Tsuge. *Kagaku (Kyoto) 41*(4), 130–131, (1986).

86HO82
Determination of Nitrogen–15 Enrichment of Nitrate and Nitrite Using Thermospray Liquid Chromatography-Mass Spectrometry. L.R. Hogge, R.K. Hynes, L.M. Nelson and M.L. Vestal. *Anal. Chem. 58*(13), 2782–2784, (1986).

86HO93
Advanced Instrumentation and Strategies for Metabolic Profiling. J.F. Holland, J.J. Leary and C.C. Sweeley. In: *Profiling of Body Fluids and Tissues* (Z. Deyl and C.C. Sweeley, eds.), *J. Chromatogr. 379*, 3–25, (1986).

86HO87
Mass Spectrometry of Inductively Coupled Plasmas. R.S. Houk. *Anal. Chem. 58*, 97A–98A, 100A–105A, (1986).

86HS99
Combined Liquid Chromatography-Mass Spectrometry for Microscale Structural Studies of Carbohydrates. F.F. Hsu, C.G. Edmonds and J.A. McCloskey. *Anal. Lett. 19*(11–12), 1259–1271, (1986).

86HS9'
Thermospray Liquid Chromatography-Mass Spectrometry of Carbohydrates. F.F. Hsu, C.G. Edmonds and J.A. McCloskey. Presented at 34th Ann. Conf. on Mass Spectrom. and Allied Topics, Cincinnati, OH, U.S.A. June 8–13, 1986. Paper FOA15. p. 919–920.

86HY76
Studies of HPLC Class Separations for Environmentally Important Compounds and Application of Those Separations to HPLC/MS Analysis. P.A. Hyldburg. *Diss. Abstr. Int. B 47*(1), 166, (1986).

86IT81
Direct Coupling of Micro High-Performance Liquid Chromatography with Fast Atom Bombardment Mass Spectrometry. II. Application to Gradient Elution of Bile Acids. Y. Ito, T. Takeuchi, D. Ishii, M. Goto and T. Mizuno. *J. Chromatogr. 385*(1), 201–209, (1986).

86IW41
Direct Coupling of Thin-Layer Chromatography [TLC] with Mass Spectrometry [MS]. Direct TLC/MS Using New Sintered TLC Plate. K. Iwatani, T. Kadono and Y. Nakagawa. *Shitsuryo Bunseki 34*(3), 181–187, (1986).

86IW49
Direct Coupling of Thin-Layer Chromatography with Mass Spectrometry: Direct TLC/SIMS and Scanning TLC/SIMS. K. Iwatani and Y. Nakagawa. *Mass Spectrosc. 34*(3), 139–196, (1986).

86JA1'
Thermospray LC/MS Analysis of Amine Salts Comparing Continuous and "Pulsed" Modes of Buffer Introduction. E.M. Jakubowski, F.E. Chou and J.R. Smith. Presented at 34th Ann. Conf. on Mass Spectrom. and Allied Topics, Cincinnati, OH, U.S.A. June 8–13, 1986. Paper MOB2. p. 1–2.

86JA05
Use of Mass Spectrometry for More Reliable Determinations of Polycyclic Aromatic Hydrocarbons in Marine Sediments. W.D. Jamieson, R. Guevremont and M.A. Quilliam. *Adv. Mass Spectrom.* *1985 (Pub. 1986), 10th* (Part B), 1255–1256, (1986).

86JO07
Soft Ionization MS/MS: Thermospray and FAB with Tandem Quadrupole MS/MS Systems. M. Johnston, W.H. McFadden and P. Christiansen. *Adv. Mass Spectrom. 1985 (Publ. 1986), 10th (Part B)*, 867–868, (1986).

86JO8'
A New Source and Probe for Thermospray Applications. D.S. Jones, R.H. Bateman, M. Lancaster, T. Liszka, S.T. Krolik and V.C. Parr. Presented at 34th Ann. Conf. on Mass Spectrom. and Allied Topics, Cincinnati, OH, U.S.A. June 8–13, 1986. Paper WPB18. p. 598–599.

86KA84
Thin Layer Chromatography-Mass Spectrometry (TLC/MS). H. Kajiura. *Seirigaku Gijutsu Kenkyukai Hokoku 8*, 14–15, (1986).

86KA81
Supercritical Fluid Extraction and Direct Fluid Injection Mass Spectrometry for the Determination of Tricothecene Mycotoxins in Wheat Samples. H.T. Kalinoski, H.R. Udseth, B.W. Wright and R.D. Smith. *Anal. Chem. 58*(12), 2421–2425, (1986).

86KA1'
Capillary Supercritical Fluid Chromatography-Mass Spectrometry for the Analysis of Labile and Nonvolatile Compounds. H.T. Kalinoski, H.R. Udseth, B.W. Wright and R.D. Smith. Presented at 34th Ann. Conf. on Mass Spectrom. and Allied Topics, Cincinnati, OH, U.S.A. June 8–13, 1986. Paper MOB12. p. 13–14.

86KA33
Ammonia and Methane Chemical Ionization Mass Spectra of Acid and Carbamate Pesticides Using Direct Supercritical Fluid Injection. H.T. Kalinoski, B.W. Wright and R.D. Smith. *Biomed. Environ. Mass Spectrom. 13*, 33–47, (1986).

86KA7'
Determination of Acrylamide in Sugar by Thermospray Multidimensional Liquid Chromatography-Mass Spectrometry. G.J. Kallos and S.S. Cutié. Presented at 34th Ann. Conf. on Mass Spectrom. and Allied Topics, Cincinnati, OH, U.S.A. June 8–13, 1986. Paper MOB5. p. 7–8.

86KA3'
Sample Holder in a Secondary-Ion-Mass Spectrometer for Decreasing Contamination by the Evaporated Fluid Matrix. H. Kambara and S. Seki. US 4,620,103 (Cl. 250–440.1; G01F21/00), 28 Oct. 1986, JP Appl. 83/226,855, 02 Dec. 1983, 5pp.

86KA6'
Secondary-Ion Mass Spectrometer. H. Kanbara and H. Hirose (Hitachi Ltd.). Jpn. Kokai Tokkyo Koho JP 61 07,556 [86 07,556] (Cl. H01J49/14), 14 Jan. 1986, Appl. 85/115,303, 30 May 1985, 4pp.

86KA5'
Studies on Size and Charge Distribution of Droplets Formed in Thermospray. V. Katta and M.L. Vestal. Presented at 34th Ann. Conf. on Mass Spectrom. and Allied Topics, Cincinnati, OH, U.S.A. June 8–13, 1986. Paper WPB16. p. 595.

86KI03
The Moving Belt Interface. A Reliable Combined Liquid Chromatography/Mass Spectrometry Interface, Which Is Capable of Performing Analyses in Three Ionization Modes. G. Kilpatrick, I.A.S. Lewis and J.F. Smith. *Adv. Mass Spectrom. 1985 (Pub. 1986), 10th* (Part B), 593–594, (1986).

86KI89
Phospholipid Molecular Species Analysis by Thermospray LC/MS. H.-Y. Kim and N. Salem, Jr. *Anal. Chem. 58*, 9–14, (1986).

86KI1'
Quantitative Analysis of Phospholipid Molecular Species by Thermospray LC/MS. H.-Y. Kim and
N. Salem, Jr. Presented at 34th Ann. Conf. on Mass Spectrom. and Allied Topics, Cincinnati,
OH, U.S.A. June 8–13, 1986. Paper TPC12. p. 361–362.

86KI034
Comparison of LC/MS Methods for the Analysis of Drugs. M.L. Kimber, D.E. Games and M.J.
Whitehouse. *Adv. Mass Spectrom. 1985 (Pub. 1986), 10th* (Part B), 583–584, (1986).

86KO88
Thermospray Interfacing for Flow Injection Analysis with Inductively Coupled Plasma Atomic
Emission Spectrometry. J.A. Koropchak and D.H. Winn. *Anal. Chem. 58*(12), 2558–2561,
(1986).

86KR5'
Simultaneous and Accurate Analysis of Trace Levels of Simple and Macrocyclic Trichothecenes by
LC/Thermospray Mass Spectrometric and Tandem Mass Spectrometric Techniques. T.
Krishnamurthy, E.W. Sarver, D.J. Beck and R. Isensee. Presented at 34th Ann. Conf. on Mass
Spectrom. and Allied Topics, Cincinnati, OH, U.S.A. June 8–13, 1986. Paper MOD12. p. 105–
106.

86KR05
Detection and Quantification of Roridins and Baccharinoids in Brazilian Plants by LC-Thermospray-
MS-MS Technique. T. Krishnamurthy, E.W. Sarver and B. Jarvis. *Adv. Mass Spectrom. 1985
(Pub. 1986), 10th* (Part B), 595–596, (1986).

86KU97
Lipids and Their Constituents. A. Kuksis and J.J. Myher. In: *Profiling of Body Fluids and Tissues*
(Z. Deyl and C.C. Sweeley, eds.), *J. Chromatogr. 279,* 57–89, (1986).

86KU07
Liquid Ionization Mass Spectrometry of Phospholipids. H. Kuwabara, I. Viden, M. Tsuchiya, J.
Sugatani and K. Saito. *J. Biochem. 100*(2), 477–484, (1986).

86LA34
Characterization and Molecular Weight Determination of Water-Soluble Polyethylene Glycol
Oligomers Using Open-Tubing Liquid Chromatography-Mass Spectrometry. S.-T. Lai. *J.
Chromatogr. 363*(2), 444–447, (1986).

86LA5'
Rapid Determination of Volatile Pollutants in Aqueous Samples via Flow Injection Mass Spectrometry. P.W. Langvardt, K.A. Brzak and P.E. Kastl. Presented at 34th Ann. Conf. on Mass Spectrom. and Allied Topics, Cincinnati, OH, U.S.A. June 8–13, 1986. Paper MPF9. p. 195–196.

86LA05
The Characterization of an Acidic Metabolite of Ranitidine in Human Urine by DLI HPLC-MS. M.S.
Lant, G.R. Manchee, L.E. Martin and J. Oxford. *Adv. Mass Spectrom. 1985 (Pub. 1986), 10th*
(Part B), 615–616, (1986).

86LE3'
Benchtop Supercritical Fluid Chromatography/Mass Spectrometry Determination of Corticosteroids.
E.D. Lee and J.D. Henion. Presented at 34th Ann. Conf. on Mass Spectrom. and Allied Topics,
Cincinnati, OH, U.S.A. June 8–13, 1986. Paper FPA13. p. 1023–1024.

86LE89
High-Performance Liquid Chromatographic Chiral Stationary Phase Separation with Filament-On
Thermospray Mass Spectrometric Identification of the Enantiomer Contaminant (S)-(+)-
Methamphetamine. E.D. Lee, J.D. Henion, C.A. Brunner, I.W. Wainer, T.D. Doyle and J. Gal.
Anal. Chem. 58(7), 1349–1352, (1986).

86LE92
Open Tubular Column Supercritical Fluid Chromatography/Mass Spectrometry on a Benchtop Mass
Spectrometer. E.D. Lee and J.D. Henion. *J. High Resolut. Chromatogr. Chromatogr. Commun.
9*(3), 172–174, (1986).

86LI0'
HPLC/MS in the Clinical Analysis of Polar Corticosteroids and Glucuronides. D. Liberato, N. Esteban and C.H.L. Shackleton. Presented at 34th Ann. Conf. on Mass Spectrom. and Allied Topics, Cincinnati, OH, U.S.A. June 8–13, 1986. Paper TPB2. p. 310–311.

86LI3'
Thermospray LC/MS in Studies of Biomedical and Clinical Interest. D.J. Liberato, N.V. Esteban and A.L. Yergey. Presented at 34th Ann. Conf. on Mass Spectrom. and Allied Topics, Cincinnati, OH, U.S.A. June 8–13, 1986. Paper FOA17. p. 923–924.

86LI86
Solvent Selection for Thermospray Liquid Chromatography/Mass Spectrometry. D.J. Liberato and A.L. Yergey. *Anal. Chem. 58*, 6–9, (1986).

86LI31
Separation and Quantification of Choline and Acetylcholine by Thermospray Liquid Chromatography/Mass Spectrometry. D.J. Liberato, A.L. Yergey and S.T. Weintraub. *Biomed. Environ. Mass Spectrom. 13*, 171–174, (1986).

86LI89
Mass Spectrometric Determination of Amines after Formation of a Charged Surface Active Derivative. W.V. Ligon, Jr. and S.B. Dorn. *Anal. Chem. 58*, 1889–1892, (1986).

86LI09
LC/MS: A Thermospray Interface and Ion Source for a Sector Field Mass Spectrometer. M. Linscheid and R. Ohlendorf. *Adv. Mass Spectrom. 1985 (Pub. 1986), 10th* (Part B), 619–620, (1986).

86LI0'
The Application of Thermospray LC/MS and Thermospray LC/MS/MS to the Analysis of Epoxy Graphite Composite Materials. J.F. Litton, S.J. Selover, M.R. Wakefield and T.M. Tuell. Presented at 34th Ann. Conf. on Mass Spectrom. and Allied Topics, Cincinnati, OH, U.S.A. June 8–13, 1986. Paper RPD15. p. 840–841.

86LL4'
Mixed-Mode Column Thermospray LC/MS. J.R. Lloyd, M.L. Cotter, D. Ohori and A. Oyler. Presented at 34th Ann. Conf. on Mass Spectrom. and Allied Topics, Cincinnati, OH, U.S.A. June 8–13, 1986. Paper WPB5. p. 574–575.

86LU3'
Supercritical Fluid Injection of Nonvolatiles with Resonant Two-Photon Ionization Detection in Supersonic Beam Mass Spectrometry. D.M. Lubman, C.H. Sin and H.M. Pang. Presented at 34th Ann. Conf. on Mass Spectrom. and Allied Topics, Cincinnati, OH, U.S.A. June 8–13, 1986. Paper TOE4. p. 383–384.

86LU37
Routine Analysis with LC-MS Coupling? I. Lüderwald. *Fresenius' Z. Anal. Chem. 323*(3), 287–288, (1986).

86LU59
Ultratrace Drug Analysis in Forensic Chemistry. I.S Lurie, J.M. Moore and D.A. Cooper. *Chem. Anal. (N.Y.) 85*, 319–352, (1986).

86MA31
Thermospray Liquid Chromatography/Mass Spectrometry of Ether Phosphocholines. A.I. Mallet and K. Rollins. *Biomed. Environ. Mass Spectrom. 13*(10), 541–544, (1986).

86MA69
Liquid Chromatography-Mass Spectrometry (LC-MS). L.E. Martin, M.S. Lant and J. Oxford. *Methodol. Surv. Biochem. Anal. 16*, 399–402, (1986).

86MA07
Coupling of Liquid Chromatography to Mass Spectrometry for the Characterization of a Complex Mixture of Chemical Intermediates. E. Martinelli, E. Santoro and V. Ambroge. *Adv. Mass Spectrom. 1985 (Pub. 1986), 10th* (Part B), 617–618, (1986).

86MA32
Development of Self-Spouting and Vacuum Nebulizing Assisted Interface for Direct Coupling of a Micro-Liquid Chromatograph with a Mass Spectrometer and Some Applications to Thermally Labile Compounds. K. Matsumoto, H.C. Lieu and S. Tsuge. *Fresenius' Z. Anal. Chem. 323*(3), 212–216, (1986).
86MA43
Application of HPLC/MS with Self-Spouting and Vacuum Nebulizing Assisted Interface to the Analysis of Triglycerides. K. Matsumoto and S. Tsuge. *Mass Spectrosc. 34*(1), 33–40, (1986).
86MA438
Analysis of Free and PTH-Amino Acids by HPLC/MS with Self-Spouting and Vacuum Nebulizing Assisted Interface. K. Matsumoto and S. Tsuge. *Mass Spectrosc. 34*(4), 243–248, (1986).
86MA17
Fundamental Conditions in Pressure-Programmed Supercritical Fluid Chromatography-Mass Spectrometry and Some Applications to Vitamin Analysis. K. Matsumoto, S. Tsuge and Y. Hirata. *Chromatographia 21*(11), 617–621, (1986).
86MA23
Development of Directly Coupled Supercritical Fluid Chromatography-Mass Spectrometry with Self-Spouting and Vacuum Nebulizing Assisted Interface. K. Matsumoto, S. Tsuge and Y. Hirata. *Anal. Sci. 2*(1), 3–7, (1986).
86MA61
Disposition of 2,4-Dinitroaniline in the Male F–344 Rat. H.B. Matthews, H.M. Chopade, R.W. Smith and L.T. Burka. *Xenobiotica 16*(1), 1–10, (1986).
86MC76
Nucleoside Modification in Archaebacterial Transfer RNA. J.A. McCloskey. *System. Appl. Microbiol. 7*, 246–252, (1986).
86MC5GA'
Experimental Approaches to the Characterization of Nucleic Acid Constituents by Mass Spectrometry. J.A. McCloskey. In: *Mass Spectrometry in Biomedical Research* (S.J. Gaskell, ed.), p. 75–95, Wiley, New York, (1986).
86MC3'
Environmental Applications of a New Thermospray LC/MS Interface for Bench-Top Mass Spectrometers. S.M. McCown and C.H. Vestal. Presented at 34th Ann. Conf. on Mass Spectrom. and Allied Topics, Cincinnati, OH, U.S.A. June 8–13, 1986. Paper WPB15. p. 593–594.
86MC22
Analysis of Herbicides by Thermospray Liquid Chromatography-Mass Spectrometry. S.M. McCown and C.H. Vestal. *Vestec Thermospray Newsletter 2*(5), 2–4, (1986).
86MC24
The Thermospray Ion Source Temperature. M.A. McLean. *Vestec Thermospray Newsletter 2*(1), 4, (1986).
86MC73
Identification of the Major Metabolites of Nafidimide in Dogs Using Thermospray Liquid Chromatography-Mass Spectrometry. M.A. McLean, M.L. Vestal, D.A. Garteiz, K. Lu and R.A. Newman. *Proc. Am. Assoc. Cancer Res. Ann. Meeting 27*(0), 423, (1986).
86MC7'
Metabolism of Nafidimide in Cancer Patients by Thermospray LC/MS. M.A. McLean, M.L. Vestal, D.A. Garteiz and R.A. Newman. Presented at 34th Ann. Conf. on Mass Spectrom. and Allied Topics, Cincinnati, OH, U.S.A. June 8–13, 1986. Paper FPA10. p. 1017–1018.
86MI01
Analysis of Intermediary Metabolites Using Thermospray LC/MS. D.S. Millington, D.A. Maltby and P. Farrow. *Adv. Mass Spectrom. 1985 (Pub. 1986), 10th* (Part B), 591–592, (1986).
86MI58
Permeable Membrane/Mass Spectrometric Measurement of Solvent Proton/Deuterium, Carbon–

12/Carbon-13 and Oxygen-16/Oxygen-18 Kinetic Isotope Effects Associated with Alpha-Chymotrypsin Deacylation: Evidence for Reaction Mechanism Plasticity. A.K. Mishra and M.H. Klapper. *Biochemistry 25*(23), 7328–7336, (1986).

86MI6'
Accurate Mass Determination of Macrocyclic Trichothecene Mycotoxins Using a LCMS Interface as an Automated Inlet. S.R. Missler, P.H. Gibbs, B.B. Jarvis and K. Wells. Presented at 34th Ann. Conf. on Mass Spectrom. and Allied Topics, Cincinnati, OH, U.S.A. June 8–13, 1986. Paper MPC2. p. 66–67.

86MI454
New Interface for Direct Liquid Introduction LC/MS. T. Mizuno, K. Matsuura and K. Azuma. *Mass Spectrosc. 34*(4), 225–234, (1986).

86MI45
Application of Direct Liquid Introduction LC/MS Negative CI Mass Spectra. T. Mizuno, K. Matsuura and K. Azuma. *Mass Spectrosc. 34*(4), 235–242, (1986).

86MI67'
The Development of Novel Sample Introduction Technique for FAB Measurement. T. Mizuno, Y. Itagaki, Y. Ito, T. Takeuchi, D. Ishii and M. Goto. Presented at 34th Ann. Conf. on Mass Spectrom. and Allied Topics, Cincinnati, OH, U.S.A. June 8–13, 1986. Paper WPB6. p. 576–577.

86MO31
Liquid Chromatographic-Mass Spectrometric Analysis of 2-Nitrofluorene and Its Derivatives. L. Moller and J.-A. Gustafsson. *Biomed. Environ. Mass Spectrom. 13*(12), 681–688, (1986).

86MO05
Studies of Groundwater Samples by Combined LC/MS. A.Van Moncur, D.E. Games and A.F. Weston. *Adv. Mass Spectrom. 1985 (Pub. 1986), 10th* (Part B), 575–576, (1986).

86NA93
Recent Advances in Analytical Instruments. 2. New Techniques in Liquid Chromatography. H. Nakanishi. *Petrotech (Tokyo) 9*(9), 823–827, (1986).

86NA21
Analysis of Drugs. Liquid Chromatography/Mass Spectrometry. T. Namba, N. Tsuneuchi and M. Hattori. *Pharm. Tech. Jpn. 2*(1), 31–39, (1986).

86NI13
Recent Developments in Mass Spectrometry. N.M.M. Nibbering. *Chem. Mag. (Rijswijk, Netherlands)* (Oct.), p. 713–715, (1986).

86NI17
A Review of Direct Liquid Introduction Interfacing for LC/MS. Part I. Instrumental Aspects. W.M.A. Niessen. *Chromatographia 21*(5), 277–287, (1986).

86NI12
A Review of Direct Liquid Introduction Interfacing for LC/MS. Part II. Mass Spectrometry and Applications. W.M.A. Niessen. *Chromatographia 21*(6), 342–354, (1986).

86OT12
An Improved Liquid Ionization Mass Spectrometer Equipped with a Three-Stage Differential Pumping System. K. Otsuka, T. Mizuno, K. Azuma, K. Musha and M. Tsuchiya. *Bunseki Kagaku 1*, 12–17, (1986).

86PA812
Characterization of Mixtures of Organic Acids by Ion-Exclusion Partition Chromatography Mass Spectrometry. F. Pacholec, D.R. Eaton and D.T. Rossi. *Anal. Chem. 58*(12), 2581–2582, (1986).

86PA8'
Design of a Pulsed Nozzle for Supercritical Fluid/Mass Spectrometry. H.M. Pang, C.H. Sin, D.M. Lubman and J. Zorn. Presented at 34th Ann. Conf. on Mass Spectrom. and Allied Topics, Cincinnati, OH, U.S.A. June 8–13, 1986. Paper RPF10. p. 888–889.

86PA81
Design of a Pulsed Source for Supercritical Fluid Injection into Supersonic Beam Mass Spectrometry. H.M. Pang, C.H. Sin, D.M. Lubman and J. Zorn. *Anal. Chem.* 58, 1581–1583, (1986).
86PA01′
Formation of (M–1)⁻ or (M+OAc)⁻ Ions in Negative Ion Thermospray Mass Spectrometry. C.E.
Parker, R.W. Smith, S.J. Gaskell and M.M. Bursey. Presented at 34th Ann. Conf. on Mass
Spectrom. and Allied Topics, Cincinnati, OH, U.S.A. June 8–13, 1986. Paper RPC14. p. 760–
761.
86PA814
Dependence of Ion Formation upon the Ionic Additive in Thermospray Liquid Chromatogra-
phy/Negative Ion Mass Spectrometry. C.E. Parker, R.W. Smith, S.J. Gaskell and M.M. Bursey.
Anal. Chem. 58(8), 1661–1664, (1986).
86PA03
Identification of Budesonide Metabolites by LC/MS. J. Paulson, C. Lindberg and S. Edsbacker.
Adv. Mass Spectrom. 1985 (Pub. 1986), 10th (Part B), 613–614, (1986).
86PE4′
An Investigation of Operational Parameters and Reproducibility of the Monodispersed Aerosol
Generation LC/MS Interface. D.D. Perkins, P.C. Winkler, L.E. Abbey and R.F. Browner.
Presented at 34th Ann. Conf. on Mass Spectrom. and Allied Topics, Cincinnati, OH, U.S.A. June
8–13, 1986. Paper WPB10. p. 584.
86PH0′
Isolation and Structure Elucidation of a Unique Hypermodified Nucleoside from *E. coli* Transfer
RNA. D.W. Phillipson, C.G. Edmonds, P.F. Crain, D.L. Smith and J.A. McCloskey. Presented
at 34th Ann. Conf. on Mass Spectrom. and Allied Topics, Cincinnati, OH, U.S.A. June 8–13,
1986. Paper FOA4. p. 900–901.
86PR07
Enzyme Kinetics of Biotransformation Reactions by a Phenoloxidase in *Mucuna pruriens* and Product Identification by LC/MS. N. Pras and A.P. Bruins. *Adv. Mass Spectrom. 1985 (Pub. 1986).
10th*(Part B), 579–580, (1986).
86QU5′
Combined LC/MS as an Aid in the Preparative Fractionation of Complex Mixtures. M.A. Quilliam,
K. Hoo, T. Berry and D.E. Games. Presented at 34th Ann. Conf. on Mass Spectrom. and Allied
Topics, Cincinnati, OH, U.S.A. June 8–13, 1986. Paper TPB15. p. 335–336.
86RA31
Use of Reversed-Phase Chromatography in Carbohydrate Analysis. E. Rajakyla. *J. Chromatogr.*
353, 1–12, (1986).
86RA2′
Thermospray LC/MS Analysis of Mycotoxins as a Function of Electrolyte Concentration and Mode
of Ionization. E. Rajakyla, K. Laasasenaho, P. Sakkers and D. Foerst. Presented at 34th Ann.
Conf. on Mass Spectrom. and Allied Topics, Cincinnati, OH, U.S.A. June 8–13, 1986. Paper
MPC5. p. 72–73.
86RI19
Generation of Hydroxyeicosatetraenoic Acids by Human Inflammatory Cells: Analysis by Thermospray Liquid Chromatography-Mass Spectrometry. R. Richmond, S.R. Clarke, D. Watson,
C.G. Chappell, C.T. Taylor and W. Graham. *Biochim. Biophys. Acta 881*(2), 159–166, (1986).
86RO8′
Analysis of Peptide Mixtures and Tryptic Digests by Gradient Liquid Chromatography/Thermospray
Mass Spectrometry. R.H. Robins and F.W. Crow. Presented at 34th Ann. Conf. on Mass Spec-
trom. and Allied Topics, Cincinnati, OH, U.S.A. June 8–13, 1986. Paper RPE5. p. 848–849.
86RO0′
Critical Parameters Effecting Thermospray Performance. R.H. Robins and F.W. Crow. Presented at

34th Ann. Conf. on Mass Spectrom. and Allied Topics, Cincinnati, OH, U.S.A. June 8–13, 1986. Paper WPB3. p. 570–571.
86RU3'
A Comparison of FAB-MS/MS and HPLC-MS/MS for the Analysis of Catecholamine Metabolites. P. Rudewicz and K. Straub. Presented at 34th Ann. Conf. on Mass Spectrom. and Allied Topics, Cincinnati, OH, U.S.A. June 8–13, 1986. Paper FPB4. p. 1033–1034.
86RU9'
Sensitivity and Selectivity in Thermospray Ionization. P. Rudewicz and K. Straub. Presented at 34th Ann. Conf. on Mass Spectrom. and Allied Topics, Cincinnati, OH, U.S.A. June 8–13, 1986. Paper MOB6. p. 9–10.
86RU88
Rapid Structure Elucidation of Catecholamine Conjugates with Tandem Mass Spectrometry. P. Rudewicz and K.M. Straub. *Anal. Chem. 58*(14), 2928–2934, (1986).
86SA03
Post-Column Chemical Derivatization in HPLC/MS Using a Moving Belt Interface. A. Sahil, C.P. Tsai, B.L. Karger and P. Vouros. *Adv. Mass Spectrom. 1985 (Pub. 1986), 10th* (Part B), 633–634, (1986).
86SA4'
Combined Liquid Chromatograph-Mass Spectrometer. M. Sakairi and H. Kanbara (Hitachi, Ltd.) Jpn. Kokai Tokkyo Koho JP 61 95,244 [86 95,244] (Cl. G01N27/62), 14 May 1986, Appl. 84/216,142, 17 Oct. 1984, 4pp.
86SA49
Direct Analysis and Differentiation of Wines by Liquid Ionization Mass Spectrometry. K. Sasakawa and M. Tsuchiya. *Mass Spectrosc. 34*(3), 159–169, (1986).
86SC59
Soft Ionization of Biomolecules: A Comparison of Ten Ionization Methods for Corrins and Vitamin B12. H.M. Schiebel and H-R. Schulten. *Mass Spectrom. Rev. 5*(3), 249–312, (1986).
86SC01
Formation and Decomposition of Charged Solid Particles in Thermospray Mass Spectrometry. G. Schmelzeisen-Redeker, L. Butfering, U. Giessmann and F.W. Röllgen. *Adv. Mass Spectrom. 1985 (Pub. 1986), 10th* (Part B), 631–632, (1986).
86SC02
Thermospray Mass Spectrometry of Diazonium and Di-, Tri- and Tetraquaternary Onium Salts. G. Schmelzeisen-Redeker, F.W. Röllgen, H. Wirtz and F. Voegtle. *Org. Mass Spectrom. 20*(10), 752–756, (1986).
86SC00
Triple-Stage Quadrupole Mass Spectrometry. R. Schubert. *Labor Praxis 10*(3), 190–191, (1986).
86SC2'
Liquid Chromatography Detectors. R.P.W. Scott. Elsevier, Amsterdam (1986), p. 272.
86SH9'
New Developments in Herbicides Analysis by Thermospray Liquid Chromatography/Mass Spectrometry. L.M. Shalaby. Presented at 34th Ann. Conf. on Mass Spectrom. and Allied Topics, Cincinnati, OH, U.S.A. June 8–13, 1986. Paper MPF6. p. 189–190.
86SH63
High-Performance Liquid Chromatography and Its Application to the Analysis of Mycotoxins. M.J. Shepherd. *Mod. Methods Anal. Struct. Elucidation Mycotoxins 1986*, 293–333, (1986).
86SH15
Direct Analysis of Nonvolatile Biomedical Compounds on Thin Layer Plates by Liquid Secondary Ion Mass Spectrometry. K. Shizukuishi, Y. Numaziri and Y. Kato. *Proc. Jpn. Soc. Med. Mass Spectrom. 11*, 85–91, (1986).
86SI6'
The Use of a Thermospray Interface to a Mass Spectrometer as a Flow Reactor for In-Situ Thermal

Degradation Studies: Applications to Beta-Lactam Antobiotics. M.M. Siegel, R.K. Isensee and D.J. Beck. Presented at 34th Ann. Conf. on Mass Spectrom. and Allied Topics, Cincinnati, OH, U.S.A. June 8–13, 1986. Paper WPB1. p. 566–567.
86SI0SA'
Recent Advances in the Interfacing of Liquid Chromatography to Mass Spectrometry. S.J. Simpson, M.A. McDowall, N. Lynaugh and D.C. Smith. In: *High Performance Liquid Chromatography in the Clinical Laboratory* (D.C. Sampson, ed.), p. 150–157, Australian Association of Clinical Biochemistry, Sydney, (1986).
86SK07
Combined LC/MS of Antibacterial and Antiparasitic Agents Isolated from Streptomyces. P.W. Skett and D.E. Games. *Adv. Mass Spectrom. 1985 (Pub. 1986), 10th* (Part B), 577–578, (1986).
86SK47
Identification of Betamethasone and a Major Metabolite in Equine Urine. D.S. Skrabalak and J.D. Henion. *J. Pharm. Biomed. Anal. 4*(3), 327–332, (1986).
86SM1RA'
Advanced Techniques in Mass Spectrometry. D.L. Smith. *Natural Products Chemistry, Proceedings of the 1st International Symposium: The Pakistan-U.S. Binational Workshop, 1984* (Publ. 1986), (A. Ur. Rahman, ed.), p. 411–456, Springer, Berlin, (1986).
86SM32'
Interfacing Methods and Performance of Capillary and Microbore Column Supercritical Fluid Chromatography-Mass Spectrometry. R.D. Smith, H.T. Kalinoski, H.R. Udseth and B.W. Wright. *Am. Chem. Soc. Abstr. 192*, ANYL0032, (1986).
86SM70
Capillary Supercritical Fluid Chromatography and Supercritical Fluid Chromatography-Mass Spectrometry. R.D. Smith, B.W. Wright and H.R. Udseth. In: *Chromatography and Separation Chemistry: Advances and Developments* (S. Ahuja, ed.), *Am. Chem. Soc. Symp. Series 297*, 260–293, (1986).
86SM71
Mass Spectrometric Analysis [of Antibiotics]. R.G. Smith. *Drugs Pharm. Sci. 27*, 141–181, (1986).
86SM07
Selective Detection and Quantification of Targeted Analytes during Liquid Chromatography-Mass Spectrometry. R.W. Smith, C.E. Parker and S.J. Gaskell. *Adv. Mass Spectrom. 1985 (Pub. 1986), 10th* (Part B), 597–598, (1986).
86SM2'
Eluent pH and Thermospray Spectra: Does the Charge on the Ion in Solution Influence the Spectrum? R.W. Smith, C.E. Parker, D.M. Johnson and M.M. Bursey. Presented at 34th Ann. Conf. on Mass Spectrom. and Allied Topics, Cincinnati, OH, U.S.A. June 8–13, 1986. Paper WPB4. p. 572–573.
86SN59
Liquid Chromatography and Liquid Chromatography-Mass Spectrometry. L.R. Snyder and S. Ajuha. *Chem. Anal. (N.Y.) 85*, 109–148, (1986).
86ST4'
Distinction of Isomeric Alpha- and Beta-Aspartame by Thermospray LC/MS. J.A. Stamp, D.A. Garteiz and T.P. Labuza. Presented at 34th Ann. Conf. on Mass Spectrom. and Allied Topics, Cincinnati, OH, U.S.A. June 8–13, 1986. Paper TPE17. p. 474–475.
86ST05
Mass Spectrometry, Possibilities of Application in Studies of Organic Compounds and Development Trends. M. Stobiecki. *Wiad. Chem. 40*(7–8), 495–532, (1986).
86ST4'
Thermospray LC-MS/MS for Rapid Profiling of Drug Metabolites. K. Straub, P. Rudewicz, L. Gutzait and B. Mico. Presented at 34th Ann. Conf. on Mass Spectrom. and Allied Topics, Cincinnati, OH, U.S.A. June 8–13, 1986. Paper FPB10. p. 1044–1045.

86ST05
On-Line Liquid Chromatography/Mass Spectrometry Employing Fast Atom Bombardment. J.G. Stroh, J.C. Cook and K.L. Rinehart, Jr. *Adv. Mass Spectrom. 1985 (Pub. 1986), 10th* (Part B), 625–626, (1986).

86ST5'
Practical Applications of On-Line Liquid Chromatography/Fast Atom Bombardment Mass Spectrometry. J.G. Stroh, J.C. Cook, K.L. Rinehart, Jr. and W.C. Snyder. Presented at 34th Ann. Conf. on Mass Spectrom. and Allied Topics, Cincinnati, OH, U.S.A. June 8–13, 1986. Paper TOE10. p. 395–396.

86ST88
Identification and Structure of Components of Leucinostatin and CC–1014 by Directly Coupled Liquid Chromatography/Fast Atom Bombardment Mass Spectrometry. J.G. Stroh, K.L. Rinehart, Jr., J. C. Cook, T. Kihara, M. Suzuki and T. Arai. *J. Am. Chem. Soc. 108*, 858–859, (1986).

86SU21
In Vitro Study of the Stereoselective Reduction of a 9-Ketocannabinoid by Thermospray LC/MS. H.R. Sullivan and D.A. Garteiz. *Vestec Thermospray Newsletter 2*, 1–3, (1986).

86SU96
Electronic Sputtering of Biomolecules. B. Sundquist, A. Hedin, P. Haakansson, G. Jonsson, M. Salehpour, G. Saeve, S. Widdiyasekera and P. Roepstorff. *Springer Proc. Phys. 9*, 6–10, (1986).

86SU89
Secondary Ion Currents in Fast Atom Bombardment of Preionized Liquids. J.A. Sunner, R. Kulatunga and P. Kebarle. *Anal. Chem. 58*, 2009–2014, (1986).

86SU5'
Analysis of Ampicillin by Thermospray High-Performance Liquid Chromatography/Mass Spectrometry in Biological Fluids. S. Suwanrumpha and M.L. Vestal. Presented at 34th Ann. Conf. on Mass Spectrom. and Allied Topics, Cincinnati, OH, U.S.A. June 8–13, 1986. Paper TPE12. p. 465–466.

86TA06
LC/MS Determination of Bromazepam, Clopenthixol and Reserpine in Serum of a Non-Fatal Case of Intoxication. A.C. Tas, J. Van Der Greef, M.C. Ten Noever De Brauw, T.A. Plomp, R.A.A. Maes, M. Höhn and U. Rapp. *J. Anal. Toxicol. 10*, 46–48, (1986).

86TA0PI'
Mass Spectrometric and Immunological Assays for the Leukotrienes. G.W. Taylor, C.G. Chappell, S.R. Clarke, D.J. Heavey, R. Richmond, N.C. Turner, D. Watson and C.T. Dollery. In: *Leukotrienes: Their Biological Significance, Biology Council Symposium 1985* (P.J. Piper, ed.), p. 67–90. Raven, New York, 1986 .

86TA65
Analysis of Lipoxygenase Products by Thermospray Liquid Chromatography-Mass Spectrometry. G.W. Taylor, D. Watson, C.T. Dollery and R. Richmond. *Top. Lipid Res. 1986*, 85–93, (1986).

86TH81
Inductively Coupled Plasma Mass Spectrometric Detection for Multielement Flow Injection Analysis and Elemental Speciation by Reversed-Phased Liquid Chromatography. J.J. Thompson and R.S. Houk. *Anal. Chem. 58*(12), 2541–2548, (1986).

86TS82
High-Performance LC/MS Determination of Volatile Carboxylic Acids Using Ion-Pair Extraction and Thermally Induced Alkylation. C.-P. Tsai, A. Sahil, J.M. McGuire, B.L. Karger and P. Vouros. *Anal. Chem. 58*, 2–6, (1986).

86UD0'
Comparison of Results from Packed and Open Tubular Capillary Column SFC/MS. H.R. Udseth and R.D. Smith. Presented at 34th Ann. Conf. on Mass Spectrom. and Allied Topics, Cincinnati, OH, U.S.A. June 8–13, 1986. Paper WPB8. p. 580–581.

86UN7'
Thermospray Liquid Chromatography/Mass Spectrometry for the Identification of Metabolites: Comparisons with Other Desorption Ionization Methods. S.E. Unger, T.J. McCormick, B.M. Warrack, S.J. Lan, B. Swanson, D. Everett and J. Mitroka. Presented at 34th Ann. Conf. on Mass Spectrom. and Allied Topics, Cincinnati, OH, U.S.A. June 8–13, 1986. Paper TOE6. p. 387–388.

86UN13
Pharmaceutical Analysis Using Thermospray Liquid Chromatography/Mass Spectrometry and Mass Spectrometry. S.E. Unger and B.M. Warrack. *Spectroscopy (Springfield, Oreg.) 1*(3), 33–38, (1986).

86VA46
Determination of Vitamin E in Seed Oils by High Performance Liquid Chromatography-Mass Spectrometry. J. Van Der Greef, A.J. Speek, A.C. Tas and J. Schrijver. *LC/GC Mag. 4*(7), 636–637, (1986).

86VA07
High Resolution MS Techniques Combined with the Moving Belt LC/MS Interface. J. Van Der Greef, A.C. Tas, M.C. Ten Noever De Brauw, U. Rapp, M. Höhn and G. Meijerhoff. *Adv. Mass Spectrom. 1985 (Pub. 1986), 10th* (Part B), 587–588, (1986).

86VA35
Characterization of Additive in Plastics by Liquid Chromatography-Mass Spectrometry. J.D. Vargo and K.L. Olson. *J. Chromatogr. 353*, 215–224, (1986).

86VE21
Environmental Applications of a Thermospray LC/MS Interface for Bench-Top Mass Spectrometers. C.H. Vestal. *Vestec Thermospray Newsletter 2*(4), 1–4, (1986).

86VE05
Automatic Thermospray Controller for Gradient Elution Liquid Chromatography/Mass Spectrometry. C.H. Vestal, G.J. Fergusson and M.L. Vestal. *Int. J. Mass Spectrom. Ion Processes 70*(2), 185–194, (1986).

86VE33
Mass Spectrometers as Detectors for Liquid Chromatography. M.L. Vestal. *Life Sci. Res. Rep. 33*, 613–629, (1986).

86VO8'
Direct Ion Evaporation of Ionic Organic and Inorganic Compounds in Thermospray HPLC/MS. R.D. Voyksner. Presented at 34th Ann. Conf. on Mass Spectrom. and Allied Topics, Cincinnati, OH, U.S.A. June 8–13, 1986. Paper RPC13. p. 758–759.

86VO05
Characterization of Organic Dyes by Thermospray HPLC/MS and FAB/MS/MS. R.D. Voyksner. *Adv. Mass Spectrom. 1985 (Pub. 1986), 10th* (Part B), 585–586, (1986).

86VO1'
Analysis of Leukotrienes, Prostaglandins and Other Metabolites of Arachidonic Acid by Thermospray HPLC/MS. R.D. Voyksner and E.D. Bush. Presented at 34th Ann. Conf. on Mass Spectrom. and Allied Topics, Cincinnati, OH, U.S.A. June 8–13, 1986. Paper MOE8. p. 114–115.

86VO3'
Analysis of Some Fusarium Mycotoxin Metabolites by Thermospray HPLC/MS. R.D. Voyksner, C.A. Haney, W.M. Hagler and S.P. Swanson. Presented at 34th Ann. Conf. on Mass Spectrom. and Allied Topics, Cincinnati, OH, U.S.A. June 8–13, 1986. Paper FOA12. p. 913–914.

86VO43
Trace Analysis of Explosives by Thermospray High-Performance Liquid Chromatography-Mass Spectrometry. R.D. Voyksner and J. Yinon. *J. Chromatogr. 354*, 393–405, (1986).

86WA8'
Analysis of Naphthenes and Organometallic Ferrocene-Type Derivatives by Normal Phase Liquid

Chromatography/Thermospray Mass Spectrometry. T.H. Wang, M. McLean and C.H. Vestal. Presented at 34th Ann. Conf. on Mass Spectrom. and Allied Topics, Cincinnati, OH, U.S.A. June 8–13, 1986. Paper WPB2. p. 568–569.

86WA99
Purification and Structural Analysis of Pyocyanin and 1-Hydroxyphenazine. D. Watson, J. MacDermot, R. Wilson, P.J. Cole and G.W. Taylor. *Eur. J. Biochem. 159*(2), 309–313, (1986).

86WA35
Steroid Glucuronide Conjugates: Analysis by Thermospray Liquid Chromatography Negative Ion Mass Spectrometry. D. Watson, G.W. Taylor and S. Murray. *Biomed. Environ. Mass Spectrom. 13*, 65–70, (1986).

86WE45
Liquid Chromatography-Mass Spectrometry, a Promising Analytic Technique. A. Wehrli. *J. Clin. Chem. Clin. Biochem. 24*(10), 755, (1986).

86WE3'
Accurate Mass HPLC-MS of a Photoaffinity Derivative of Colchicine. S.T. Weintraub, L.J. Floyd, L.D. Barnes and R.F. Williams. Presented at 34th Ann. Conf. on Mass Spectrom. and Allied Topics, Cincinnati, OH, U.S.A. June 8–13, 1986. Paper FPA8. p. 1013–1014.

86WH7'
Further Adventures with an Electrospray Ion Source. C.M. Whitehouse, F. Levin, C.K. Meng and J.B. Fenn. Presented at 34th Ann. Conf. on Mass Spectrom. and Allied Topics, Cincinnati, OH, U.S.A. June 8–13, 1986. Paper WOE1. p. 507–508.

86WI6'
Studies with Thermospray Liquid Chromatography-Mass Spectrometry. R. Willoughby, F. Poeppel, J. Buchner and V. Pizzitola. Presented at 34th Ann. Conf. on Mass Spectrom. and Allied Topics, Cincinnati, OH, U.S.A. June 8–13, 1986. Paper WPB17. p. 596–597.

86WI2'
Studies of Beta-Adrenergic Blocker Drugs with Liquid Chromatography-Mass Spectrometry. R.C. Willoughby, F. Poeppel, J.D. Buchner, V. Pizzitola, A.J. Pirone and S.P. Assenza. Presented at 34th Ann. Conf. on Mass Spectrom. and Allied Topics, Cincinnati, OH, U.S.A. June 8–13, 1986. Paper FPB3. p. 1032–1033.

86WR95
Capillary Supercritical Fluid Chromatgraphy-Mass Spectrometry. B.W. Wright, H.T. Kalinoski, H.R. Udseth and R.D. Smith. *J. High Resolut. Chromatogr. Chromatogr. Commun. 9*(3), 145–153, (1986).

86WR8'
Supercritical Fluid Chromatography for High Molecular Weight Organic Analysis. B.W. Wright, H.R. Udseth and R.D. Smith. Report 1986, EPA/600/2–86/092; Order No. PB87–110524/GAR, 58p. , (1987).

86YA31
Stable Isotope Dilution Quantification of Mutagens in Cooked Foods by Combined Liquid Chromatography/Thermospray Mass Spectrometry. Z. Yamaizumi, H. Kasai, S. Nishimura, C.G. Edmonds and J.A. McCloskey. *Mutation Res. 173*, 1–7, (1986).

86YA11
Application of Secondary Ion Mass Spectrometry to the Quantitative Analysis of Acetylcarnitine and Propionylcarnitine in Urine. S. Yamamoto, H. Kakinuma, T. Nishimuta and K. Mori. *Iyo Masu Kenkyukai Koenshu 11*, 151–156, (1986).

86YA66
1. Application and Optimization of Thermospray HPLC/MS with Direct Electrical Heating of the Capillary Vaporizer. 2. A Moving Belt Transport Detector for HPLC Based on Thermospray Vaporization. L. Yang. *Diss. Abstr. Int. B 46*(7), 2286, (1986).

86YA8'
Studies of Protein Sequencing by On-Line Enzymatic Hydrolysis and LC-Thermospray MS. L.

Yang, K. Stachowiak and D.F. Dyckes. Presented at 34th Ann. Conf. on Mass Spectrom. and Allied Topics, Cincinnati, OH, U.S.A. June 8–13, 1986. Paper RPE12. p. 858–859.

86YE01
Novel Metabolites of Docosahexaenoic Acid (C22:6) Determined by Thermospray LC/MS. J.A. Yergey, H.-Y. Kim and N. Salem, Jr. *Adv. Mass Spectrom. 1985 (Pub. 1986), 10th* (Part B), 601–602, (1986).

86YE84
High-Performance Liquid Chromatography/Thermospray Mass Spectrometry of Eicosanoids and Novel Oxygenated Metabolites of Docosahexaenoic Acid. J.A. Yergey, H.Y. Kim and N. Salem, Jr. *Anal. Chem. 58*(7), 1344–1348, (1986).

86YE87
Detectors for Liquid Chromatography. E.S. Yeung and R.E. Synovec. *Anal. Chem. 58*(12), 1237A–1256A, (1986).

86YI09
Metabolic Studies of Explosives in Laboratory Animals by LC/MS. J. Yinon and D.-G. Hwang. *Adv. Mass Spectrom. 1985 (Pub. 1986), 10th* (Part B), 569–570, (1986).

86YI54
Metabolic Studies of Explosives. IV. Determination of 2,4,6-Trinitrotoluene and Its Metabolites in Blood of Rabbits by High-Performance Liquid Chromatography/Mass Spectrometry. J. Yinon and D.-G. Hwang. *J. Chromatogr. 375*, 154–158, (1986).

86YI0'
Determination of 2,4,6-Trinitrotoluene and Its Metabolites in Urine of Munition Workers by Micro-LC/MS. J. Yinon and D.-G. Hwang. Presented at 34th Ann. Conf. on Mass Spectrom. and Allied Topics, Cincinnati, OH, U.S.A. June 8–13, 1986. Paper RPC9. p. 750–751.

86YI13
Metabolic Studies of Explosives. 5. Detection and Analysis of 2,4,6-Trinitrotoluene and Its Metabolites in Urine of Munition Workers by Micro Liquid Chromatography/Mass Spectrometry. J. Yinon and D.-G. Hwang. *Biomed. Chromatogr. 1*(3), 123–125, (1986).

86ZA3'
Modification of a Finnigan Thermospray Interface to Allow Discharge Chemical Ionization of the Thermospray Vapor. D. Zakett, G.J. Kallos and P.J. Savickas. Presented at 34th Ann. Conf. on Mass Spectrom. and Allied Topics, Cincinnati, OH, U.S.A. June 8–13, 1986. Paper MOB3. p. 3–4.

87AB47
Direct Analysis of the Major Human Seminal Prostaglandins by Thermospray Liquid Chromatography-Mass Spectrometry. J. Abian and E. Gelpi. *J. Chromatogr. 394*(1), 147–154, (1987).

87AB15
Measurement of Ochratoxin A in Barley Extracts by Liquid Chromatography-Mass Spectrometry. D. Abramson. *J. Chromatogr. 391*(1), 315–320, (1987).

87AG69
Identification of 12-Keto–5,8,10-heptadecatrienoic Acid as an Arachidonic Acid Metabolite Produced by Human HL–60 Leukemia Cells. A.P. Agins, M.J. Thomas, C.G. Edmonds and J.A. McCloskey. *Biochem. Pharmacol. 36*(11), 1799–1805, (1987).

87AL45
Micro Liquid Chromatography-Mass Spectrometry with Low Flow Gradient Elution. Studies of Electrostatic Nebulization and Fused-Silica Column Design. H. Alborn and G. Stenhagen. *J. Chromatogr. 394*(1), 35–49, (1987).

87AL03
Micro Liquid Chromatography-Mass Spectrometry Combination. Application to Allelochemical Compounds. H. Alborn and G. Stenhagen. *Am. Chem. Soc. Symp. Ser. 330*, 313–327, (1987).

87AL6'

Characterisation of Radiation-Induced Damage to Polyadenylic Acid Using HPLC-MS-MS. A.J. Alexander, A.F. Fuciarelli, P. Kebarle and J.A. Raleigh. Presented at 35th Ann. Conf. on Mass Spectrom. and Allied Topics, Denver, CO, U.S.A. May 24–29, 1987. p. 1106–1107.

87AL94

Characterization of Radiation-Induced Damage to Polyadenylic Acid Using High-Performance Liquid Chromatography/Tandem Mass Spectrometry. A.J. Alexander, P. Kebarle, A.F. Fuciarelli and J.A. Raleigh. *Anal. Chem.* 59(20), 2484–2491, (1987).

87AN6'

Electrophoresis-Mass Spectrometry Probe. B.D. Andresen and R. Fought. US 4,705,616 (Cl. 204/299R; GO1N24/00), 10 Nov 1987, Appl. 906,847, 15 Sept 1986, 15pp.

87AP0'

Development of a MAGIC LC/MS System. A. Apffel and B. Nordman. Presented at 35th Ann. Conf. on Mass Spectrom. and Allied Topics, Denver, CO, U.S.A. May 24–29, 1987. p. 440.

87AP9'

The Application of FAB and Thermospray to the Structure Determination of omega-Carboxy-alpha-Amino Acids. R.T. Aplin, R.R. Dixon and G.J.W. Fleet. Presented at 35th Ann. Conf. on Mass Spectrom. and Allied Topics, Denver, CO, U.S.A. May 24–29, 1987. p. 429–430.

87AP1'

The Application of Thermospray to Cyanogenic Glycosides. R.T. Aplin and S.G. Errington. Presented at 35th Ann. Conf. on Mass Spectrom. and Allied Topics, Denver, CO, U.S.A. May 24–29, 1987. p. 431.

87AR19

Application of a Combined Supercritical Fluid Chromatograph-Mass Spectrometer to the Investigation of Gas Phase Ion-Molecule Reactions. P.J. Arpino and J. Cousin. *Rapid Commun. Mass Spectrom.* 1(2), 29–33, (1987).

87AR69

Supercritical Fluid Chromatography-Mass Spectrometry Coupling. P.J. Arpino, J. Cousin and J. Higgins. *Trends Anal. Chem.* 6(3), 69–73, (1987).

87AR43

Evaluation of the Potential of Thermospray Liquid Chromatography-Mass Spectrometry in Neurochemistry. F. Artigas and E. Gelpi. *J. Chromatogr.* 394(1), 123–134, (1987).

87AS24

Micro- and Microbore Liquid Chromatography-Fast Atom Bombardment-Mass Spectrometry (LC-FAB-MS) Analysis of Peptides and Antibiotics. A.E. Ashcroft. *Org. Mass Spectrom.* 22(11), 754–757, (1987).

87AS4'

Continuous Flow Fast Atom Bombardment: A New LC-MS Technique. A.E. Ashcroft and J.R. Chapman. Presented at 35th Ann. Conf. on Mass Spectrom. and Allied Topics, Denver, CO, U.S.A. May 24–29, 1987. p. 1094–1095.

87AS45

Continuous-Flow Fast Atom Bombardment Mass Spectrometry. A.E. Ashcroft, J.R. Chapman and J.S. Cottrell. *J. Chromatogr.* 394(1), 15–20, (1987).

87BA6'

Application of High Resolution Mass Measurements in Thermospray LC/MS. L. Baczynskyi and G.E. Bronson. Presented at 35th Ann. Conf. on Mass Spectrom. and Allied Topics, Denver, CO, U.S.A. May 24–29, 1987. p. 1076–1077.

87BA21

Polyethylene Glycols as Mass Calibrants in Thermospary LC/MS. M.A. Baldwin and G.J. Langley. *Org. Mass Spectrom.* 22(8), 561–563, (1987).

87BA2'

Application of Continuous Flow FAB Mass Spectrometry for the Study of Non-Polar, Thermally

Labile Molecules. M. Barber, D.J. Bell, L.W. Tetler, A. Ashcroft, R.S. Brown and G.J. Elliot. Presented at 35th Ann. Conf. on Mass Spectrom. and Allied Topics, Denver, CO, U.S.A. May 24–29, 1987. p. 642–643.
87BA27
Applications of a Continuous Flow Probe in FAB Mass Spectrometry. M. Barber, L.W. Tetler, D. Bell, A.E. Ashcroft, R.S. Brown and C. Moore. *Org. Mass Spectrom.* 22(9), 647–650, (1987).
87BA71
Analysis of Pesticides by On-Line Coupling Liquid Chromatography-Mass Spectrometry with a Direct Liquid Introduction Interface. D. Barcelo and J. Albaiges. *Rev. Agroquim. Tecnol. Aliment.* 27(4), 471–483, (1987).
87BA45
Comparison between Positive, Negative and Chloride-Enhanced Negative Chemical Ionization of Organophosphorus Pesticides in On-Line Liquid Chromatography-Mass Spectrometry. D. Barcelo, F.A. Maris, R.B. Geerdink, R.W. Frei, G.J. De Jong and U.A.Th. Brinkman. *J. Chromatogr.* 394(1), 65–76, (1987).
87BA87
Packed Capillary Liquid Chromatography-Mass Spectrometry Using Both Direct-Coupling and Moving-Belt Interfaces. A.C. Barefoot and R.W. Reiser. *J. Chromatogr.* 398, 217–226, (1987).
87BA67'
An Evaluation of the Continuous-Flow FAB Probe as an LC-MS Interface. R.H. Bateman, D. Jones and S.T. Krolik. Presented at 35th Ann. Conf. on Mass Spectrom. and Allied Topics, Denver, CO, U.S.A. May 24–29, 1987. p. 1096–1097.
87BA5'
LC-MS Accurate Mass Measurement Using Plasmaspray. R.H. Bateman, D. Jones and H. Major. Presented at 35th Ann. Conf. on Mass Spectrom. and Allied Topics, Denver, CO, U.S.A. May 24–29, 1987. p. 415–416.
87BA23
Use of Fast Atom Bombardment Mass Spectrometry to Identify Materials Separated on High-Performance Thin-Layer Chromatography Plates. K.J. Bare and H. Read. *Analyst (London)* 122(4), 433–436, (1987).
87BE24'
Thermospray Analysis of Achromaphoric and Chromaphoric Glutathione Conjugates. M. Bean, S. Pallante-Morell, D. Duik and C. Fenselau. Presented at 35th Ann. Conf. on Mass Spectrom. and Allied Topics, Denver, CO, U.S.A. May 24–29, 1987. p. 1104–1105.
87BE2'
Liquid Chromatography-Mass Spectrometry of Pesticides and Butyltin Compounds. T.A. Bellar and W.L. Budde. Presented at 35th Ann. Conf. on Mass Spectrom. and Allied Topics, Denver, CO, U.S.A. May 24–29, 1987. p. 1072–1073.
87BE9'
Characterization of Mixtures of Organic Acids by Glow-Discharge Ionization Thermospray LC-MS. F.A. Bencsath and F.H. Field. Presented at 35th Ann. Conf. on Mass Spectrom. and Allied Topics, Denver, CO, U.S.A. May 24–29, 1987. p. 29–30.
87BE4'
Forensic Comparison of Explosives by Thermospray Liquid Chromatography-Mass Spectrometry. D.W. Berberich and D.D. Fetterolf. Presented at 35th Ann. Conf. on Mass Spectrom. and Allied Topics, Denver, CO, U.S.A. May 24–29, 1987. p. 1074–1075.
87BE0'
Studies of Natural Products by Packed Column SFC/MS. A.J. Berry, D.E. Games, I.C. Mylchreest, J.R. Perkins and S. Pleasance. Presented at 35th Ann. Conf. on Mass Spectrom. and Allied Topics, Denver, CO, U.S.A. May 24–29, 1987. p. 1100–1101.
87BE8'
Interfaces for Packed and Capillary Column SFC/MS. A.J. Berry, D.E. Games, I.C. Mylchreest,

J.R. Perkins and S. Pleasance. Presented at 35th Ann. Conf. orr Mass Spectrom. and Allied Topics, Denver, CO, U.S.A. May 24–29, 1987. p. 648–649.

87BE3'

The Utility of Thermospray Liquid Chromatography/Mass Spectrometry for the Direct Analysis of Rat Bile for Acetaminophen and Two of Its Conjugates. L.D. Betowski, W.A. Korfmacher, J.O. Lay, Jr., D.W. Potter and J.A. Hinson. Presented at 35th Ann. Conf. on Mass Spectrom. and Allied Topics, Denver, CO, U.S.A. May 24–29, 1987. p. 453–454.

87BE45

Direct Analysis of Rat Bile for Acetaminophen and Two of Its Conjugated Metabolites via Thermospray Liquid Chromatography/Mass Spectrometry. L.D. Betowski, W.A. Korfmacher, J.O. Lay Jr., D.W. Potter and J.A. Hinson. *Biomed. Environ. Mass Spectrom.* *14*(12), 705–709, (1987).

87BE43

Thermospray LC/MS/MS Analysis of Wastewater for Disperse Azo Dyes. L.D. Betowski, S.M. Pyle, J.M. Ballard and G.M. Shaul. *Biomed. Environ. Mass Spectrom.* *14*(1), 343–354, (1987).

87BI97

Membrane Inlet for Selective Introduction of Volatile Compounds Directly into the Ionization Chamber of a Mass Spectrometer. M.E. Bier and R.G. Cooks. *Anal. Chem.* *59*(4), 597–601, (1987).

87BI45

Characterization of Benzo[*a*]pyrene Metabolites by High Performance Liquid Chromatography-Mass Spectrometry with a Direct Liquid Introduction Interface and Using Negative Chemical Ionization. R.H. Bieri and J. Greaves. *Biomed. Environ. Mass Spectrom.* *14*(10), 555–561, (1987).

87BL41

Structure Eludication of Drug Metabolites Using Thermsopray Liquid Chromatography-Mass Spectrometry. T.J.A. Blake. *J. Chromatogr.* *394*(1), 171–182, (1987).

87BL01

Application of High-Performance Liquid Chromatography and Thermospray High-Performance Liquid Chromatography-Mass Spectrometry to the Analysis and Identification of 2',3'-Dideoxyadenosine and Its Metabolite in Biological Media. P.A. Blau, J.W. Hines and R.D. Voyksner. *J. Chromatogr.* *420*(1), 1–12, (1987).

87BO37

Pharmaceutical and Toxicological Analyses. L. Boniforti, M.G. Quaglia and M. Terracciano. *Ann. Ist. Super. Sanita* *23*(1), 97–104, (1987).

87BO99

Hyphenated MS at Pittcon. S. Borman. *Anal. Chem.* *59*(11), 769A–774A, (1987).

87BR9'

Transport Processes, Loss Mechanisms and Signals in MAGIC-LC/MS. R.F. Browner, J.D. Kirk, D.A. Garrett and D.D. Perkins. Presented at 35th Ann. Conf. on Mass Spectrom. and Allied Topics, Denver, CO, U.S.A. May 24–29, 1987. p. 409–410.

87BR92

Ion Spray Interface for Combined Liquid Chromatography/Atmospheric Pressure Ionization Mass Spectrometry. A.P. Bruins, T.R. Covey and J.D. Henion. *Anal. Chem.* *59*(22), 2642–2626, (1987).

87BR97

Determination of Sulfonated Azo Dyes by Liquid Chromatography/Atmospheric Pressure Ionization Mass Spectrometry. A.P. Bruins, L.O.G. Weidolf, J.D. Henion and W.L. Budde. *Anal. Chem.* *59*(22), 2647–2652, (1987).

87BU5'

Ion Source Design for Thermospray LC-MS. J.D. Buchner, R.C. Willoughby, S.N. Ketkar and R.J. Madden. Presented at 35th Ann. Conf. on Mass Spectrom. and Allied Topics, Denver, CO, U.S.A. May 24–29, 1987. p. 245–246.

87BU65
Direct Analysis of Thin-Layer Chromatograms and Electrophoretograms by Secondary Ion Mass Spectrometry. K.L. Busch. *Trends Anal. Chem. (Pers. Ed.)* 6(4), 95–100, (1987).
87BU2′
Direct Analysis of Organometallic Compounds by Chromatography/SIMS. K. Busch and S.J. Doherty. *Am. Chem. Soc. Abstr.* 193, INOR0462, (1987).
87BU49
Studies with a Dual-Beam Thermospray Interface in High-Performance Liquid Chromatography-Mass Spectrometry. L. Butfering, G. Schmelzeisen-Redeker and F.W. Röllgen. *J. Chromatogr.* 394(1), 109–118, (1987).
87CA46
Identification of Prescription Drugs in Adulterated Chinese Herbal Medications. T. Cairns, E.G. Siegmund and B.R. Rader. *Pharm. Res.* 4(2), 126–129, (1987).
87CA67
Enzymes and Mass Spectrometry: A Dynamic Combination. R.M. Caprioli. *Mass Spectrom. Rev.* 6(2), 237–287, (1987).
87CA52
Applications of Mass Spectrometry to Enzymic Reactions. R.M. Caprioli. *Biochem. Soc. Trans.* 15(1), 162–164, (1987).
87CA3′
Following Dynamic Biochemical Systems "On-Line" Using FABMS. R.M. Caprioli. Presented at 35th Ann. Conf. on Mass Spectrom. and Allied Topics, Denver, CO, U.S.A. May 24–29, 1987. p. 273–274.
87CA0′
Microbore HPLC/MS Using a Continuous Flow FAB Probe Interface. R.M. Caprioli, B. DaGue, T. Fan and W. Moore. Presented at 35th Ann. Conf. on Mass Spectrom. and Allied Topics, Denver, CO, U.S.A. May 24–29, 1987. p. 640–641.
87CA61
Microbore HPLC/Mass Spectrometry for the Analysis of Peptide Mixtures Using a Continuous Flow Interface. R.M. Caprioli, B. DaGue, T. Fan and W.T. Moore *Biochem. Biophys. Res. Commun.* 146(1), 291–299, (1987).
87CA39
Analytical Chemistry and the Games of the XXIIIrd Olympiad in Los Angeles, 1984. D.H. Catlin, R.C. Kammerer, C.K. Hatton, M.H. Sekera and J.L. Merdink. *Clin. Chem. (Winston-Salem, N.C.)* 33(2), 319–327, (1987).
87CH41
Improvements in Instrumentation for Thermospray Operation on a Magnetic Sector Mass Spectrometer. J.R. Chapman and J.A.E. Pratt. *J. Chromatogr.* 394(1), 231–238, (1987).
87CH62
Thermospray High-Performance Liquid Chromatographic-Mass Spectrometric Analysis of the Degradation Products of Piroximone. T.M. Chen, J.E. Coutant, A.D. Sill and R.R. Fike *J. Chromatogr.* 396, 382–388, (1987).
87CH8′
Application of Supercritical Fluid Chromatography-Mass Spectrometry in the Analysis of Fossil Fuels. E.K. Chess, H.T. Kalinoski, B.W. Wright, H.R. Udseth and R.D. Smith. Presented at 35th Ann. Conf. on Mass Spectrom. and Allied Topics, Denver, CO, U.S.A. May 24–29, 1987. p. 1068–1069.
87CH5′
LC/MS and MS/MS Analyses of Carbamate Pesticides. K.S. Chui, C. Tanaka and A. Van-Langenhove. Presented at 35th Ann. Conf. on Mass Spectrom. and Allied Topics, Denver, CO, U.S.A. May 24–29, 1987. p. 115–116.

87CO851
Construction of a Supercritical Fluid Chromatograph-Mass Spectrometer Instrument System Using Capillary Columns, and a Chemical Ionization Source Accepting High Flow-Rates of Mobile Phase. J. Cousin and P.J. Arpino. *J. Chromatogr. 398*, 125–141, (1987).

87CO7'
Effects of Source Design on the Nature of Thermospray Spectra. J.E. Coutant, B.L. Ackermann and M.L. Vestal. Presented at 35th Ann. Conf. on Mass Spectrom. and Allied Topics, Denver, CO, U.S.A. May 24–29, 1987. p. 417–418.

87CO85
Development of Rapid, Specific, and Sensitive Methods for the Determination of Drugs Based on Mass Spectrometry. T.R. Covey. *Diss. Abstr. Int. B 48*(4), 1025, (1987).

87CR1'
Heat-Induced Dissociation in LC/MS Analyses. J.L. Crawford, J.L. Little and D.R. Wilder. Presented at 35th Ann. Conf. on Mass Spectrom. and Allied Topics, Denver, CO, U.S.A. May 24–29, 1987. p. 411–412.

87CR01'
Determination of Impurities in Renin Inhibitor Peptide by Thermospray LC/MS. F.W. Crow, K.D. Evans, K.S. Manning and R.H. Robins. Presented at 35th Ann. Conf. on Mass Spectrom. and Allied Topics, Denver, CO, U.S.A. May 24–29, 1987. p. 1080–1081.

87CR0'
Investigation of Guanine Adducts by Positive and Negative Thermospray Mass Spectrometry. J. Crowley, M. O'Hara, P. Valladier, L.J. Marnett and R. Hood. Presented at 35th Ann. Conf. on Mass Spectrom. and Allied Topics, Denver, CO, U.S.A. May 24–29, 1987. p. 700.

87DA1'
Application of Methanol Thermospray Ionization to Compounds of Pharmaceutical Interest. P.R. Das, R.D. Malchow and B.N. Pramanik. Presented at 35th Ann. Conf. on Mass Spectrom. and Allied Topics, Denver, CO, U.S.A. May 24–29, 1987. p. 711–712.

87DA0'
HPLC-Thermospray Mass Spectrometric Analysis of DNA Bases. W.C. Davidson, R.J. Greathead, T.L. Walden, Jr., W.F. Blakley, E. Holwitt and M. Dizdaroglu. Presented at 35th Ann. Conf. on Mass Spectrom. and Allied Topics, Denver, CO, U.S.A. May 24–29, 1987. p. 900–901.

87DE24
Hybridized Techniques: Hyphenated and Slashed Instrumental Methods. J.A. De Haseth. *Spectroscopy (Eugene, Oregon) 2*(10), 14–16, (1987).

87DE90
Direct Coupling of Open-Tubular Liquid Chromatography with Mass Spectrometry. J.S. de Wit, C.E. Parker, K.B. Tomer and J.W. Jorgenson. *Anal. Chem. 59*(19), 2400–2404, (1987).

87DE27
Studies of Metalloprotein Species by Directly Coupled High-Performance Liquid Chromatography-Inductively Coupled Plasma Mass Spectrometry. J.R. Dean, S. Munro, L. Ebdon, H.M. Crews and R.C. Massey. *J. Anal. At. Spectrom. 2*(6), 607–610, (1987).

87DE55
Focal Points in Mass Spectrometry. W.N. Delgass and R.G. Cooks. *Science 235*, 545–552, (1987).

87DE3'
Evaluation of a Disposable Moving Belt LC-MS Interface. W. DeMaio. Presented at 35th Ann. Conf. on Mass Spectrom. and Allied Topics, Denver, CO, U.S.A. May 24–29, 1987. p. 423–424.

87DE85
Development and Evaluation of a Disposable Moving Ribbon Liquid Chromatography-Laser Desorption Mass Spectrometry Interface. W. DeMaio. *Diss. Abstr. Int. B 48*(4), 1025, (1987).

87DE2'
Analysis of Glutathione Conjugates and Related Compounds by TSP MS. J.S.M. de Wit, C.E.

Parker, R.W. Smith, M.B. Gopinathan, O. Hernandez and J.R. Bend. Presented at 35th Ann. Conf. on Mass Spectrom. and Allied Topics, Denver, CO, U.S.A. May 24–29, 1987. p. 512–513.

87DE9′
Development of a New Interface for Combined Open-Tubular LC/MS. J.S.M. de Wit, C.E. Parker, K.B. Tomer and J.W. Jorgenson. Presented at the 35th Ann. Conf. on Mass Spectrom. and Allied Topics, Denver, CO, U.S.A. May 24–29, 1987. p. 419–420.

87DU8′
Mass Spectrometry Analysis of Esperamicins. G.R. Dubay, G.S. Groenewold and J. Golik. Presented at 35th Ann. Conf. on Mass Spectrom. and Allied Topics, Denver, CO, U.S.A. May 24–29, 1987. p. 438–439.

87ED79
Structural Characterization of Four Ribose-Methylated Nucleosides from the Transfer RNA of Extremely Thermophilic Archaebacteria. C.G. Edmonds, P.F. Crain, T. Hashizume, R. Gupta, K.O. Stetter and J.A. McCloskey. *J. Chem. Soc., Chem. Commun. 1987*, 909–910, (1987).

87ES6′
Separation and Identification of Acetylated Monoglycerides by Reversed-Phase High Performance Liquid Chromatography/Mass Spectrometry. R. Espino, B. Nelson and R. Weinkam. Presented at 35th Ann. Conf. on Mass Spectrom. and Allied Topics, Denver, CO, U.S.A. May 24–29, 1987. p. 1086–1087.

87ES94
Stable Isotope Dilution Thermospray Liquid Chromatography/Mass Spectrometry Method for Determination of Sugars and Sugar Alcohols in Humans. N.V. Esteban, D.J. Liberato, J.B. Sidbury and A.L. Yergey. *Anal. Chem. 59*(13), 1674–1677, (1987).

87EV49
Analysis of Native Neutral Glycosphingolipids by Combined High Performance Liquid Chromatography/Mass Spectrometry. J.E. Evans and R.H. McCluer. *Biomed. Environ. Mass Spectrom. 14*(4), 149–153, (1987).

87FA4′
The Characterization of Carotenoids and Retinoids by HPLC-MS-MS. P.E. Farrow, R.F. Taylor and L.M. Yelle. Presented at 35th Ann. Conf. on Mass Spectrom. and Allied Topics, Denver, CO, U.S.A. May 24–29, 1987. p. 1084–1085.

87FI8′
Fast Atom Bombardment HPLC-MS of Peptides Using a VG Moving Belt Interface. D. Fisher, H. Nguyen and R. Weinkam. Presented at 35th Ann. Conf. on Mass Spectrom. and Allied Topics, Denver, CO, U.S.A. May 24–29, 1987. Paper No. FPA 18.

87FL7′
Environmental Applications of the Thermospray LC/MS Interface: Qualitative Analysis of Sulfonated Azo Dyes. D.A. Flory, M.A. McLean, M.L. Vestal and L.D. Betowski. Presented at 35th Ann. Conf. on Mass Spectrom. and Allied Topics, Denver, CO, U.S.A. May 24–29, 1987. p. 157–158.

87FL0′
Instrumental Modifications to a Modular Chromatograph/SIMS Instrument. R.A. Flurer, G.C. Di Donato, K.L. Duffin and K.L. Busch. Presented at 35th Ann. Conf. on Mass Spectrom. and Allied Topics, Denver, CO, U.S.A. May 24–29, 1987. p. 1140–1141.

87FR40
Fourth Symposium on Liquid Chromatography-Mass Spectrometry and Mass Spectrometry-Mass Spectrometry. [1986; Montreux, Switzerland] R.W. Frei (ed.). *J. Chromatogr. 394*(1), 1–270, (1987).

87FR9′
Supercritical Fluid Chromatography-Mass Spectrometry of Carotenoids and Other Polar Lipids. N.M. Frew and C.G. Johnson. *Am. Chem. Soc. Abstr. 193*, AGFD0019, (1987).

87GA3GI'
Applications of High Performance Liquid Chromatography/Mass Spectrometry (LC/MS) in Food Chemistry. D.E. Games. In: *Applications of Mass Spectrometry in Food Science* (J. Gilbert, ed.), p. 193–237, Elsevier, London, 1987.

87GA41
Supercritical Fluid Chromatography-Mass Spectrometry. D.E. Games, A.J. Berry, I.C. Mylchreest, J. Perkins and S. Pleasance. *Anal. Proc. (London)* 24(21), 371–372, (1987).

87GA65
Supercritical Fluid Chromatography Combined with Mass Spectrometry. D.E. Games, A.J. Berry, I.C. Mylchreest, J.R. Perkins and S. Pleasance. *Lab. Pract.* 36(2), 49–50, (1987).

87GA14
What Ever Became of Vitamin B12? D.A. Garteiz. *TSP Report* 1(1), 4, (1987).

87GA11
TSP-CID/MS: A New Dimension of Structural Analysis in LC/MS. D.A. Garteiz and W.H. McFadden. *TSP Report* 1(1), 1–3, (1987).

87GA47
Determination of Serum Cortisol by Thermospray Liquid Chromatography/Mass Spectrometry: Comparison with Gas Chromatography/Mass Spectrometry. S.J. Gaskell, K. Rollins, R.W. Smith and C.E. Parker. *Biomed. Environ. Mass Spectrom.* 14(12), 717–722, (1987).

87GE41
Halogenated Mobile Phase Additives for Improved Detection Performance in Liquid Chromatography-Negative Chemical Ionization Mass Spectrometry. R.B. Geerdink, F.A. Maris, G.J. de Jong, R.W. Frei and U.A.Th. Brinkman. *J. Chromatogr.* 394(1), 51–64, (1987).

87GE3'
The Utility of Solution Electrochemistry MS for Enhancing Thermospray Spectra and Elucidating Electrochemical Redox Processes. T.A. Getek. Presented at 35th Ann. Conf. on Mass Spectrom. and Allied Topics, Denver, CO, U.S.A. May 24–29, 1987. p. 443–444.

87GO87
Metabolism of the Mutagen MeIQx In Vivo: Metabolite Screening by Liquid Chromatography-Thermospray Mass Spectrometry. N.J. Gooderham, D. Watson, J.C. Rice, S. Murray, G.W. Taylor and D.S. Davies. *Biochem. Biophys. Res. Commun.* 148(3), 1377–1382, (1987).

87GO2'
The Use of Thermospray LC/MS for the Characterization of Surfactants. P.C. Goodley. Presented at 35th Ann. Conf. on Mass Spectrom. and Allied Topics, Denver, CO, U.S.A. May 24–29, 1987. p. 1092–1093.

87GR4'
Direct Liquid Introduction (DLI) High Performance Liquid Chromatography-Mass Spectrometry (HPLC-MS) and Thermsopray HPLC-MS of Conjugated Benzo(a)pyrene (B(a)P) Metabolites. J. Greaves and R.H. Bieri. Presented at 35th Ann. Conf. on Mass Spectrom. and Allied Topics, Denver, CO, U.S.A. May 24–29, 1987. p. 1034–1035.

87HA69
Isotope Ratio Measurements in Nutrition and Biomedical Research. D.L. Hachey, W.W. Wong, T.W. Boutton and P.D. Klein. *Mass Spectrom. Rev.* 6(2), 289–328, (1987).

87HA28
Transfer RNA(5-Methylaminomethyl–2-Thiouridine)-Methyltransferase from *Escherichia coli* K–12 Has Two Enzymatic Activities. T.G.Hagervall, C.G. Edmonds, J.A. McCloskey and G.R. Bjork. *J. Biol. Chem.* 262(18), 8488–8495, (1987).

87HA01
Qualitative Determination of Polymethoxylated Flavones in Valencia Orange Peel Oil and Juice by LC-UV/VIS and LC/MS Techniques. M. Hakj-Mahammed and B.Y. Meklati. *Lebensm.-Wiss. Technol.* 20(3), 111–114, (1987).

87HA17
Determination of Volatile Organic Compounds in Aqueous Systems by Membrane Inlet Mass Spectrometry. B.J. Harland, P.J.D. Nicholson and E. Gillings. *Water Res. 21*(1), 107–113, (1987).

87HA93
Drug Metabolism, Pharmacokinetics and Toxicity. D.J. Harvey. *Mass Spectrom. 9*, 303–327, (1987).

87HA87
Analysis of Commercial Waxes Using Capillary Supercritical Fluid Chromatography/Mass Spectrometry. S.B. Hawthorne and D.J. Miller. *J. Chromatogr. 388*(2), 397–409, (1987).

87HA2'
Identification of Urinary Rat Metabolites of CGS 15855A Utilizing LC/MS, GC/MS and a Stable Isotope Labile Tracer. M. Hayes, D. Gaudry, H. Egger, J. Miotto, M. Bax, R. Iannucci and D. Alkalay Presented at 35th Ann. Conf. on Mass Spectrom. and Allied Topics, Denver, CO, U.S.A. May 24–29, 1987. p. 502.

87HA89
Enantiomeric Composition Analysis of Amphetamine and Methamphetamine by Chiral Phase High-Performance Liquid Chromatography-Mass Spectrometry. S.M. Hayes, R.H. Liu, W-S. Tsang, M.G. Legendre, R.J. Berni, D.J. Pillion, S. Barnes and M.H. Ho. *J. Chromatogr. 398*, 239–246, (1987).

87HE6'
Determination of Anabolic Steroids by API SFC/MS and SFC/MS/MS Using Packed-Columns. J. Henion, L. Weidolf, E.D. Lee and T.R. Covey. Presented at 35th Ann. Conf. on Mass Spectrom. and Allied Topics, Denver, CO, U.S.A. May 24–29, 1987. p. 16–17.

87HE65
Comparison of Coal Tars by Liquid Chromatography-Mass Spectrometry. A.A. Herod, W.R. Ladner, B.J. Stokes, A.J. Berry, D.E. Games and M. Höhn. *Fuel 66*(7), 935–946, (1987).

87HOO'
The Analysis of Desulfo Glucosinolates Using Thermospray LC/MS. L.R. Hogge, D.W. Reed and E.W. Underhill. Presented at 35th Ann. Conf. on Mass Spectrom. and Allied Topics, Denver, CO, U.S.A. May 24–29, 1987. p. 1090–1091.

87HS78
New Techniques for Microscale Structural Studies by Mass Spectrometry. F.F. Hsu. Univ. Microfilms Int., Order No. DA8626624. *Diss. Abstr. Int. 47*(8), 3328, (1987).

87HS4'
Direct Interfacing of Capillary Supercritical Fluid Chromatography with Mass Spectrometry Electron and Chemical Ionization. S.H. Hsu, E.D. Lee and J.D. Henion. Presented at 35th Ann. Conf. on Mass Spectrom. and Allied Topics, Denver, CO, U.S.A. May 24–29, 1987. p. 434–435.

87HU57
Advances in Supercritical Fluid Chromatography. D. Huang. *Sepu 5*(5), 287–296, (1987).

87HU24
Analysis of Peptide Digests Using High Performance Liquid Chromatography Coupled with Continuous Flow Fast Atom Bombardment Mass Spectrometry (FAB-MS). D.W. Hutchinson, A.R. Wollfitt and A.E. Ashcroft. *Org. Mass Spectrom. 22*, 304–306, (1987).

87HU8'
Continuous Flow FAB-MS Analysis of Peptide Digests. D.W. Hutchinson, A.R. Wollfitt, K.R. Jennings and A.E. Ashcroft. Presented at 35th Ann. Conf. on Mass Spectrom. and Allied Topics, Denver, CO, U.S.A. May 24–29, 1987. p. 638–639.

87IT4'
Recent Developments of Frit-FAB LC/MS Interface. Y. Itagaki, T. Kobayashi, T. Mizuno, E. Kubota and D. Ishii. Presented at 35th Ann. Conf. on Mass Spectrom. and Allied Topics, Denver, CO, U.S.A. May 24–29, 1987. p. 644.

87IT16
Direct Coupling of Micro High-Performance Liquid Chromatography with Fast Atom Bombardment Mass Spectrometry. III. Application to Oligosaccharides. Y. Ito, T. Toyohide, D. Ishii, M. Goto and T. Mizuno. *J. Chromatogr. 391*(1), 296–302, (1987).

87IW49
Application of Molecular-Secondary-Ion Mass Spectrometry for Drug Metabolism Studies. I. Direct Analysis of Conjugates by Thin-Layer Chromatography Secondary-Ion Mass Spectrometry. H. Iwabuchi, A. Nakagawa and K.-I. Nakamura. *J. Chromatogr. 414*, 139–148, (1987).

87JO17
Analysis of Nonderivatized Peptides by Thermospray Using a Magnetic Sector Mass Spectrometer. D.S. Jones and S.T. Krolik. *Rapid Commun. Mass Spectrom. 1* (4), 67–68, (1987).

87KA43
Capillary Supercritical Fluid Chromatography-Mass Spectrometry. H.T. Kalinoski, H.R. Udseth, E.K. Chess and R.D. Smith. *J. Chromatogr. 394*(1), 3–14, (1987).

87KA07
Analytical Applications of Capillary Supercritical Fluid Chromatography-Mass Spectrometry. H.T. Kalinoski, H.R. Udesth, B.W. Wright and R.D. Smith. *J. Chromatogr. 400*, 307–316, (1987).

87KA6'
New Approaches to the Supercritical Fluid Chromatography-Mass Spectrometry Interface for Analysis of Nonvolatile Materials. H.T. Kalinoski, H.R. Udseth, J.A. Olivares and R.D. Smith. Presented at 35th Ann. Conf. on Mass Spectrom. and Allied Topics, Denver, CO, U.S.A. May 24–29, 1987. p. 646–647.

87KA1'
Field Limit for Ion Evaporation from Mobilities of Charged Thermospray Droplets. V. Katta and A.L. Rockwood. Presented at 35th Ann. Conf. on Mass Spectrom. and Allied Topics, Denver, CO, U.S.A. May 24–29, 1987. p. 441–442.

87KA86
Studies in Ion Evaporation From Charged Droplets in Thermospray. V. Katta. *Diss. Abstr. Int. 48/10B*, 2986, (1987).

87KE92
Developments and Trends in Instrumentation. T.R. Kemp. *Mass Spectrom. 9*, 122–142, (1987).

87KE6'
Mass Spectrometry of New Dideoxynucleosides and 4-Thiodideoxynucleosides. M.B. Kempff, R. Hood, J.P. Horowitz and E. Palomino. Presented at 35th Ann. Conf. on Mass Spectrom. and Allied Topics, Denver, CO, U.S.A. May 24–29, 1987. p. 896–897.

87KIE6'
The Formation of Ions in Thermospray Liquid Chromatography/Mass Spectrometry. D.A. Kidwell, J.H. Callahan and M.M. Ross. Presented at 35th Ann. Conf. on Mass Spectrom. and Allied Topics, Denver, CO, U.S.A. May 24–29, 1987. p. 436–437.

87KI45
The Reduction of Collision Induced Dissociation Effects in Thermospray Sources on Sector Instruments. G. Kilpatrick, I.A.S. Lewis and J.F. Smith. *Biomed. Environ. Mass Spectrom. 14*, 155–159, (1987).

87KI92
Application of Thermospray High-Performance Liquid Chromatography/Mass Spectrometry for the Determination of Phospholipids and Related Compounds. H.Y. Kim and N. Salem, Jr. *Anal. Chem. 59*(5), 722–726, (1987).

87KI8'
Thermospray LC/MS Analysis of the Effect of Alcohol on Polyunsaturated Phospholipid Molecular Species Levels. H.Y. Kim and N. Salem, Jr. Presented at 35th Ann. Conf. on Mass Spectrom. and Allied Topics, Denver, CO, U.S.A. May 24–29, 1987. p. 1108–1109.

87KI45
Determination of Eicosanoids, Phospholipids and Related Compounds by Thermospray Liquid Chromatography-Mass Spectrometry. H.Y. Kim, J.A. Yergey and N. Salem, Jr. *J. Chromatogr. 394*(1), 155–170, (1987).

87KI1'
The Analysis of Microbial Natural Products by Thermospray LC-MS. G.S. King, L. Gordon and N. Runnalls. Presented at 35th Ann. Conf. on Mass Spectrom. and Allied Topics, Denver, CO, U.S.A. May 24–29, 1987. p. 401–402.

87KI34'
The Identification of a Sulfate Conjugate of a Pyrethroid Metabolite by Negative Ion Soft Ionisation Techniques. G.S. King, J. Heath, N. Runnalls and I. Hammond. Presented at 35th Ann. Conf. on Mass Spectrom. and Allied Topics, Denver, CO, U.S.A. May 24–29, 1987. p. 403–404.

87KO8'
Thermospray/Mass Spectrometry and Thermospray/MS/MS of Doxylamine and Pyrilamine Metabolites. W.A. Korfmacher, L.D. Betowksi, C.L. Holder and R.K. Mitchum. Presented at 35th Ann. Conf. on Mass Spectrom. and Allied Topics, Denver, CO, U.S.A. May 24–29, 1987. p. 1088–1089.

87KR45
Combined Thin-Layer Chromatography/Mass Spectrometry without Substance Elution: Application to the Rapid Analysis of Dipeptide Mixtures. R. Kraft, D. Buttner, P. Franke and G. Etzold. *Biomed. Environ. Mass Spectrom. 14*(1), 5–7, (1987).

87KR41
Combined Thin-Layer Chromatography/Mass Spectrometry without Substance Elution: Use for Direct Identification of Amino Acids as Dansylated Esters. R. Kraft, A. Otto, H.-J. Zopfl and G. Etzold. *Biomed. Environ. Mass Spectrom. 14*(1), 1–4, (1987).

87KR49
Direct Electrically Heated Spray Device for a Moving Belt Liquid Chromatography-Mass Spectrometry Interface. G.M. Kresbach, T.R. Baker, R.J. Nelson, J. Wronka, B.L. Karger and P. Vouros. *J. Chromatogr. 394*(1), 89–100, (1987).

87KU73
LC/MS of Natural Glycerolipids, Sterols and Steryl Esters. A. Kuksis, L. Marai, J.J. Myher and S. Pind. *J. Chromatogr. Libr. 37*, 403–440, (1987).

87LA0'
Methods and Apparatus for Mass Spectrometric Analysis of Fluids. W.M. Lagna. US 4,667,100 (Cl. 250–585; B01D59/44), 19 May 1987, Appl. 724,166, 17 Apr. 1985, 12pp.

87LA43
Automated Sample Preparation On-Line with Thermospray High-Performance Liquid Chromatography-Mass Spectrometry for the Determination of Drugs in Plasma. M.S. Lant, J. Oxford and L.E. Martin. *J. Chromatogr. 394*(1), 223–230, (1987).

87LA54
Detection Techniques for Capillary Supercritical Fluid Chromatography. D.W. Later, D.J. Bornhop, E.D. Lee, J.D. Henion and R.C. Wieboldt. *LC-GC 5*(9), 804–806, (1987).

87LA93
Supercritical Fluid Chromatography Interface for a Differentially Pumped Dual-Cell Fourier Transform Mass Spetrometer. D.A. Laude, Jr., S.L. Pentoney, Jr., P.R. Griffiths and C.L. Wilkins. *Anal. Chem. 59*(18), 2283–2288, (1987).

87LE99
Supercritical Fluid Chromatography/Fourier Transform Mass Spectrometry. E.D. Lee, J.D. Henion, R.B. Cody and J.A. Kinsinger. *Anal. Chem. 59*(9), 1309–1312, (1987).

87LE4'
Ion Spray Liquid Chromatography/API Tandem Mass Spectrometry Determination of Peptides. E.D.

Lee, L.O.G. Weidolf and J.D. Henion. Presented at 35th Ann. Conf. on Mass Spectrom. and Allied Topics, Denver, CO, U.S.A. May 24–29, 1987. p. 654–655.
87LE1'
Fundamental Studies of Thermospray Ionization: Investigation of the Thermospray Plume. M.S. Lee and R.A. Yost. Presented at 35th Ann. Conf. on Mass Spectrom. and Allied Topics, Denver, CO, U.S.A. May 24–29, 1987. p. 421–422.
87LE5'
High Mass Ion Production Using Electrohydrodynamic Ionization. T.D. Lee, K. Leggesse, J.F. Mahoney and J. Perel. Presented at 35th Ann. Conf. on Mass Spectrom. and Allied Topics, Denver, CO, U.S.A. May 24–29, 1987. p. 35–36.
87LE2'
Mass Spectrometer Having a Thermospray Ion Source Suitable for Analysis of Liquids. I.A. Lewis and D.C. Smith. US 4,647,772 (Cl. 250–288; D01D59/44), 03 Mar. 1987, GB Appl. 84/4,683, 22 Feb. 1984, 14 pp.
87LE45
The Reduction of Collision Induced Dissociation Effects in Thermospray Sources on Sector Instruments. I.A.S. Lewis and J.F. Smith. *Biomed. Environ. Mass Spectrom. 14*(4), 155–159, (1987).
87LI71
Thermospray HPLC/MS: A New Mass Spectrometric Technique for the Profiling of Steroids. D.J. Liberato, A.L. Yergey, N. Esteban, C.E. Gomez-Sanchez and C.H.L. Shackleton. *J. Steroid Biochem. 27*(1–3), 61–70, (1987).
87LI47
Optimization of Thermospray Conditions. Effect of Repeller Potential and Vaporizer Temperature. C. Lindberg and J. Paulson. *J. Chromatogr. 394*(1), 117–121, (1987).
87LI45
The Use of On-Line Liquid Chromatography/Mass Spectrometry and Stable Isotope Techniques for the Identification of Budesonide Metabolites. C. Lindberg, J. Paulson and S. Edsbacker. *Biomed. Environ. Mass Spectrom. 14*(10), 535–541, (1987).
87LI0'
Quantitation by LCMS of Neuropeptides Isolated by Immunoaffinity Chromatography from Spinal Cord. C.A. Lisek, J. Bailey, T.L. Yaksh, I. Jardine, M.A. McDowall and D.C. Smith. Presented at 35th Ann. Conf. on Mass Spectrom. and Allied Topics, Denver, CO, U.S.A. May 24–29, 1987. p. 1110–1111.
87LL93
Mixed-Mode Column Thermospray Liquid Chromatography/Mass Spectrometry. J.R. Lloyd, M.L. Cotter, D. Ohori and A.R. Dyler. *Anal. Chem. 59*(20), 2533–2534, (1987).
87LU6'
Supercritical Fluid Injection of Small Polar Molecules Using Derivatization with Detection by Two-Photon Ionization in Supersonic Beam-Mass Spectrometry. D.M. Lubman, C.H. Sin and H.M. Pang. Presented at 35th Ann. Conf. on Mass Spectrom. and Allied Topics, Denver, CO, U.S.A. May 24–29, 1987. p. 56–57.
87MA99
Mechanism for High-Mass Sample Ion Desolvation in Electrohydrodynamic Mass Spectrometry. J.F. Mahoney, J. Perel and T.D. Lee. *Int. J. Mass Spectrom. Ion Processes 79*(3), 249–266, (1987).
87MA55
Analysis of Non-Ionic Surface Active Agents by Supercritical Fluid Chromatography/Mass Spectrometry. K. Matsumoto, S. Tsuge and Y. Hirata. *Mass Spectrosc. 35*(1), 15–22, (1987).
87MA58
Disposition of 8-Methoxypsoralem in the Rat: Induction of Metabolism In Vivo and In Vitro and Identification of Urinary Metabolites by Thermospray Mass Spectrometry. D.C. Mays, S.G. Hecht, S.E. Unger, C.M. Pacula, J.M. Climie, D.E. Sharp and N. Gerber. *Drug Metab. Dispos. 15*(3), 318–328, (1987).

87MC53
Structure Determination of a New Fluorescent Tricyclic Nucleoside from Archaebacterial tRNA. J.A. McCloskey, P.F. Crain, C.G. Edmonds, R. Gupta, T. Hashizume, D.W. Phillipson and K.O. Stetter. *Nucleic Acids Res.* 15(2), 683–693, (1987).

87MC41
Thermospray Collisionally Induced Dissociation with Single and Multiple Mass Analyzers. W.H. McFadden, D.A. Garteiz and E.G. Siegmund. *J. Chromatogr.* 394(1), 101–108, (1987).

87MC51
Techniques for Increased Use of Thermospray Liquid Chromatography-Mass Spectrometry. W.H. McFadden and S.A. Lammert. *J. Chromatogr.* 385, 201–211, (1987).

87MC6'
Thermospray Analysis: Structural Identification Using Repeller CID with a Triple Stage Quadrupole to Obtain CID/MS/MS Data. W.H. McFadden, P.J. Rudewicz and D.A. Garteiz. Presented at 35th Ann. Conf. on Mass Spectrom. and Allied Topics, Denver, CO, U.S.A. May 24–29, 1987. p. 636–637.

87MC31
Analysis of Trichothecene Mycotoxins by Thermospray Liquid Chromatography-Mass Spectrometry. M.A. McLean and C.H. Vestal. *Vestec Thermospray Newsletter* 3(2), 1–4, (1987).

87ME49
Thermospray Liquid Chromatography-Mass Spectrometry in Food and Agricultural Research. F.A. Mellon, J.R. Chapman and J.A.E. Pratt. *J. Chromatogr.* 394(1), 209–222, (1987).

87ME3'
Solute Aggregation in Electrospray Ionization of Arginine Solutions. C.K. Meng and J.B. Fenn. Presented at 35th Ann. Conf. on Mass Spectrom. and Allied Topics, Denver, CO, U.S.A. May 24–29, 1987. p. 723–724.

87ME34'
Multiple Charging in Electrospray Ionization of Polyethlyene Glycols. C.K. Meng, S.F. Wong and J.B. Fenn. Presented at 35th Ann. Conf. on Mass Spectrom. and Allied Topics, Denver, CO, U.S.A. May 24–29, 1987. p. 33–34.

87MI0'
Supercritical Fluid Chromatography/Mass Spectrometric (SFC/MS) Analysis of Waxes. D.J. Miller and S.B. Hawthorne. Presented at 35th Ann. Conf. on Mass Spectrom. and Allied Topics, Denver, CO, U.S.A. May 24–26, 1987. p. 1070–1071.

87MI416
Characterization of New Diagnostic Acylcarnitines in Patients with Beta-Ketothiolase Deficiency and Glutaric Aciduria Type I Using Mass Spectrometry. D.S. Millington, C.R. Roe and D.A. Maltby. *Biomed. Environ. Mass Spectrom.* 14(12), 711–716, (1987).

87MI41
Thermospray Liquid Chromatographic-Mass Spectrometric Analysis of Mutagenic Substances Present in Tryptophan Pyrolysates. H. Milon, H. Bur and R. Turesky. *J. Chromatogr.* 394(1), 201–208, (1987).

87MI7'
Thermospray Spectra of Quaternary Ammonium Salts. D. Mitchell, K.R. Jennings and J.H. Scrivens. Presented at 35th Ann. Conf. on Mass Spectrom. and Allied Topics, Denver, CO, U.S.A. May 24–29, 1987. p. 427–428.

87MI59
Application of a New Sample Introduction System for FAB Mass Spectrometry. T. Mizuno, T. Kobayashi, Y. Ito and D. Ishii. *Mass Spectrosc.* 35(1), 9–13, (1987).

87MO6'
Determination of Enzymic Reaction Parameters by On-Line FABMS. W.T. Moore, T. Fan and R.M. Caprioli. Presented at 35th Ann. Conf. on Mass Spectrom. and Allied Topics, Denver, CO, U.S.A. May 24–29, 1987. p. 566–567.

87MO67
Separation and Characterization of Oleanene-Type Pentacyclic Triterpenes from *Glypsophilia arrositii* by Liquid Chromatography-Mass Spectrometry. H.B. Mostad and J. Doehl. *J. Chromatogr. 396*, 157–168, (1987).

87MU83
Mass Spectrometric Studies of Enzymatic Reactions. S.L. Murawski. *Diss. Abstr. Int. B 48*(2), 413–414, (1987).

87MU6'
Thermospray LC/MS Identification of Mitomycin C and Porfiromycin DNA Adducts. S.M. Musser, P.S. Callery and S.-S. Pan. Presented at 35th Ann. Conf. on Mass Spectrom. and Allied Topics, Denver, CO, U.S.A. May 24–29, 1987. p. 496–497.

87NI51
Open-Tubular Liquid Chromatography-Mass Spectrometry with a Capillary-Inlet Interface. W.M.A. Niessen and H. Poppe. *J. Chromatogr. 385*, 1–15, (1987).

87NI41
Problems in Interfacing Open-Tubular Liquid Chromatography and Mass Spectrometry. W.M.A. Niessen and H. Poppe. *J. Chromatogr. 394*(1), 21–34, (1987).

87NI93
Conversion of Liquids to Cluster Beams by Adiabatic Expansion of Liquid Jets: Mass Spectrometric Analysis of Molecular Association in Aqueous Solution Systems. N. Nishi and K. Yamamoto. *J. Am. Chem. Soc. 109*(24), 7353–7361, (1987)

87OH33
Recent Progress in Mass Spectrometry. M. Ohashi and H. Miyazaki. *Farumashia 23*(10), 1033–1038, (1987).

87OL90
On-Line Mass Spectrometric Detection for Capillary Zone Electrophoresis. J.A. Olivares, N.T. Nguyen, C.R. Yonker and R.D. Smith. *Anal. Chem. 59*(8), 1230–1232, (1987).

87OS4'
Adaption of Thermospray LC-MS for Micro LC. R.M. Osterman, C.R. Blakley, G.J. Fergusson and M.L. Vestal. Presented at 35th Ann. Conf. on Mass Spectrom. and Allied Topics, Denver, CO, U.S.A. May 24–29, 1987. p. 634–635.

87OT98
On-Line Thermospray-LC-MS of Nonionic Surfactants. K.H. Ott, W. Wagner-Redeker and W. Winkle. *Fett Wiss. Technol. 89*(5), 208–213, (1987).

87PA81
Trace Analysis of Polar Agrochemicals Using Mass Spectrometry Techniques. J.A. Page. *Pestic. Sci. 18*(4), 291–300, (1987).

87PA2'
Ion Spray API LC/MS and LC/MS/MS Determination of Sulfoconjugated Anabolic Steroids in the Horse. T.M. Pawlowski, L. Weidolf, E. Lee and J. Henion. Presented at 35th Ann. Conf. on Mass Spectrom. and Allied Topics, Denver, CO, U.S.A. May 24–29, 1987. p. 432–433.

87PE55
3,4-Dichlorobenzyloxyacetic Acid Is Extensively Metabolized to a Taurine Conjugate in Rats. R.C. Peffer, D.J. Abraham, M.A. Zermaitis, L.K. Wong and J.D. Alvin. *Drug Metab. Dispos. 15*(3), 305–311, (1987).

87PH22
Isolation and Structure Elucidation of an Epoxide Derivative of the Hydermodified Nucleoside Queosine for *Escherichia coli* Transfer RNA. D.W. Phillipson, C.G. Edmonds, P.F. Crain, D.L Smith, D.R. Davis and J.A. McCloskey. *J. Biol. Chem. 262*(8), 3462–3471, (1987).

87PI6'
Supercritical Fluid Chromatography-Mass Spectrometry: Design and Operating Considerations, and Industrial Applications Involving a "High-Mass" Quadrupole and Mass Spectrometry/Mass Spectrometry. J.D. Pinkston, G.D. Owens, B.R. DeMark, R.L. Dobson and D.S. Millington. *Am. Chem. Soc. Abstr. 194*, ANYL0106, (1987).

87PI0'

Supercritical Fluid Chromatography-Mass Spectrometry Using a "High Mass" Quadrupole Mass Spectrometer and Splitless Injection. J.D. Pinkston, G.D. Owens, D.S. Millington, D.A. Maltby, L.J. Burkes and T.E. Delaney. Presented at 35th Ann. Conf. on Mass Spectrom. and Allied Topics, Denver, CO, U.S.A. May 24–29, 1987. p. 650–651.

87RA93

On-Line Microbore High-Performance Liquid Chromatography-Capillary Gas Chromatography-Mass Spectrometry. II. Application to Analysis of Solvent Refined Coal. T.V. Raglione, J.A. Troskosky and R.A. Hartwick. *J. Chromatogr.* *409*, 213–221, (1987).

87RA95

On-Line Microbore High-Performance Liquid Chromatography-Capillary Gas Chromatography-Mass Spectrometry. I. Development of Isotachic Eluent Splitters. T.V. Raglione, J.A. Troskosky and R.A. Hartwick. *J. Chromatogr.* *409*, 205–212, (1987).

87RA41

Determination of Mycotoxins in Grain by High-Performance Liquid Chromatography and Thermospray Liquid Chromatography-Mass Spectrometry. E. Rajakyla, K. Laasasenaho and P.J.D. Sakkers. *J. Chromatogr.* *384*, 391–402, (1987).

87RE3'

Packed Capillary LC/MS Using a Moving Belt Interface. R.W. Reiser and A.C. Barefoot. Presented at 35th Ann. Conf. on Mass Spectrom. and Allied Topics, Denver, CO, U.S.A. May 24–29, 1987. p. 413–414.

87RO1'

On-Line Thermospray (TSP) Analysis of the Cephalosporin Ceftiofur and Related Compounds in the Bulk Drug and in Rat Urine. R.H. Robins and P.S. Jaglan. Presented at 35th Ann. Conf. on Mass Spectrom. and Allied Topics, Denver, CO, U.S.A. May 24–29, 1987. p. 961–962.

87RO12'

Does the Charging of a Droplet Alter Its Vapor Pressure? Application to Thermospray. A.L. Rockwood and V. Katta. Presented at 35th Ann. Conf. on Mass Spectrom. and Allied Topics, Denver, CO, U.S.A. May 24–29, 1987. p. 31–32.

87RO98

Ion Mobility Spectrometer after Supercritical Fluid Chromatography. S. Rokushika, H. Hatano and H.H. Hill, Jr. *Anal. Chem.* *59*(1), 8–12, (1987).

87RO94

Analysis of Mixtures by Mass Spectrometry. Part II: Liquid Chromatography/Mass Spectrometry and Supercritical Fluid Chromatography/Mass Spectrometry. M.E. Rose. *Mass Spectrom.* *9*, 264–284, (1987).

87RO16

Direct HPLC/MS Using a Fused Silica Capillary Interface. R.T. Rosen and J.E. Dziedzic. *Chem. Anal. (N.Y.) 91*, 176–186, (1987).

87RO31

Initial Observations Using a Vestec Thermospray HPLC Interface on a VG 7070 Mass Spectrometer. R.T. Rosen and T.G. Hartman. *Vestec Thermospray Newsletter 3*(1), 1, (1987).

87RU6'

Analysis of Ethylene Oxide/Propylene Oxide Bis-Phenol A Adducts by Thermospray LC/MS/MS. P.J. Rudewicz, T.R. Covey and W.H. McFadden. Presented at 35th Ann. Conf. on Mass Spectrom. and Allied Topics, Denver, CO, U.S.A. May 24–29, 1987. p. 536–537.

87RU2BE'

Rapid Metabolic Profiling Using Thermospray LC-MS/MS. P. Rudewicz and K. Straub. In: *Drug Metabolism from Molecules to Man [10th European Drug Metabolism Workshop]* (D. Benford, J.W. Bridges and G.G. Gordon, eds.), p. 208–212. Taylor and Franis, London, 1987.

87SA7'

Characteristics of Liquid Chromatograph/Atmospheric Pressure Ionization Mass Spectrometer (LC/API-MS). M. Sakairi and H. Kambara. Presented at 35th Ann. Conf. on Mass Spectrom. and Allied Topics, Denver, CO, U.S.A. May 24–29, 1987. p. 407–408.

87SA21
Thermospray-Liquid Chromatography/Mass Spectrometry of Eicosanoids and Phospholipids. N. Salem, Jr., H.Y. Kim and J.A. Yergey. *Inst. Natl. Santé Rech. Med. [Colloq.]* *152*, 151–161, (1987).

87SA2′
Thermospray LC/MS of Triglycerides. R.A. Sanders, B.A. Charpentier and C.H. Vestal. Presented at 35th Ann. Conf. on Mass Spectrom. and Allied Topics, Denver, CO, U.S.A. May 24–29, 1987. p. 1082–1083.

87SA61
Oligosaccharide Structural Studies by On-Line HPLC-MS Using Fast Atom Bombardment Ionization. S. Santikarn, G.R. Her and V.N. Reinhold. *J. Carbohydr. Chem.* *6*(1), 141–154, (1987).

87SA59
The Formation of a Glutathione Conjugate Derived form Propranolol. H.A. Sasame, D.J. Liberato and J.R. Gillette. *Drug Metab. Dispos.* *15*(3), 349–355, (1987).

87SC49
Experience with Routine Applications of Liquid Chromatography-Mass Spectrometry in the Pharmaceutical Industry. K.H. Schellenberg, M. Linder, A. Groeppelin and F. Erni. *J. Chromatogr.* *394*(1), 239–252, (1987).

87SE5′
A Direct Liquid Fluid Inlet for Time-of-Flight Mass Spectrometry Using Resonance Enhanced Multiphoton Ionization/Fragmentation. K.R. Segar, G.R. Kinsel and M.V. Johnston. Presented at 35th Ann. Conf. on Mass Spectrom. and Allied Topics, Denver, CO, U.S.A. May 24–29, 1987. p. 405–406.

87SE65
High-Performance Liquid Chromatographic Analysis of Phytoestrogens in Soy Protein Preparations with Ultraviolet, Electrochemical and Thermospray Mass Spectrometric Detection. K.D.R. Setchell, M.B. Welsh and C.K. Lim. *J. Chromatogr.* *386*, 315–323, (1987).

87SE57
Rapid High-Performance Liquid Chromatography Assay for Salivary and Serum Caffeine Following an Oral Load. K.D.R. Setchell, M.B. Welsh, M.J. Klooster, W.F. Balistreri and C.K. Lim. *J. Chromatogr.* *385*, 267–274, (1987).

87SH11
On-Line LC/MS as a Problem-Solving Tool for the Analysis of Thermally Labile Herbicides. L.M. Shalaby. *Chem. Anal. (N.Y.)* *91*, 161–175, (1987).

87SH2′
Variable Molecular Separator. D. Sharp. US 4,654,052 (Cl. 55–67; B01D15/08), 31 Mar. 1987, Appl. 784,000, 24 June 1985, 17 pp.

87SI99
Thermospray Mass Spectrometer Interface Used as a Flow Reactor for In Situ Thermal Degradation Studies: Applications to Beta-Lactam Anitbiotics. M.M. Siegel, R.K. Isensee and D.J.Beck. *Anal. Chem.* *59*(7), 989–995, (1987).

87SI11
TSP-(CID/MS): Translating Daughter Ions into Chemical Structures. E. Siegmund, W.H. McFadden and D.A. Garteiz. *TSP Report* *1*(2), 1–4, (1987).

87SI45
A Comparison of Chromatographic and Chromatographic/Mass Spectrometric Techniques for the Determination of Polycyclic Aromatic Hydrocarbons in Marine Sediments. P.G. Sim, R.K. Boyd, R.M. Gershey, R. Guevremont, W.D. Jamieson, M.A. Quilliam and R.J. Gergely. *Biomed. Environ. Mass Spectrom.* *14*(8), 375–381, (1987).

87SM65
Fundamentals and Practice of Supercritical Fluid Chromatography-Mass Spectrometry. R.D. Smith, H.T. Kalinoski and H.R. Udseth. *Mass Spectrom. Rev.* *6*(4), 445–496, (1987).

87SM3'
Capillary Zone Electrophoresis with On-Line Mass Spectrometric Detection. R.D. Smith, J.A. Olivares and N.T. Nguyen. Presented at 35th Ann. Conf. on Mass Spectrom. and Allied Topics, Denver, CO, U.S.A. May 24–29, 1987. p. 283–284.

87SM93
Mass Spectrometer Interface for Microbore and High Flow Rate Capillary Supercritical Fluid Chromatography with Splitless Injection. R.D. Smith and H.R. Udseth. *Anal. Chem. 59*(1), 13–22, (1987).

87SM41
Eluent pH and Thermospray Mass Spectra; Does the Charge on the Ion in Solution Influence the Mass Spectrum? R.W. Smith, C.E. Parker, D.M. Johnson and M.M. Bursey. *J. Chromatogr. 394*(1), 261–270, (1987).

87SO8'
GC/FTIR/MS and LC/MS Analysis of Nitrosobenzene Decomposition Products. R.T. Solsten, H. Fujiwara and S.J. Wratten. Presented at 35th Ann. Conf. on Mass Spectrom. and Allied Topics, Denver, CO, U.S.A. May 24–29, 1987. p. 1078–1079.

87SP1'
An Apparatus and Method for Mass Analysis of a Chemical Sample. Spectros Ltd. Jpn. Kokai Tokkyo Koho JP 62 85,851 [87 85,851] (Cl. G01N23/225), 20 Apr. 1987, US Appl. 767,819, 21 Aug. 1985, 9pp.

87ST49
Positive Secondary-Ion Mass Spectra and Thin-Layer Chromatography/Mass Spectrometry of Phenothiazine Drugs. M.S. Stanley and K.L. Busch. *Anal. Chim. Acta 194*, 199–209, (1987).

87ST9'
Application of Chromatography/SIMS. M.S. Stanley, S.J. Doherty, K.L. Duffin and K.L. Busch. Presented at 35th Ann. Conf. on Mass Spectrom. and Allied Topics, Denver, CO, U.S.A. May 24–29, 1987. p. 19–20.

87ST99
Simultaneous Qualitative Detection of Alkylated Ammonium Ions by Thermospray Mass Spectrometry. M. Stein. *Vom Wasser 69*, 39–47, (1987).

87ST79
On-Line Liquid Chromatography/Fast Atom Bombardment Mass Spectrometry. J.G. Stroh. *Diss. Abstr. Int. B 47*(9), 3739, (1987). Univ. Microfilms Int. Order No. DA8701628.

87ST52
Liquid Chromatography-Fast Atom Bombardment Mass Spectrometry: Recent Developments. J.G. Stroh and K.L. Rinehart. *LC/GC Mag. 5*(7), 562–570, (1987).

87SU7'
Thermodynamic and Kinetic Control of Ion Intensities in Atmospheric Pressure Ionization Mass Spectrometry. J. Sunner, M. Ikonomou and P. Kebarle. Presented at 35th Ann. Conf. on Mass Spectrom. and Allied Topics, Denver, CO, U.S.A. May 24–29, 1987. p. 37–38.

87SU98
Dominance of Gas-Phase Basicities over Solution Basicities in the Competition for Protons in Fast Atom Bombardment Mass Spectrometry. J. Sunner, A. Morales and P. Kebarle. *Anal. Chem. 59*(10), 1378–1383, (1987).

87SU1'
The Identification of Ampicillin and Its Metabolites by Thermospray LC/MS. S. Suwanrumpha, D.A. Flory and R.B. Freas. Presented at 35th Ann. Conf. on Mass Spectrom. and Allied Topics, Denver, CO, U.S.A. May 24–29, 1987. p. 471–472.

87TA2'
Thin Layer Chromatography Mass Spectrometer. S. Takahashi. PCT Int. Appl. WO 87 01,452 (Cl. G01N23/225), 12 Mar. 1987, JP Appl. 85/190,377, 29 Aug. 1985, 17pp.

87TA45
Biomedical Applications of Thermospray Liquid Chromatography-Mass Spectrometry. G.W. Taylor and D. Watson. *J. Chromatogr. 394*(1), 135–146, (1987).

87TE58
Application of Mass Spectrometry to Peptide Sequencing. L.W. Tetler. *Biochem. Soc. Trans. 15*(1), 158–162, (1987).

87TS20
Directly Coupled System of LC/MS (Liquid Chromatography/Mass Spectrometry). S. Tsuge. *Kagaku (Kyoto) 42*(2), 140–141, (1987).

87UN92
Interchangeable Insert Thermospray Probe. S.E. Unger, T.J. McCormick, M.S. Bolgar and J.B. Hunt. *Anal. Chem. 59*(8), 1242–1243, (1987).

87UN95
Comparison of Desorption Ionization Methods for the Analysis of Neutral Seven-Coordinate Technetium Radiopharmaceuticals. S.E. Unger, T.J. McCormick, E.N. Treher and A.D. Nunn. *Anal. Chem. 59*, 1145–1149, (1987).

87VA47
Identification and Quantification of Diketopiperazines by Liquid Chromatography-Mass Spectrometry Using a Moving Belt Interface. J. Van Der Greef, A.C. Tas, L.M. Nijssen and J. Jetten. *J. Chromatogr. 394*(1), 77–88, (1987).

87VE32
Thermospray Liquid Chromatography-Mass Spectrometry Determination of Quaternary Ammonium Herbicides. C.H. Vestal. *Vestec Thermospray Newsletter 3*(1), 2–4, (1987).

87VO16
Thermospray HPLC/MS for Monitoring the Environment. R.D. Voyksner. *Chem. Anal. (N.Y.) 91*, 146–160, (1987).

87VO29
Characteristics of Ion Evaporation Ionization in Thermospray High Performance Liquid Chromatography/Mass Spectrometry. R.D. Voyksner. *Org. Mass Spectrom. 22*(8), 513–518, (1987).

87VO43
Determination of Prostaglandins and Other Metabolites of Arachidonic Acid by Thermospray HPLC/MS Using Post Column Derivatization. R.D. Voyksner and E.D. Bush. *Biomed. Environ. Mass Spectrom. 14*(5), 213–220, (1987).

87VO431
Derivatization to Improve Thermospray HPLC/MS Sensitivity for the Determination of Prostaglandins and Thromboxane B2. R.D. Voyksner, E.D. Bush and D. Brent. *Biomed. Environ. Mass Spectrom. 14*(9), 523–531, (1987).

87VO430
Analysis of Some Metabolites of T–2 Toxin, Diacetoxyscirpenol and Deoxynivalenol by Thermospray High-Performance Liquid Chromatography-Mass Spectrometry. R.D. Voyksner, W.M. Hagler, Jr. and S.P. Swanson. *J. Chromatogr. 394*(1), 183–200, (1987).

87VO17
Application of Thermospray HPLC/MS/MS for Determination of Triazine Herbicides. R.D. Voyksner, W.H. McFadden and S.A. Lammert. *Chem. Anal. (N.Y.) 91*, 247–258, (1987).

87VO5'
Determination of the Photodegradation Products of Basic Yellow 2 by Thermospray HPLC/MS and GC/MS. R.D. Voyksner, T.W. Pack, H.S. Freeman, W.N. Hsu and C.A. Haney. Presented at 35th Ann. Conf. on Mass Spectrom. and Allied Topics, Denver, CO, U.S.A. May 24–29, 1987. p. 155–156.

87VO23'
Analysis of Cyclodextrins by FAB/MS, FAB/MS/MS and Thermospray HPLC/MS. R.D. Voy-

ksner, F.P. Williams, D.L. Koble and H.H. Seltzman. Presented at 35th Ann. Conf. on Mass Spectrom. and Allied Topics, Denver, CO, U.S.A. May 24–29, 1987. p. 1032–1033.
87VO8'
Comparison of Moving Belt/FAB, Thermospray and Continuous Flow FAB Probe HPLC/MS Techniques for the Analysis of Peptides. R.D. Voyksner, F.P. Williams and K. Tyczkowska. Presented at 35th Ann. Conf. on Mass Spectrom. and Allied Topics, Denver, CO, U.S.A. May 24–29, 1987. p. 1098–1099.
87WA95
LC/MS and SFC/MS: Will They Replace GC/MS? M. Warner. *Anal. Chem.* 59(13), 855A–858A, (1987).
87WA29
Identification of Steroids in Rat Adrenal Glands by Liquid Chromatography Thermospray Mass Spectrometry. D. Watson, G.W. Taylor, S. Laird and G.P. Vinson. *Biochem. J.* 242(1), 109–114, (1987).
87WE1'
Thermospray HPLC/MS Studies of a Prostaglandin Derivative Luprositol and Its Metabolites. C.A. Weerasinghe, L.A. Locke and R. Wang. Presented at 35th Ann. Conf. on Mass Spectrom. and Allied Topics, Denver, CO, U.S.A. May 24–29, 1987. p. 451–452.
87WE11
Thermospray Mass Spectrometry and Accurate Mass Measurement of Morphine and a Cephalosporin. J. Welby and P. Watkins. *TSP Report* 1(3), 1–3, (1987).
87WH11
Measurement of Prostaglandin Production by Isolated Tissues and Cells in Culture. A.R. Whorton. *Methods Enzymol.* 141(Cell Regul., Pt. B), 341–350, (1987).
87WI6'
Liquid Chromatography-Mass Spectrometry Analyses of Non-Volatile Organic Environmental Pollutants. R.C. Willoughby, S. Mitrovich, J.G. Dulak, F. Poeppel and D. Sauter. Presented at 35th Ann. Conf. on Mass Spectrom. and Allied Topics, Denver, CO, U.S.A. May 24–29, 1987. p. 136–137.
87WI9'
Particle Beam-Liquid Chromatography-Mass Spectrometry (PB-LC-MS): Advantages and Applications. R.C. Willoughby and F. Poeppel. Presented at 35th Ann. Conf. on Mass Spectrom. and Allied Topics, Denver, CO, U.S.A. May 24–29, 1987. p. 289.
87WI72
Development and Applications of the MAGIC LC/MS Interface. P.C. Winkler. *Diss. Abstr. Int. B* 47(8), 3332, (1987).
87WR9'
Capillary Supercritical Fluid Chromatography (SFC) and SFC/MS. B.W. Wright and R.D. Smith. *Am. Chem. Soc. Abstr.* 194, GEOC0029, (1987).
87YA3'
Analysis of PTH Amino Acids by Thermospray Liquid Chromatography-Mass Spectrometry. L. Yang and C.H. Vestal. Presented at 35th Ann. Conf. on Mass Spectrom. and Allied Topics, Denver, CO, U.S.A. May24–29, 1987. p. 1103–1104.
87YA0'
Analysis of Insecticides by Thermospray Liquid Chromatography-Mass Spectrometry. L. Yang and C.H. Vestal. Presented at 35th Ann. Conf. on Mass Spectrom. and Allied Topics, Denver, CO, U.S.A. May 24–29, 1987. p. 130–131.
87YE43
Metabolic Kinetics and Quantitative Analysis by Isotope Dilution Thermospray LC/MS. A.L. Yergey, N.V. Esteban and D.J. Liberato. *Biomed. Environ. Mass Spectrom.* 14(11), 623–625, (1987).
87YI0'
Forensic Mass Spectrometry. J. Yinon (ed.). CRC, Boca Raton, Fla., 1987, 240pp.

87YI43
Application of Liquid Chromatography-Mass Spectrometry in Metabolic Studies of Explosives. J. Yinon and D.-G. Hwang. *J. Chromatogr.* *394*(1), 253–260, (1987).

87YO77
Conduct Studies of Supercritical Fluids Relevant to Chromatography and Mass Spectrometry. C.R. Yonker and R.D. Smith. *Govt. Rep. Announce. Index (U.S.)* *87*(7), Abstr. No. 711,484, (1987).

87ZA00
Simplified Interface for Electron Ionization Supercritical Fluid Chromatography/Mass Spectrometry. S.D. Zaugg, S.J. Deluca, G.U. Holzer and K.J. Voorhees. *J. High Resolut. Chromatogr. Chromatogr. Commun.* *10*(2), 100–101, (1987).

88AN32
Thermospray Cartridge Rejuvenation. Anonymous. *Ion Notes* *3*(1), 5–6, (1988).

88BE55
Packed Column Supercritical Fluid Chromatography/Mass Spectrometry Using a Thermospray Source in the Filament-On Mode. A.J. Berry, D.E. Games, I.C. Mylchreest, J.R. Perkins and S. Pleasance. *Biomed. Environ. Mass Spectrom.* *15*(2), 105–110, (1988).

88CA73
Analysis of Biochemical Reactions with Molecular Specificity Using Fast Atom Bombardment Mass Spectrometry. R.M. Caprioli. *Biochemistry* *27*(2), 521–525, (1988).

88DA53
Comparison of Methanol Thermospray (Filament On) and Direct Chemical Ionization Mass Spectrometry to the Study of Biologically Active Steroids. P.R. Das, B.N. Pramanik, R.D. Malchow and K.J. Ng. *Biomed. Environ. Mass Spectrom.* *15*(5), 253–256, (1988).

88GA59
Continuous Flow Fast Atom Bombardment Liquid Chromatography/Mass Spectrometry: Studies Involving Conventional Bore Liquid Chromatography with Simultaneous Ultraviolet Detection. D.E. Games, S.Pleasance, E.D. Ramsey and M.A. McDowall. *Biomed. Environ. Mass Spectrom.* *15*(3), 179–182, (1988).

88HU21
Analyses of Testosterone and Related Compounds by Supercritical Fluid Chromatography/Mass Spectrometry (SFC/MS) and SFC/MS/MS. J. Hurst. *TSP Report* *2*(1), 1–3, (1988).

88JE71
A Comparison of Mass Spectrometry Methods for Structural Determination and Analysis of Phospholipids. N.J. Jensen and M.L. Gross. *Mass Spectrom. Rev.* *7*(1), 41–70, (1988).

88KA09
Pressure Programmed Microbore Column Supercritical Fluid Chromatography-Mass Spectrometry for the Determination of Organophosphorous Insecticides. H.T. Kalinoski and R.D. Smith. *Anal. Chem.* *60*(6), 529–535, (1988).

88KA86
Studies on Ion Formation from Charged Droplets in Thermospray. V. Katta. *Diss. Abstr. Int. B* *48*(10), 2986, (1988).

88SA04
Characteristics of a Liquid Chromatograph/Atmospheric Pressure Ionization Mass Spectrometer. M. Sakairi and H. Kambara. *Anal. Chem.* *60*(8), 774–780, (1988).

88SM06
Capillary Zone Electrophoresis-Mass Spectrometry Using an Electrospray Ionization Interface. R.D. Smith, J.A. Olivares, N.T. Nguyen and H.R. Udseth. *Anal. Chem.* *60*(5), 436–441, (1988).

88ST08
Rapid Protein Sequencing by the Enzyme-Thermospray Liquid Chromatographic/Mass Spectrometric Method. K. Stachowiak, C. Wilder, M.L. Vestal and D.F. Dyckes. *J. Am. Chem. Soc.* *110*(6), 1758–1765, (1988).

88ST7MC'
Peptide and Protein Sequencing by Thermospray LC/MS. K. Stachowiak, D.F. Dyckes and M.L. Vestal. In: *The Analysis of Peptides and Proteins by Mass Spectrometry* (C.J. McNeal, ed.), p. 167–178, Wiley, New York, (1988).

88WE53
Determination of Boldenone Sulfoconjugate and Related Steroid Sulfates in Equine Urine by High-Performance Liquid Chromatography/Tandem Mass Spectrometry. L.O.G. Weidolf, E.D. Lee and J.D. Henion. *Biomed. Environ. Mass Spectrom. 15*(5), 283–290, (1988).

88WI29
Liquid Chromatography/Mass Spectrometry. A Promising Coupling Technique. H.M. Widmer. *Chimica 42*(1), 29–32, (1988).

88WI09
Performance of an Improved Monodisperse Aerosol Generation Interface for Liquid Chromatography/Mass Specctrometry. P.C. Winkler, D.D. Perkins, W.K. Williams and R.F. Browner. *Anal. Chem. 60*(5), 489–493, (1988).

88WO26
Multiple Charging in Electrospray Ionization of Poly(ethylene glycols). S.F. Wong, C.K. Meng and J.B. Fenn. *J. Phys. Chem. 92*(2), 546–550, (1988).

9.3. AUTHOR INDEX

Arpino, P.J. (*cont.*)
75MC73	76AR61	77AR9'
78AR6'	78DA51	78LO3'
78TA79	79AR1KL'	79AR12
79AR6'	79AR59	80AR84
81AR37	82AR13	82AR3'
82AR96	82AR51	82AR14
82DE13	82DE42	82DE5'
83AR33	83AR13	83AR4'
83GU13	84AR9'	85AR33
85AR45	86AR01	86AR86
86CO45	86DE2'	87AR69
87AR19	87CO851	

Artigas, F. 87AR43
Ashcroft, A.E. 83ST15 85AS79
87AS45 87AS4' 87AS24
87BA2' 87BA27
87HU8'
Ashraf, J. 86DE23
Assenza, S.P. 84AS89 86WI2'
Augustin, D.J. 85DI4'
Aylott, I.C. 81AY1
Azoulay, M. 84AZ32
Azuma, K. 80YA87 86MI45
86MI454 86OT12

Baczynskyi, L. 87BA6'
Bailey, J. 87LI0'
Baillie, T.A. 80BU24 86BU85
Baker, T.R. 87KR49
Baldwin, M.A. 73BA71 74AR10
76MC9' 87BA21
Balistreri, W.F. 87SE57
Ballard, J.M. 84BE64 85BE8'
86BA15 86BE5' 87BE43
Barafsky, D.F. 84BA7'
Baram, G.I. 84AL00 85AL15
85AL10 86AL05
Barber, M. 84BA45 87BA2'
87BA27
Barber, W.E. 86BA81
Barcelo, D. 87BA45 87BA71
Barchas, J.D. 83KE58
Bare, K.J. 87BA23
Barefoot, A.C. 87BA87 87RE3'
Barnard, D. 78DY9'
Barnes, L.D. 86WE3'
Barnes, S. 87HA89
Barth, H.G. 82MA43 84MA60
86BA81
Bateman, R.H. 86JO8' 87BA67'
87BA5'

Baty, J.D. 79BA9' 81BA90
82BA91 85BA25
Bax, M. 86HA9' 87HA2'
Bean, M. 87BE24'
Beattie, I.G. 84GA5' 85BE26
86BE03
Beaugrand, C. 83AR4' 83BE63
84AR9' 85AR45 86AR01
Beavis, R.C. 85BE2' 86BE99
Beck, D.J. 86KR5' 86SI6'
87SI99
Beck, O. 86FA92
Beck, W. 81ER64
Beck, W.R. 80ER78
Bell, D.J. 87BA2' 87BA27
Bellar, T.A. 87BE2'
Belmans, M. 86ES1'
Bencsath, F.A. 86BE5' 87BE9'
Bend, J.R. 87DE7'
Benninghoven, A. 79BE6' 80BE59
Berberich, D.W. 87BE4'
Bermond, A. 84EL97
Berni, R.J. 87HA89
Berry, A.J. 86BE37 86BE1'
86BE31 86HE11 87BE0'
87BE8' 87GA41 87GA65
87HE65 88BE55
Berry, T. 86QU5'
Berry, V. 86BE86
Berthou, F. 86DE2'
Bertino, D.J. 85BE9'
Betowski, L. 84BE64 85BE8'
86BA15 86BE5' 87BE43
87BE3' 87BE35 87FL7'
87KO8'
Beuhler, R.J. 82BE79
Beydon, P. 81LA5'
Bieber, L.L. 86BI34
Biemann, K. 85BI23
Bier, M.E. 87BI97
Bieri, R.H. 87BI45 87GR4'
Bill, J.C. 82KE6'
Bjork, G.R. 87HA28
Blaas, W. 85TI31 85TI83
Blais, J.C. 84IN13
Blake, T.J.A. 87BL41
Blakley, C.R. 77BL8' 77MC6'
78BL81 78MC8' 79BL2'
79MC88 79MC8' 80BL0'
80BL67 80BL26 80BL21
80BL2' 80BL86 81BL5'
81CA6' 81DE83 81DV3'

Callery, P.S.	85LA2'	85LA29		Co, R.P.	85JE49	
87MU6'				Cocbelli, L.	83GA61	
Calvo, K.C.	83CA55			Cocksedge, M.J.	82GA95	
Canada, D.C.	82CA4'			Cody, R.B.	86HE4'	87LE99
Caprioli, R.M.	86CA1'	86CA89		Cohen, A.	82YI4'	83YI87
86CA18	87CA67	87CA52		Cohen, S.A.	84YU15	
87CA0'	87MO6'	87CA61		Colby, B.N.	74CO65	74SI51
87CA3'	88CA73			Cole, P.J.	86WA99	
Carmody, J.J.	80BL0'	80BL2'		Cole-Hamilton, D.J.	85CO55	
80BL26	80BL21	80BL67		Colin, H.	77AR9'	
81CA6'	81DV3'			Compson, K.R.	83CH4'	83CH8'
Carroll, D.I.	74CA66	74CA3'		Cook, J.C.	84ST9'	85ST4'
74HO25	74HO93	75CA79		85ST75	86ST88	86ST5'
76HO95	77HO33	78HO03		86ST05		
81CA77				Cook, K.D.	82CH41	82CH5'
Catlin, D.H.	87CA39			83CH56	83CH2'	84MU65
Catlow, D.A.	84GA4'	85CA33		85CO2'	85CO12'	85MA75
Cerbulis, J.	85CE20			86CO57		
Champlin, P.	79HI5'			Cooks, R.G.	81UN36	82BU87
Chan, K.W.S.	82CH5'	82CH41		83CO23	85BR73	87BI97
83CH2'	83CH56			87DE55		
Chan, K.Y.	86CO9'			Cooper, D.A.	86LU59	
Chang, C.	85CH79			Corbelli, L.	82AL99	
Chang, T.T.	85CH7KA'			Corey, T.	84CO6'	84CR61
Chapman, J.R.	83CH61	83CH8'		Cos, R.P.	84CO10	
83CH4'	84CH6'	85AS79		Cottee, F.H.	85CO55	
85CH47	85CH3'	85CH7'		Cotter, M.L.	86LL4'	87LL93
85CH01	86CH33	86CH5'		Cotter, R.J.	83FE1'	83FE13
86GR7'	87AS45	87AS4'		84FE69	84LA4'	85CO1'
87CH41	87ME49			85FE78		
Chappell, C.G.	86RI19	86TA0PI'		Cottrell, J.	86CA89	87AS45
Charpentier, B.A.	87SA2'			86CO45	87AR69	
Chedda, G.B.	81DE83	81ED9'		87CO851		
Chen, T.M.	87CH62			Cousin, J.	87AR19	
Chess, E.K.	85SM5'	87CH8'		Coutant, J.E.	85CO0'	86CO9'
87KA43				87CO7'	87CH62	
Chizov, O.S.	80BU24			Covey, T.R.	83CO55	84HE1'
Chopade, H.M.	86MA61			84HE3'	84HE0'	84SK59
Chou, F.E.	86JA1'			85CO3'	85CO74	85CR30
Chrisman, R.	81UN36			85HE5'	85HE01	85HE7'
Christensen, R.G.	79CH0'			86BR5'	86BR9'	86CO03
80CH1RE'	80CH8RE'	80HE08		86CO7'	86CO81	86CO83
81CH31	81CH7KI'	81WH6'		86HE09	87BR92	86CR92
82CH7'	83CH11	83CH7'		87CO85	87HE6'	87RU6'
84CH8'	85CH336			Cox, H.L., Jr.	72GI6'	78DO53
Christiansen, P.	86JO07			Cox, R.P.	85JE49	
Christman, R.F.	85BE9'			Crain, P.F.	86PH0'	87ED79
Chui, K.S.	87CH5'			87MC53	87PH22	
Chung, H.L.	85BL9'			Crawford, J.L.	87CR1'	
Clarke, S.R.	86RI19	86TA0PI'		Crews, H.M.	87DE27	
Clegg, G.A.	71CL01			Crosby, D.G.	79SK1'	80SK88
Clench, M.R.	86GR7'			Crow, F.W.	86RO8'	86RO0'
Climie, J.M.	87MA58			87CR01'		
				Crowley, J.	87CR0'	

Hirata, Y. (*cont*.)			Hwang, D.-G.	83YI85	84YI6′	
78TA09	79TS87	80TS01	85YI97	85YI65	85YI0′	
82TS1′	82YO14	86HI10	85YI7′	86YI87′	85YI09	
86MA17	87MA55		86YI54	86YI0′	86YI13	
Hiroaka, K.	79TS4′		87YI43			
Hirose, H.	86KA6′		Hyanek, J.	86DE97		
Hirschfeld, T.	80HI27		Hyldburg, P.A.	86HY76		
Hirter, P.	79GA7′	80GA17	Hynes, R.K.	85HO9′	86HO82	
81GA31	85HI39					
Hitachi Ltd.	82HI64′	82HI4′	Iannucci, R.	87HA2′		
84HI8′	84HI7′	85HI3′	Iida, Y.	81DA97	81II29	
Ho, M.H.	87HA89		81OK97			
Hoffman, P.A.	78LO3′		Ikonomou, M.	87SU7′		
Hoffmann, J.J.	82LU02		Imasaka, T.	86FU85		
Hogge, L.R.	85HO9′	86HO82	Inchaouh, J.	84IN13		
87HO0′			Iribarne, J.V.	76IR47	81IR9′	
Höhn, M.	85HO2′	85VA31	81IR4′	82TH9′	82TH49	
85VA37	86TA06	86VA07	83IR01			
86VO46	87HE65		Isensee, R.K.	86KR5′	86SI6′	
Holder, C.L.	87KO8′		87SI99			
Holland, J.F.	86HO93		Ishibashi, N.	86FU85		
Holwitt, E.	87DA0′		Ishii, D.	82TA51	82TA0′	
Holzer, G.U.	85DE0′	85HO88	85IT61	86MI67′	86IT81	
85HO9SA′	87ZA00		87IT16	87IT4′	87MI59	
Hoo, K.	86QU5′		Itagaki, Y.	86MI67′	87IT4′	
Hood, R.	87CR0′	87KE6′	Ito, Y.	85IT61	86IT81	
Horakova, M.	86DE97		86MI67′	87IT16	87MI59	
Horman, I.	84GA13		Itoh, K.	82YO14		
Horning, E.C.	74CA3′	74CA66	Iwabuchi, H.	87IW49		
74HO93	74HO25	75CA79	Iwatani, K.	86IW41	86IW49	
76HO95	77HO33	78HO03				
81CA77			Jackson, M.D.	82WR80		
Horning, M.G.	74CA3′	74CA66	Jaglan, P.S.	87RO1′		
74HO25	76HO93	74HO95	Jakubowski, E.M.	86JA1′		
Horowitz, J.P.	87KE6′		Jamieson, W.D.	80RA4′	81RA5′	
Horton, R.L.	79HO1′		82RA7′	83LE6′	83RA55	
Houghton, E.	81HO88		84QU7′	86JA05	87SI45	
Houk, R.S.	81HO7′	82HO5′	Japan Spectroscopic Co. Ltd.	82JA8′		
82HO70	86HO87	86TH81	Jardine, I.	87LI0′		
Howard, C.C.	80RO5′		Jarvis, B.	86KR05	86MI6′	
Hsu, F.F.	85ED6′	85HS14	Jarvis, D.E.	80ER78	81ER64	
86ED07	86HS9′	86HS99	84ER16			
87HS78			Jenkins, E.E.	81DE83	81ED9′	
Hsu, S.H.	87HS4′		Jennings, K.R.	87HU8′	87MI7′	
Hsu, W.N.	87VO5′		Jensen, B.B.	84CO10	85JE49	
Huang, D.	87HU57		Jensen, N.J.	88JE71		
Huang, Z.	85ST75	85ST4′	JEOL Ltd.	81JE3′	81JE5′	
Huber, J.F.K.	83HU17		82JE6′	82JE3′	82JE2′	
Hunt, D.F.	80HU0′		83JE0′	84JE4′	84JE8′	
Hunt, J.B.	87UN92		85JE8′	85JE0′		
Hurst, J.	88HU21		Jetten, J.	87VA47		
Hutchinson, D.W.	87HU24	87HU8′	JGC Corp.	84JG1′		

86WH7′	87ME34′	87ME3′	Murawski, S.L.	84MU65	87MU83
88WO26			Murray, S.	85WA34	85WA20
Merdink, J.L.	87CA39		86WA35	87GO87
Meresz, O.	83FO7′	83FO08	Musha, K.	84MU8′	85MU46
83GA8′	84GA14	84GA48	86OT12
Meyer, W.	81BR0′		Musser, S.M.	87MU6′
Meyerhoff, G.	83DO65	85VA37	Myers, M.N.	70GI49
Michl, J.	81OR0′	82ST5′	Myher, J.J.	83KU05	83MA10
Mico, B.	86ST45′		84MY39	84PI21	85CE20
Milberg, R.M.	85ST75		85KU27	85KU22	86KU97
Miller, D.J.	87HA3′	86HA87	87KU73
87MI0′			Mylchreest, I.C.	87BE8′	87BE0′
Millington, D.S.	79YO4′	80MI89	87GA41	87GA65	88BE55
84LI6′	84YE98	85LI43
85MI59	86MI01	85MI0′	Nabichvrishvili, D.S.		82OR14
87MI416	87PI0′	87PI6′	Nagano, T.	84SM10′
Milon, H.	83MI13	84MI25	Nakagawa, A.	87IW49
85MI09	87MI41		Nakagawa, Y.	86IW41	86IW49
Min, Z.		83WH68	Nakajima, F.	78MI4′
Miotto, J.	87HA2′		Nakamae, K.	81NA7′	84GI19
Mirgrodskaya, O.A.	85AL15	85AL10	Nakamura, K.-I.	87IW49
Mishra, A.K.	86MI58		Nakanishi, H.	86NA93
Misler, S.R.	86MI6′		Namba, T.	84HA20	86NA21
Mitchell, D.	87MI7′		Nearing, M.E.	81RA5′	83RA55
Mitchum, R.K.	83MI55	87KO8′	Nelson, B.	87ES6′
Mitroka, J.	86UN7′		Nelson, D.	80BR65	83NE07
Mitrovich, S.	87WI6′		Nelson, L.M.	86HO82
Mittelman, A.	81DE83		Nelson, R.J.	87KR49
Miyagi, H.	78MI4′		Neukermans, A.	80ME0′
Miyazaki, H.	76TA91	87OH33	Newman, R.A.	85LU12	86MC7′
Mizuno, T.	80YA87	84HA20	86MC73
84MI256	86IT81	86MI67′	Ng, K.J.	88DA53
86MI45	86MI454	86OT12	Ngo, A.	83TH5′
87IT16	87IT4′	87MI59	Nguyen, H.	87FI8′
Mobley, R.C.	68DO16	68DO90	Nguyen, N.T.	87OL90	87SM3′
Mochizuki, K.	80YO36	82YO14	88SM06
Moll, H.	85SH71		Nibbering, N.M.M.	81OT66	82NI13
Moller, L.	86MO31		86NI13
Moller, J.	86BO89		Nicholson, P.J.D.	87HA17
Montaudo, G.	86GA53		Nicolaides, N.	78MC2′	79MC78
Moore, C.	87BA27		Niessen, W.M.A.	85NI37	86NI17
Moore, J.M.	86LU59		86NI12	87NI51	87NI41
Moore, L.E.	83CH61		Nijssen, L.M.	84VA4AD′	85VA03
Moore, W.T.	87CA61	87CA0′	87VA47
87MO6′			Nikolaev, V.I.	84AL00	84AL96
Morales, A.	87SU98		84GA49	85AL15	85AL10
Morgan, D.J.	82GA02		86AL05	86AL07
Mori, K.	86YA11		Nishi, N.	87NI93
Morita, M.	86HA17		Nishimura, S.	84ED1′	85ED4′
Morrissey, M.A.	86HI1′		86ED75	86YA31
Mostad, H.B.	87MO67		Nishimuta, T.	86YA11
Munro, S.	87DE27		Nishishita, T.	76NI95
Munroe, M.	74SC95	74SC65	Nishizawa, Y.	86HA17

Noda, T.	84OK67	
Noel, S.F.	83SM87	
Noeller, H.G.	80NO9'	
Nonaka, T.	82TS05	82TS6'
83TS65		
Nordman, B.	87AP0'	
Novak, F.P.	85NO83	
Novotny, M.	81NO34	84NO0'
Numaziri, Y.	86SH15	
Nunn, A.D.	87UN95	
Nyssen, L.M.	84GA499	
O'Brien, I.	86GO09	
Oehme, M.	85OE9FA'	
Oertli, C.U.	77HO33	
Oestvold, G.	81OE14	
O'Hara, M.	87CR0'	
Ohashi, M.	87OH33	
Ohki, T.	82TA51	
Ohlendorf, R.	86LI09	
Ohori, D.	86LL4'	87LL93
Ohta, J.	85MA07	
Ohta, K.	82TS1'	
Okada, S.	81II29	81OK97
Okumura, Y.	78TA09	
Okutani, T.	84OK67	
Olivares, J.A.	82HO5'	87KA6'
87OL90	87SM3'	88SM06
Olson, K.L.	85OL5'	85OL37
85VA72	86VA35	
Ordzhonikidze, K.G.	82OR14	
Orth, R.G.	78UD9'	81OR0'
82ST5'		
Osei, Y.P.	82OR14	
Oseitwum, E.Y.	80QU2'	
Osterman, R.M.	86GA1'	87OS4'
Oswald, E.O.	78WR7'	
Otake, W.	81TS15	
Otsuka, K.	85TS8'	85TS69
86OT12		
Ott, K.H.	81OT66	87OT98
Otto, A.	87KR41	
Owens, G.D.	87PI0'	87PI6'
Oxford, J.	81MA11	82MA15
83DI2'	84MA41	85LA33
86LA05	86MA69	87LA43
Oyler, A.	86LL4'	
Pacholec, F.	86PA812	
Pack, T.W.	85VO3'	87VO5'
Pacula, C.M.	87MA58	
Page, J.A.	85CO55	87PA81

Pallante-Morell, S.	87BE24'	
Palmer, A.G., III.	82LE9CO'	
Palmer, M.J.	85DI4'	
Palomino, E.	87KE6'	
Pan, S.-S.	87MU6'	
Pang, H.	82ED6'	
Pang, H.M.	86LU3'	86PA8'
86PA81	87LU6'	
Parillo, R.J.	86GR7'	
Parker, C.E.	81VO7'	82PA73
82PA27	82PA75	82VO43
83PA6'	83VO6'	84SM10'
85BU77	85PA48	85PA93
85PA6'	85PA71	86PA814
86PA01'	86SM07	86SM2'
87DE9'	87DE2'	87DE90
87GA47		
Parks, O.W.	85CE20	
Parr, I.	84SM10'	
Parr, V.C.	86JO8'	
Pastorino, A.M.	85FR7BR'	
Patel, A.C.	84PA42	
Paulson, J.	86PA03	87LI47
87LI45		
Pavlenko, V.A.	84AL96	84AL00
Pawlowski, T.M.	87PA2'	
Peaden, P.A.	84NO0'	
Peffer, R.C.	87PE55	
Pellizzari, E.D.	83VO67'	84VO5'
84VO21	84V067	84VO16
84VO3'		
Pentoney, S.L., Jr.	87LA93	
Perel, J.	83GO4'	84MA4'
87LE5'	87MA99	
Perkins, D.D.	86AB2'	86BR45
86PE4'	87BR9'	88WI09
Perkins, J.R.	86BE1'	86BE37
86BE31	87BE0'	87BE8'
87GA41	87GA65	88BE55
Pesce, A.J.	86HA21	
Pesch, R.	83DO65	
Peters, H.	80BE59	
Petersen, B.	80KI4'	81KI39
Petty, J.D.	80ST87	
Pfluger, K.H.	85DA22	85DA51
Philips, C.R.	79KE1'	
Phillips, F.C.	84ER16	
Phillips, F.O.	82ER98	
Phillipson, D.W.	86PH0'	87MC53
87PH22		
Phinney, M.J.	80PH10	
Pilipenko, A.T.	83KU59	

9.3. SUBJECT INDEX

—with sector MS

83YI7'	83YI83	83YI85
84YI6'	85YI65	85YI97
85YI0'	85YI7'	86YI54
86YI0'	86YI13	

DLI (unsplit)

—with FAB

85IT61	86CA1'	86CA18
86CA89	86IT81	86KA3'
86MI67'	87AN6'	87AS24
87AS45	87AS4'	87BA67'
87BA27	87BA2'	87CA61
87HU24	87IT16	87IT4'
87MI59	87SA61	
88CA73	88GA59	

—with gas nebulizer

83AP50	87SE5'

—with laser vaporizer

78LO3'	79LO2'	80LO4'
80LO84		

—with micro-LC

78AR6'	78HE07	80BR65
80HE3TO'	80HE75	80HE81
80HE815	80PR0'	80YA87
81HE4'	81HE8'	81HE33
81HE97	81SC65	81TI85
82BR44	82DE5'	82EC2'
82EC82	82HE99	82KR19
82SM43	82SM71	83BR63
83EC65	83GA61	83GO1'
83HE3'	83HE8'	83HE5'
83LE69	83SM67	83SU47
84AL00	84AP23	84AR9'
84BR1'	84BR31	84BR62
84CO6'	84HA20	84HE8'
84SK56	84SK59	85AR45
85CR30	85ES8'	85ES21
85HI39	85MA33	85NI37
85PR3'	85SK11	85TI83
85TS70	86ES5'	86LA34
86MI45	86SK47	86YI13
87AL03	87AL45	87BA87
87DE9'	87GR4'	87NI41

Review of —

84HE80	84WE18	85BR39
85HE03	85LE33	

—with plasma chromatography

73KA63	79HA26

—with sector MS

83CH4'	83CH61	83CH8'
83CH4'	83ST15	84CH6'

Moving band (*cont.*)

80EG23	80GA6'	80GA8'
80GA72	80GA73	80GA83
80GA86	80HA6'	80HU0'
80KA81	80KI4'	80KI95
80MA95	80MI89	80QU2'
80RA80	80RO5'	80SK88
80SM0'	80SM97	80ST87
80TH4'	81BA90	81BR0'
81EC79	81GA4'	81GA6'
81GA31	81HA32	81HO88
81KE6'	81KI39	81MA11
81SM2'	81SM5'	81SM07
81SM30	81SM33	81SM39
81TH87	81TW8'	81UN36
81WE43	81WR85	81YU89
82AB47	82AL15	82AL99
82CA46	82CA43	82EC92
82EC93	82GA02	82GA95
82HA0'	82KI25	82MA15
82MA34	82SC33	82TA0'
82VO6'	82VO15	82WE91
82WR01	82WR80	83CA03
83CA04	83DO65	83EV0'
83EV3'	83EV94	83GA7'
83GA61	83GA93	83LA67
83LA69	83MC1'	83MC87
83RO87	83SM87	83WH68
84GA499	84GA5'	84GA4'
84GA17	84JO45	84JU58
84JU59	84LE18	84MA41
84MA76	84MC2'	84QU7'
84ST3'	84VO62	85BA25
85BE26	85BU41	85DA7'
85HO2'	85KI7'	85KR73
85KR72	85LE35	85MC6'
85OL5'	85QU89	85ST73
85VA03	85VA37	86HE11
86KI034	86QU5'	86TA06
86VA46	87EV49	87HA89
87VA47		

—with aerosol deposition

80HA6'	81HA9'	81HA32
81SM5'	81SM30	81SM39
81YU89	82HA0'	83YU03
84HA5'	84HA69	84KA2'
84SC59	84TH44	84YU15
85HA53	85OL37	86KI03
86MO31	86PA03	

—with thermospray deposition

83HA5'	84FA0'	84FA60
84HA62	84YA62	85FA66
85FA1'	86YA66	87KR49

Band broadening
82GA02	83HA1′	83HA55
83VO3′		

—with chemical derivatization
80QU2′	81AM53	81YU89
82MC33	82VO15	82VO6′
86TS82		

—with chemical reactor
77ER27	78PR11	80ER78
81ER64	81PR26	83VO99
84ER16	85ER26	

—with continuous extractor
79KA6′	79KA14	80KA81
80KI95	80KI4′	81KI39
82KI25	82VO15	82VO6′
83LA67	85VO9′	

—with DCI
81BR0′

—with FAB
82KE6′	83DO65	83JE0′
83LE0′	83LE20′	84ST9′
85DE6′	85ST4′	85ST75
86AR86	86BO4′	86HE34′
86HI01	86KI03	86SK07
86ST45′	86ST5′	86ST88
87ST79	87ST52	87VO23′

—with HRMS
82BA91	85VA03	86MI6′

—with isotopically labeled standards
77GA5′	78GA6′	78GA5RE′
80GA83	80MI89	82MA15
85VA37	86PA03	

—with laser desorption
80HA6′	80HU0′	80NO9′
80RE2′	80SM0′	80WE5′
81HA32	81HA9′	82HA0′
83HA5′	83HU17	84FA0′
84FA60	84HA62	85FA66
85FA1′	86DE8′	86FA66
87DE3′	87DE85	

—with micro-LC
81GA6′	82AL99	82GA02
82GA95	82TA0′	82WE91
83AL83	83FO7′	83FO08
83GA61	83GA84	83GA3′
83LA69	83MC87	83VO3′
84EV3′	84FR8′	84HA5′
84HA69	84KA2′	84PA42
84SC59	84WE18	85GA37
85HA53	85OL37	86AB0′
86HY76	86MO05	86PA03
87BA87	87RE3′	

—with nonvolatile buffers
   81KE6'

—with SIMS

79BE6'	80BE59	80SM0'
80SM97	81SM2'	81SM5'
81SM07	81SM30	81SM33
81UN36	83SM22	84FA0'
84FA60	85BU8'	85FA1'
85FA66	86BE99	86DE8'
86FA66	87DE3'	

—with supercritical fluid chromatography

86BE1'	86BE31	86BE37
87BE9'		

Review of—

86HI10	87AR69	87GA65
87KA43		

Moving wire

73KA63	74SC65	74SC95
77SC13	79SC95	

Review of —

82AL15	84GA13	84GA14
85BR39	85KA33	

Thermospray

80BL21	80BL67	81CA6'
81VE09	82BL7'	82ED6'
82YE8'	83BL50	83CO55
83ED6'	83ED9'	83KI9HR'
83LI7'	83LI51	83MC93
83PI8'	83VE5'	83VE8'
83VE63	83VO67'	83YE0'
84BE64	84ED1'	84GA5'
84HE3'	84JO45	84KI64
84KI3'	84LE3'	84MC0'
84PO8'	84SC83	84SM10'
84ST97	84TA59	84VO7'
84VO5'	84VO78'	84VO67
84VO3'	84WE9'	84YE0'
84YE98	84YE4'	85AR45
85AS79	85BE9'	85BL3'
85BU3'	85CO74	85CR30
85DE6'	85DI4'	85ED6'
85ED4'	85ED37	85ED45'
85ES3'	85FO9'	85GA37
85GA114	85GA34	85GA11
85GO5'	86GO09	85HO9'
85JO28	85KI7'	85KI60
85KI8'	85LI43	85LU12
85MI0'	85MI59	85PA6'
85PA48	85RO4'	85SH3'
85SH21	85SH5'	85SU8'
85TA59	85VE4'	85VE49
85VO77	85VO70	85VO0'
85VO3'	85VO5'	85VO86'

—with micro-LC
   83HE2'   85BR39   87OS4'
—with MS/MS
   84BE64   85BE8'   85COO'
   85HE7'   85MC6'   85SH2'
   85SI2'   85UN12   86BA15
   86BE5'   86CO9'   86CO83
   86JO07   86KR5'   86RU3'
   86RU88   86ST4'   86UN13
   87BE43   87BE4'   87MC41
   87MC51   87RU6'   87RU2BE'
   87VO17
—with sector MS
   84CH6'   84LA4'   84VE60
   85CA33   85CH7'   85CH3'
   85CH4'   85FE7GE' 86GR7'
   86LI09   86WE3'   87BA6'
   87CH41   87RO31
Vaporizer
   85HS14   85SC3'   85VE73
   86VE05
Universal LC detector
   84AO43
Other Interfaces
  Heated-wire concentrator
   79CH0'   80CH1RE' 80CH8RE'
   81CH31   81WH6'   83CH7'
   83CH11   84CH8'   85CH336
  Polymer membrane
   75JO70   79WE08   85BR73
OTHER LC/MS METHODS (BY MODE OF IONIZATION)
Atmospheric Pressure Ionization (API)
   74CA3'   74HO25   74HO93
   75CA79   76HO95   78HO03
   79TS4'   81CA77   82HE41
   83SA17   83TH9'   83TH2'
   84HE1'   84HI8'   85CO3'
   85HE5'   85LA2'   86BR9'
   86CO83   86HE1'   87AL94
   87CO85   87HE6'   87SA7'
   88SA04
Review of—
   83MI55
Discharge Ionization
   82AR3'   84HE1'   84PO8'
   85LA2'   85MA03   85VE1'
   86BR9'   86ZA3'
Electrohydrodynamic Ionization (EHD)
   82JE3'   84JE8'   84JG1'
Electrospray
   80FI6'   84WH2'   85WH2'
   85WH75   86BR5'   87LE4'

—with capillary zone electrophoresis

87OL90	87SM3'	88SM06

Field Desorption (FD)

80MA9'	82JE2'

Inductively Coupled Plasma (ICP)

82HO5'	82HO70	86KO88
86TH81	87DE27	

Ion Evaporation

81IR4'	81IR9'	82TH9'
82TH49	85BR39	86BR9'
86HE1'	87BR92	87BR97
87PA2'	88WE53	

Laser Desorption (LD)

78LO3'	80HU0'	83HU17

Liquid Ionization

85TS8'

Plasma Desorption (PD)

82JU76	82JU79	82SC85
83DA23	83JU49	83JU15
83JU67	84TH44	85DA22
85DA51	86SU96	

Secondary Ion Mass Spectrometry (SIMS)

85BE2'	86BE99

LC/GC/MS

82LU02	87RA93	87RA95

LC/MS/MS

80SM0'	80SM97	81SM2'
81SM5'	81SM07	81SM30
81SM33	82FE33	82HE41
82SM71	82SM43	82TH9'
82TH49	82VO2'	82VO43
82VO45	82VO51	83SM67
83VO6'	83VO65	84BE64
85BE8'	85CO0'	85HE7'
86HE09	86JO07	85MC6'
85MC3'	85SH2'	85SI2'
85SM40	85UN12	86BA15
86BE5'	86CO9'	86CO83
86KR5'	86RU3'	86SC00
86ST45'	86UN13	87MC51

TLC/MS

80RA4'	81RA5'	81UN36
82BU87	82RA7'	83RA55
84BA7'	85BU8'	85CH7KA'
85DA22	85DA51	85KA36
85KU85	85NO83	85TA52
86DI7'	86DI81	86FI74
86HA17	86IW41	86IW49
86KA84	86SH15	86SU96
86YA11	87BA23	87BU2'
87BU65	87FL0'	87IW49
87KR41	87KR45	87ST9'
87ST49	87TA2'	

RELATED PHYSICS AND MASS SPECTROMETRY
Atmospheric Pressure Ionization (API)

77HO33	81CA77	82KA43
83MI55	87SU7'	

Bibliography of—
81FR61

Chemical Ionization (CI)

79MC95	81DA97	81II29
81OK97	82BE79	82CA46
82DU2'	82VO43	82VO45
82YI4'		

Direct Chemical Ionization (DCI)

80RA80	82SU19

DLI Systems

68TA22	68TA28	69TA38
69TA43	72SK60	81TI85
82CH41	83GU13	

Direct fluid injection (DFI) of supercritical fluids

83SM1'	83SM85	83UD8'
84SM0'	84SM95	84UD2'
85SA8'	85SM31	85SM40
86KA81	86KU07	86LU3'
86PA8'	86PA81	87LU6'

—with MS/MS

83SM26	83SM85	83SM07
83SM56		

Electron Ionization (EI)

80KI95	81KI39

Electrohydrodynamic Ionization (EHD)

74CO65	74SI51	78ST20
78ST52	80SK84	82CH5'
83CH56	83CH2'	84AL96
84GA49	84JE8'	84JG1'
84MU65	84OK67	85CO2'
85CO12'	85MA75	86CO57
87LE4'	87MA99	

Electrospray

68DO16	68DO90	70MA1'
70MA27	71CL01	72GI6'
73GI2'	75DO4'	78DO53
80FI6'	81NA7'	84GI19
84LA2'	84WH8'	84WH2OG'
84YA81	84YA819	84YA93
86WH7'	87ME3'	87ME34'
88WO26		

Fast Atom Bombardment (FAB)

82RI84	84FE69	84IN13
84LI13	85CO55	85FE41
85LI73	86BO89	86CA18
86CA89	86SU89	87CA0'
87HU8'	87LI0'	87MO6'
87MU83	87SU98	

Field Ionization/Field Desorption (FI/FD)

| 81GI87 | 81OT66 | 82TS05 |
| 82TS6' | 83SC68 | 84SU91 |

Glow Discharge

81FO20

Inductively Coupled Plasma (ICP)

45RA0'	58LO0'	81GR62
81HO7'	82HO5'	82HO70
86HO87	86KO88	87PA2'

Ion Evaporation

76IR47	81IR4'	81IR9'
82ZO23	82ZO25	83IR01
84TH0'	86DE46	

Ion Mobility Spectrometry

87RO98

Liquid Ionization

79TS4'	80TS85	81TS15
82TS05	82TS6'	83TS3'
83TS6'	83TS65	84KU23
84MU8'	84TS23	84TS00
85MU46	85TS69	86OT12
86SA49		

Membrane inlet mass spectrometry

83CA55	83HE11	84CO10
84RI48	85BR73	85HE78
85HI41	85JE49	85LL19
86DE25	86LA5'	86MI58
87BI97	87HA17	

Mass Spectrometry/Mass Spectrometry (MS/MS)

78MC4'	80MC19	81GA4'
81MC39	81MC40	82TH9'
83AL83	83BE63	83GA61
6JO07	86VO05	87SI11
87SI99		

Molecular Beams

| 70GI49 | 85JO90 | 86FU85 |

Plasma Desorption (PD)

| 82JU76 | 85SU41 |

Reviews of—

80MC19	81MC40	82AR13
82BU87	82NI13	82RI84
82TS73	83DE33	83VE56
83VE27	83ZA87	84BA45
84KA59	85FE41	85KA36
85SU41	86SC59	87CA52
87DE55		

Secondary Ion Mass Spectrometry (SIMS)

| 81OR0' | 82ST5' | 84IN13 |
| 84LI13 | 84OK67 | 86LI89 |

Solution Introduction

84DO67

Spark Source

| 82OR14 | 87NI93 |

Spectral Deconvolution
    80HA84

Thermospray

53DO43	80BL0'	80BL2'
81CA6'	81DV3'	82VE9'
83FE13	83FE1'	83FE9'
83VE27	83VE63	84BL1'
84FE69	84GO8'	84PI3'
84PI13	84SC47	84TH0'
85AL9'	85BU77	85CO1'
85DE8'	85KA9'	85LA29
85SC02	85SC3'	85SC00
85SC37	85VE1'	85VE7'
85VE73	85VE0'	86AL7'
86AL81	86BE5'	86DE23
86GE6'	86HA87	86HA9'
86HA21	86JA1'	86KA7'
86KA5'	86LI86	86PA01'
86PA814	86RU9'	86SC01
86SC02	86SI6'	86SM2'
86VO1'	87BU5'	87CH5'
87GA14	87KA1'	87KE6'
87KI459	87KO8'	87LE45
87RO12'	87SI99	87SI11
87UN92	87WE11	88AN32
88KA86		

REVIEWS
    LC/MS

76MC63	78DA9TS'	78GA5RE'
78HA38	78KE33	78PR11
78SU15	78TA79	79AR12
79AR1KL'	79TS14	79ZE19
80GA17	80GA70	80GA72
80MC87	80TS01	81GA31
81GA84	81GA4M0'	81MA33
81OE14	81PR26	81QI28
81TS25	81TS66	82AL15
82AR13	82AR96	82DE03
82ER11	82GA27	82GA93
82LE39	82MC98	82NI13
82WI29	82YO29	83AR33
83ES98	883FE13	83GA02
83GA93	83GA11	83GA8'
83GA84	83LE12	83MA18
84BR98	84BU28	84DE61
84DE77	84GA18	84GA13
84GA90	84GA14	84GA48
84HE80	84KA59	84LE23
84SC83	84VE65	85AL45
85AR33	85BR39	85EC25
86GA03	85GO57	85HE03
85KA33	85SU41	85VO15
85YI05	86AN21	86AR86

86BE86	86BR45	86CO45
86CO81	86CR92	86LE37
86MA69	86NA21	86NI12
86NI17	86SC00	86SI0SA'
86SM70	86SN59	86VE33
86WE45	87AR69	87BO99
87GA65	87KU73	87OH33
87RO94	87TS20	87WA95
88WI29		

Other Topics With LC/MS

76AR61	76TA91	77HO33
79MC72	80AJ02	80BU24
80HI27	80LE87	80MC19
81GA92	81LA5'	81MC40
81MC39	81MC63	81ME66
81NO34	82AR51	82BO47
82BU43	82GR4'	82KN96
82MA43	82MC33	82PO31
82TS73	83CO23	83DE33
83KU59	83NE07	83RA7PA'
83VE27	83VE56	83WI21
83ZA87	84AS89	84BA45
84HI91	84MA60	84MC61
84RO76	84WH93	84WO81
85BI23	85EG47	85FE1'
85FI26	85HA84	85HE01
85JU1EI'	85KA36	85KE82
85KU27	85KU22	85RO80
85TA96	85WH5FA'	85YU83
86BA81	86BI34	86BU85
86CA43	86DE97	86FA92
86FE99	86FI01	86HI10
86HO93	86KU97	86LU59
86MC76	86NA93	86NI13
86SC2'	86SC59	86SH63
86SM71	86SM1RA'	86ST05
86YE87	87BO37	87CA39
87CA67	87DE24	87DE55
87GA3GI'	87HA69	87HA93
87KE92	87TE58	87WH11
87YI0'	88JE71	

ABSTRACTS

    81AY1

BOOKS

    86DE2'

PATENTS

68TA5'	76MC9'	77MC87'
78BR7'	78MI4'	78RY0'
79BE6'	79HO1'	80FI6'
80MA9'	80ME0'	80NO9'
80RE2'	80WE5'	81BR8'
82HI64'	82JA8'	82JE3'
81IR4'	81JE3'	81JE5'

PATENTS (*cont.*)

81WH6'	82DE47'	82HI4'
82JE2'	82JE6'	83JE0'
84BL1'	84HI7'	84HI8'
84JE4'	84JE8'	84JG1'
84LA2'	84NO0'	85FI4'
85HI3'	85JE8'	85JE0'
86KA6'	86SA4'	87AN6'
87LA0'	87LE2'	87SH2'
87SP1'		

APPLICATIONS

Active Hydrogen Determination

81DA97	81HE97	83TS3'
84HE80	84TS00	

Alkaloids

79ME3'	80EC86	80GA8'
80HU0'	81DI5'	81GA4'
81KE56	81UN36	82DI03
82EC92	82EC93	82TH9'
82TH49	83CH4'	83IR01
86BE1'	86BE37	86JA1'
87BE0'	87ME49	88BE55

Amino Acids

74HO93	75GR9'	76MC63
77MC6'	79HI46	79TS14
80BE59	80BL0'	80BL2'
80BL26	80BL67	80HA6'
80SC39	80SM0'	80TS41
80TS85	81BL5'	81DV3'
81HA9'	81HA32	81IR9'
81OK97	81SM33	82DI03
82HO70	82TH9'	82TS1'
82YO14	83IR01	83SA17
83TS65	84AL96	84AO43
84FA0'	84FA60	84GA49
84HA62	84PI3'	84ST97
84TA59	84TH0'	84TS00
84YA62	84YE0'	85DE8'
85FA1'	85GA37	85GO5'
85LA29	85LI61	85MA03
85TA59	85TS07	85WH2'
86CO7'	86CO9'	86DE8'
86DE97	86FO09	86GA1'
86MI45	86OT12	86SM2'
86WI6'	86YA8'	87AS4'
87AS24	87LE1'	87ME3'
87SE5'		

Derivatives

Dansyl

84AL00	85AL15	87AS45
87KR41		

Drugs and Metabolites (*cont.*)

82MA15	82SC85	82VO15
83AP50	83BE63	83CH4'
83CH11	83CO55	83HE2'
83IR01	83MC93	83TH2'
83VO67'	83YI83	84CH6'
84CR61	84FE69	84HA20
84HE0'	84HE3'	84HE1'
84JO45	84MA41	84PA42
84PO8'	84SC83	84ST3'
84VO16	85AS79	85CA33
85CH33	85CH3'	85CO0'
85CO74	85CR3'	85CR30
85CR71	85DE8'	85EC25
85FA1'	85FE7GE'	85FR7BR'
85GA34	85GA11	85HA84
85HE5'	85HE7'	85HE03
85HI39	85LA33	85LE33
85LU12	85SH2'	85ST73
85SU8'	85VO15	85VO77
85VO0'	85YA5'	86BE1'
86BE37	86BR5'	86BR9'
86CO83	86DI81	86GA21
86GE6'	86GO09	86HA9'
86HA21	86HE09	86KI034
86KI1'	86LE89	86LI3'
86LL4'	86MI45	86PA814
86ST45'	86SU21	86TA06
86UN13	86WI2'	86YA66
87AP0'	87BE45	87BL41
87CA46	87CH62	87CO85
87GA11	87HA89	87HA2'
87HS4'	87IT4'	87IW49
87KI6'	87KO8'	87LA43
87LA93	87LI47	87NI41
87NI51	87PE55	87RU2BE'
87SA59	87SC49	87SI11
87ST49	87WE11	87WE1'

Chemotherapeutics

83JU15	83JU67	84VO78'
85DA22	85VO33	86MC7'
86MC73	86WE3'	

Dexamethasone/betamethasone

83CA03	83CO55	83EC65
83HE3'	83SU47	83TH2'
84SK56	84SK59	85SK11
86SK47	87BA5'	

Radiopharmaceuticals

87UN95

Elements

86TH81

Environmental Analysis
  Dyes

83ST15	84BE64	84MU65
85BE8'	85HA53	85NO83
85VO70	85VO3'	86BA15
86BE5'	86BR5'	86BR9'
86FL8'	86HE1'	86KA1'
86VO05	87BE43	87BR97
87BU5'	87VO5'	

  Fulvic acids
    85BE9'

  Halogenated and polyhalogenated compounds

73KA63	77AR9'	77BL8'
78DY9'	79DY95	80DY4'
80PH10	80ST87	83VO99
83VO49	85CO1'	

  Herbicides

77MC6'	79AR1KL'	79GA0'
79MC8'	79MC88	79SK1'
80SK88	81EV86	82PA27
83AP50	83EV94	83PA6'
85BL3'	85MA33	85MC3'
85RO4'	85RO6'	85SH21
85SH3'	86BR45	86MC22
87BA87	87DE9'	87RE3'
87SH11	87VA47	87VE32
87VO17		

  Pesticides and metabolites

73KA63	76MC29	77AR9'
77HE7'	77SC13	78HO03
78WR7'	79KA14	79TS14
80GA6'	80GA86	80KA81
80MC80	80YO36	81EC79
81GA31	81GA84	82AL99
82CA43	82GA93	82KI25
82PA73	82WR01	82WR80
83CA03	83LE69	83LE11
83PA6'	83WH68	84AP23
84SM61	84VO7'	84VO62
84VO21	84VO67	84VO3'
85BU3'	85FO9'	85KA5'
85MC16'	85SH3'	85SH5'
85SI2'	85SM5'	85TA96
85TS70	85VO71	85VO5'
85WI3'	86BE37	86BE1'
86BR45	86CA43	86FO09
86FR1'	86GA1'	86KA33
86KA1'	86MC3'	86SM70
86VE21	86WR95	87BA45
87BA71	87BE9'	87CH5'
87DE9'	87DE90	87LA93

Organic acids

84AP23	84YA81	85BR4'
85VO6'	85VO9'	86PA812
86SM2'	86TS82	87BE9'

Mycotoxins

77MC56	83SM07	83SM56
83TH2'	85SM5'	85SM32
85TI31	85TI83	85UD3'
85VO89	86KA81	86KR05
86KR5'	86MI6'	86RA2'
86SH63	86SM70	86VO3'
87AB15	87KR49	87MC31
87RA41	87VO430	87VO431

Natural Products

77SC13	78GA51	78GA5RE'
79GA7'	80EC86	80GA72
81BR0'	81GA4MO'	81GA84
81WE43	82AL99	82GA93
82LU02	83AL83	83GA7'
83GA3'	84BR1'	84BR31
84BR62	84GA90	84MA76
85ST4'	86GR7'	86SK07
86ST45'	86WA99	87AL03
87AL45	87BE9'	87FR9'
87HA01	87HO0'	87KI1'
87LI45	87MA55	87MA58
87MO67	87SE65	87TA45

Pungent oleoresins

84GA499	84VA4AD'	85VA03

Neurotransmitters

83HU17	83KE58	83MI13
84MI256	87AR43	87ST9'

Nucleic Acid Constituents

Nucleobase

77MC6'	78BL81	78MC8'
79CH0'	79VE97	80BL2'
80BL26	80PH10	81ED9'
82ZO25	83TH5'	84IN13
84SM10'	85CH47	86LL4'
86RU9'		

—Adducts

87DA0'		

Nucleoside

77MC6'	78BL81	78MC8'
79ME3'	79TS14	79VE97
80BL0'	80BL2'	80BL21
80BL26	80BL86	80CH1RE'
80GA8'	80GA73	80HA6'
80ME87	80PH10	80QU2'
81BL5'	81BR0'	81ED9'
81GA31	81GA84	81HA9'
81HA32	81IR9'	81OK97

Peptides (*cont.*)

81HA32	81YU89	82KE5'
82KI25	82TH9'	82TH49
82TS1'	82YE8'	82YO14
83BL50	83CH4'	83CH61
83DO65	83FE13	83HA5'
83KE05	83KI9HR'	83PI8'
83RO87	83SA17	83ST15
83YU03	84GA4'	84GI19
84HA62	84HA69	84KI64
84KI3'	84LA4'	84MI256
84PI13	84PI66	84PI3'
84ST9'	84ST97	84TA59
84WH2'	84WH8'	84YA93
84YA62	84YU15	85AL15
85AL10	85BU8'	85CH3'
85GO5'	85KI60	85LI61
85MI09	85ST75	85TA52
85TA59	85TS07	85VA31
85VE49	85WH2'	85WH75
86BO4'	86CA89	86CA18
86DE8'	86DI7'	86FO09
86GR7'	86IW49	86JO07
86RO0'	86RO8'	86ST45'
86ST4'	86ST88	86YA8'
87AS24	87AS45	87AS4'
87BA67'	87BE24'	87CA0'
87CA61	87CH41	87CR01'
87DA1'	87HU24	87HU8'
87KR45	87LE45	87LE1'
87LE4'	87LI0'	87MI59
87SE5'	87ST79	87ST52
87VO8'	88CA73	88GA59
88SM06	88ST08	

Glucopeptides
83FE13

Phenols

74HO93	79MC8'	79MC88
79WR2'	80BL86	80CH8RE'
80MC80	80TH4'	81CH31
81EV86	81SC65	81TH87
81WR85	82GA95	82WE91
83DO65	84FR8'	84GA13
84GA17	84HA5'	84TH44
86PR3'	85PR07	87ME49

Phthalates

79HI46	79TS87	80MC80
80TS41	83PO7'	84EL97

Porphyrins

78MC2'	79MC78	80EG23
85AR45	85EG47	87SE5'

Sugars, Polysaccharides, and Derivatives (*cont.*)

83HA5′	83TS65	83YE7′
84AO43	84DO67	84FE69
84GA17	84LE18	84MI256
84SC47	84TA59	85AL37
85AL8′	85AR45	85BE9′
85KI7′	85LI61	85SA8′
85SH71	85TA59	85TS8′
85TS07	86BU1′	86CO7′
86CO9′	86DI7′	86GA1′
86HE34′	86HS9′	86HS99
86IW49	86JO8′	86KA1′
86KA7′	86LI3′	86MA32
86MI454	86RA31	86WR95
87BA2′	87BU49	87DA1′
87ES94	87HS78	87IT16
87PI0′	87SA61	87YE43

Surfactants

83SC68	85HI39	85LE35
85LI73	85MU46	87BA23
87GO2′	87MA55	87MI7′
87OT98	87RU6′	87SM93

Vitamins

74SC95	75GR9′	76HE4′
80BL0′	80BL21	80DY4′
80SC39	80YA87	82DE42
82DE5′	83KE58	83SA17
83TH5′	84AO43	84CH6′
84GA17	84MI25	84TA59
85CH33	85TA59	85VA31
85WH2′	86MA32	86MA17
86VA46	86VO1′	87BA27
87DA1′	87GA14	87LE1′
87YE43		

Xanthines

83BL50	83CO55	83DO65
83HE2′	83SA17	83ST15
83TH2′	85AS79	85CR3′
85CR71	85RO6′	86BE1′
86HE4′	86LE92	87KR49
87LE99	87LI47	87NI51
87SA7′	87SE57	

Index